Progress in Mathematics
Volume 105

Series Editors
J. Oesterlé
A. Weinstein

New Developments in Lie Theory and Their Applications

Juan Tirao
Nolan Wallach

Editors

Birkhäuser

Boston · Basel · Berlin

Juan Tirao
Facultad de Matematica
Astronomia y Fisica
Universidad Nacional de Córdoba
5000 Córdoba
Argentina

Nolan R. Wallach
Dept. of Mathematics
University of California
 at San Diego
La Jolla, CA 92093
USA

Library of Congress in-Publication Data

New Developments in Lie theory and their applications / edited by Juan
 Tirao, Nolan Wallach.
     p.   cm. -- (Progress in mathematics : 105)
     Papers from the Third Workshop on Representation Theory of Lie
Groups and Their Applications, held Aug.-Sept. 1989 in Córdoba,
Argentina.
     Includes bibliographical references.
     ISBN 0-8176-3916-6 (alk. paper). -- ISBN 3-7643-3619-6 (alk.
paper.
     1. Lie groups--Congresses.   2. Representations of groups-
-Congresses. I. Tirao, Juan 1942 -   . II. Wallach, Nolan R.
III. Series : Progress in mathematics (Boston, Mass.) ; vol. 105.
QA387.N48   92-1255
  512'.55--dc20   CIP

Printed on acid-free paper

ISBN 0-8176-3619-6
ISBN 3-7643-3619-6

Camera-ready copy prepared by the Authors in TeX.
Printed and bound by Quinn-Woodbine, Woodbine, NJ.
Printed in the U.S.A.

9 8 7 6 5 4 3 2 1

Table of Contents

Introduction

Representation theory, and more generally Lie theory, has played a very important role in many of the recent developments of mathematics and in the interaction of mathematics with physics. In August-September 1989, a workshop (Third Workshop on Representation Theory of Lie Groups and its Applications) was held in the environs of Córdoba, Argentina to present expositions of important recent developments in the field that would be accessible to graduate students and researchers in related fields. This volume contains articles that are edited versions of the lectures (and short courses) given at the workshop.

Within representation theory, one of the main open problems is to determine the unitary dual of a real reductive group. Although this problem is as yet unsolved, the recent work of Barbasch, Vogan, Arthur as well as others has shed new light on the structure of the problem. The article of D. Vogan presents an exposition of some aspects of this problem, emphasizing an extension of the orbit method of Kostant, Kirillov. Several examples are given that explain why the orbit method should be extended and how this extension should be implemented.

Another active direction of research is the (not related) problem of giving geometric realizations of admissible (and hopefully unitary) representations of real reductive groups. This is the subject of the article by W. Schmid. In light of the various classifications of admissible representations of real reductive groups (Langlands, Beilinson-Bernstein, Vogan-Zuckerman, Vogan) the structure of these representations is well understood from the perspective of analysis and algebra ($(\mathfrak{g}, K)$-modules). The main algebraic constructions are cohomological induction and $\mathcal{D}$-module theoretic induction. The article of W. Schmid describes his work (and that of his coworkers) on a geometric approach to the construction of admissible representations of real reductive groups (the Hecht, Miličić, Schmid, Wolf construction) which is used to relate the various classifications. The article also contains an introduction to hyperfunction theory and a survey of results on functorial globalizations of $(\mathfrak{g}, K)$-modules.

The representation theory of reductive groups has been intimately connected with number theory (in particular the theory of automorphic forms) since the basic work of Gelfand, Piatetski-Shapiro and Selberg in the 1950s and 1960s. In the form of the Langlands program, number theory has suggested important new approaches to problems in representation theory (the flavor of the influence of the Langlands philosophy can be seen in the article of Vogan). The article of N. Wallach is an introduction to the theory of automorphic forms with emphasis on the classical theory (from the perspective of representation theory). Wallach emphasizes the theory for $SL(2)$ and shows how the classical theory can be reformulated into a representation theoretic language. The article also contains some exposition of the general theory.

Lie theory has almost from it earliest beginnings been ultimately connected with physics. Certainly, general relativity and quantum mechanics have made substantial use of Lie theory. The article of B. Kostant contains research that he and his coworkers have done on a remarkable representation of $SO(4, 4)$ which is analogous to the metaplectic (or Weil) representation of the two-fold covering group of the symplectic group. In his article Kostant shows how this representation is intimately connected with general relativity on the one hand and to the triality associated with $SO(8)$ on the other.

The Penrose twistor theory relates complex analysis to general relativity and Maxwell's equations. This theory has influenced and has been influenced by the geometric methods of constructing representations of reductive groups (see the article of W. Schmid). Twistor theory is related to the group $SU(2, 2)$ and is a four-dimensional theory. In the article of J. Rawnsley, an extension of this theory is given to arbitrary even dimensions. The twistor space for a compact symmetric space of inner type is determined.

Since the late 1960s there have been several important generalizations of (finite-dimensional) reductive groups and their representations. Although Kač-Moody algebras were first studied for purely mathematical reasons (work of Kač on finite order automorphisms of simple Lie algebras over $\mathbb{C}$, the proof of the Dyson-McDonald identities using the Kač character formula, the use of the normal operators of Lepowsky-Wilson to construct affine Kač-Moody algebras), the physics of the 1970s and 1980s influenced and was influenced by the development of this theory (notably through the theory of vortex operators and string theory). Related to this development is the so-called theory of quantum groups (called quantum deformations of enveloping algebras) in the article of G. Lusztig. These generalizations of reductive Lie groups are not even groups; they are actually Hopf algebras that deform in a parameter the universal enveloping algebra of a simple Lie algebra (or even a Kač-Moody algebra). These objects were initially studied to analyze explicitly solvable Hamiltonian systems. The article of G. Lusztig gives an introduction to this theory and in particular to the important problem of constructing "good" bases for these algebras and analyzing the specializations of the deformation parameter to a root of unity and descent to a finite characteristic.

One of the most fertile areas in which representation theory has played an important role is the study of eigenfunction expansions of elliptic differential operators. The classical functions of analysis are for the most part derived by separation of variables along the orbits of an action of a reductive Lie group from an appropriate Laplacian. Also, the theory of partial differential equations has played a basic role in the analytic theory of representations (see the article of Schmid). The article of C. Gordon is more differential geometric in nature than the others and gives an exposition of the extent to which the spectrum of the Laplace-Beltrami operator determines the topology and or geometry of a Riemannian manifold. Since the early example of Milnor on isospectral manifolds, a virtual "zoo" of examples has been found. A significant number involve homogenous spaces or Lie theory. In Gordon's article, there is a survey of some of the most fruitful techniques of constructing isospectral manifolds.

Several shorter contributions have been also included in this volume. For the theory of automorphic forms, there are articles by R. Miatello and F. Williams, respectively, on the Kuznetsov formulas and the Selberg zeta function for discrete subgroups of reductive groups; J. Rohlfs and B. Speh discuss the use of the Lefschetz fixed point formula to study automorphic representations. The article of J. Tirao studies the important question of determining the structure of the centralizer of a maximal compact subgroup in the enveloping algebra of a real reductive group, and N. Andruskiewitsch considers the related question of polynomial invariants. The article of J. Pantoja and J. Soto-Andrade examines a generalization of the Weil representation to $SL(n, k)$, where k is a finite field. The contribution of A. S. Dubson gives an exposition of several results on local intersection multiplicities for holonomic modules.

Automorphic Forms

Nolan R. Wallach

Notes by Roberto Miatello

Introduction.

The purpose of these lectures is to give a relatively easily accessible introduction to the theory of automorphic forms. For most of the lectures we will confine our attention to the least complicated (but still important in its own right) case of $SL(2, \mathbf{R})$. The lectures begin with a discussion of the relationships between the η-function, the θ-function and classical Eisenstein series. The emphasis here is on their q-expansions which we will see have interesting group and representation theoretic interpretations. Here we give some implications of a combinatorial nature to the explicit calculations of the expansions. These functions were singled out for study for two reasons. The first is that they are the simplest automorphic forms that can be explicitly written. The second is that they are intimately related to the representation theory of loop groups and the Virasoro algebra (precisely through their q-expansions).

In the second section we relate the explicit material in the first with the classical theory of holomorphic automorphic forms. The main constructions are (holomorphic) Eisenstein series and Poincaré series. The modern theory of automorphic forms begins with the introduction of the Maass wave forms (now just called Eisenstein series). The general notion of automorphic form (in the case of $SL(2, \mathbf{R})$) is given in section 3. The rest of the section involves first showing that classical automorphic forms are automorphic forms in the general sense and showing that Eisenstein series are automorphic forms. In section 4, the advertised interpretation of q-expansions is given. Also, the notion of cusp form and a basic property of cusp forms is given. In section five, we leave the cozy environs of $SL(2, \mathbf{R})$ for the general arithmetic case describing how the concepts for $SL(2, \mathbf{R})$ generalize. Here we give a description of the basic theorems of Langlands on the decomposition of $L^2(\Gamma \backslash G)$. We end the section with a rapid discussion of the trace formula.

Most of the emphasis in the modern theory of automorphic forms is to its relationship with the Langlands program (see the article [A] and the related material in the volume that contains it). Also, the general structure theory of arithmetic groups is alluded to in section 5. A good reference for this material is [Z].

1. Some examples of automorphic forms.

In this section we will discuss several important functions that arise in analytic number theory. We will emphasize their functional equations and use them as concrete examples of automorphic forms. The main reference for this section is [Kn] where most of the omitted proofs can be found.

We begin with an important function in additive number theory. Set

$$\mu(q) = \Pi_{n=1}^{\infty}(1 - q^n)$$

where $q = e^{2\pi i z}$, $z \in H = \{z \in \mathbb{C}|\ \mathrm{Im} z > 0\}$. The product is absolutely convergent for $|q| < 1$; hence μ defines a non-vanishing holomorphic function on H.

If we use the expansion

$$\frac{1}{1 - q^n} = \sum_{m \geq 0} q^{nm}$$

for $|q| < 1$ and formally combine terms we find that

$$\mu(q)^{-1} = \sum_{n=0}^{\infty} p(n)q^n$$

where $p(n)$ is the classical partition function, that is, $p(n)$ counts the number of ways in which n can be written as a sum of positive integers. For instance, $p(0) = 0$, $p(1) = 1$, $p(4) = 5$, $p(5) = 7$.

Euler's famous pentagonal number theorem says that

$$\mu(q) = \sum_{n=-\infty}^{+\infty} (-1)^n q^{(3n+1)n/2}$$

that is

$$\mu(q) = 1 + \sum_{n=1}^{\infty}(-1)^n\{q^{\frac{3n^2-n}{2}} + q^{\frac{3n^2+n}{2}}\}.$$

If we multiply this identity times the expansion above for $\mu(q)^{-1}$ we have

$$1 = \sum_{n=0}^{\infty} p(n)q^n + \sum_{n=0}^{\infty}\left\{\sum_{m=1}^{\infty}(-1)^m[p(n - \frac{3m^2 - m}{2}) + p(n - \frac{3m^2 + m}{2})]\right\}q^n.$$

These yield a reasonably efficient recurrence relation for $p(n)$:

$$p(n) = \sum_{m=1}^{\infty}(-1)^{m+1}[p(n - \frac{3m^2 - m}{2}) + p(n - \frac{3m^2 + m}{2})].$$

For example, if $n = 6$,

$$p(6) = p(5) + p(4) - p(1) = 7 + 5 - 1 = 11.$$

A remarkable classical function closely related to $\mu(q)$ is the Dedekind η-function. Set

$$\eta(z) = q^{\frac{1}{24}}\mu(q), \quad q = e^{2\pi i z}, z \in H.$$

One can show (using a combination of the pentagonal number theorem and Poisson summation) that η satisfies the following functional equations:

$$\eta(z+1) = e^{\frac{\pi i}{12}}\eta(z), \quad \eta\left(-\frac{1}{z}\right) = e^{-\frac{\pi i}{4}}\eta(z)z^{\frac{1}{2}} \tag{1}$$

(here we choose the branch of the square root, holomorphic in $\mathbf{C}-\{t \in \mathbf{R} \mid t \le 0\}$, so that $\sqrt{i} = e^{\frac{\pi i}{4}}$).

If $g = \begin{bmatrix} a & b \\ c & d \end{bmatrix} \in SL(2,\mathbf{R})$, $z \in H$, we write $g \cdot z = \frac{az+b}{cz+d}$ (the usual action of $SL(2,\mathbf{R})$ on the upper half plane by linear fractional transformations) and $j(g,z) = cz + d$ (the so-called classical automorphy factor). A direct calculation shows that

$$j(g_1 g_2, z) = j(g_1, g_2 \cdot z)j(g_2, z).$$

Since

$$\begin{bmatrix} 0 & 1 \\ -1 & 0 \end{bmatrix} \text{ and } \begin{bmatrix} 1 & 1 \\ 0 & 1 \end{bmatrix}$$

generate the full modular group $\Gamma = SL(2,\mathbf{Z})$, (1) is equivalent with

$$\eta(\gamma z) = j(\gamma, z)^{\frac{1}{2}}\chi(\gamma)\eta(z), \quad \gamma \in \Gamma, z \in H \tag{2}$$

with χ a function on Γ such that

$$\chi\left(\begin{bmatrix} 0 & 1 \\ -1 & 0 \end{bmatrix}\right) = e^{-\frac{\pi i}{4}}, \quad \chi\left(\begin{bmatrix} 1 & 1 \\ 0 & 1 \end{bmatrix}\right) = e^{\frac{\pi i}{12}}$$

and

$$\chi(\gamma_1\gamma_2)j(\gamma_1\gamma_2, z)^{\frac{1}{2}} = \chi(\gamma_1)j(\gamma_1, \gamma_2 \cdot z)^{\frac{1}{2}}\chi(\gamma_2)j(\gamma_2, z)^{\frac{1}{2}}$$

(see the remark above about $z^{\frac{1}{2}}$).

In general, if $\chi : \Gamma \to S^1$ is a function such that

$$\chi(\gamma_1\gamma_2)j(\gamma_1\gamma_2, z)^r = \chi(\gamma_1)j(\gamma_1, \gamma_2 \cdot z)^r\chi(\gamma_2)j(\gamma_2, z)^r$$

then χ is called a multiplier of order r for Γ. If r is an integer then χ is a character of Γ.

Notice that if χ is the multiplier for η then $\chi^{24} = 1$. We therefore set

$$\Delta(z) = \eta(z)^{24} = q\Pi_{n=1}^{\infty}(1 - q^n)^{24}.$$

This function has the functional equation

$$\Delta(\gamma z) = j(\gamma, z)^{12}\Delta(z), \quad \gamma \in \Gamma.$$

This, together with the q-expansion at ∞ says that Δ is a cusp form of weight 12 (and $\chi = 1$) in the sense of Section 2.

We note that if we write $\delta = \eta^8$, then

$$\delta(-\frac{1}{z}) = z^4\delta(z),$$

$$\delta(z + 3) = \delta(z).$$

In the terminology of Section 3, $\delta(z)$ is a cusp form of weight 4 for the group generated by

$$\begin{bmatrix} 0 & 1 \\ -1 & 0 \end{bmatrix} \text{ and } \begin{bmatrix} 1 & 3 \\ 0 & 1 \end{bmatrix}$$

which has index 3 in $SL(2, \mathbf{Z})$

Another important classical example of automorphic form is the θ-function which is defined by the following expansion.

$$\theta(z) = 1 + 2\sum_{n=1}^{\infty} e^{\pi i n^2 z}.$$

We note that $\theta(z + 2) = \theta(z)$. The Poisson summation implies that

$$\theta(-\frac{1}{z}) = e^{-\frac{\pi i}{4}}z^{\frac{1}{2}}\theta(z).$$

Also, the Jacobi triple formula ([Kn])

$$\Pi_{n=0}^{\infty}(1 - q^{2n+2})(1 + q^{2n+1}z)(1 + q^{2n+1}z^{-1}) = \sum_{m=-\infty}^{+\infty} z^{m^2}q^m$$

can be used to prove that

$$\theta(z) = \eta(\frac{z+1}{2})^2/\eta(z + 1).$$

Let Γ_θ denote the subgroup of $SL(2, \mathbf{Z})$ generated by

$$\begin{bmatrix} 0 & 1 \\ -1 & 0 \end{bmatrix} \text{ and } \begin{bmatrix} 1 & 2 \\ 0 & 1 \end{bmatrix}$$

that is,

$$\Gamma = \left\{ \gamma \in SL(2, \mathbf{Z}) \mid \gamma \equiv I \text{ or } \gamma \equiv \begin{bmatrix} 0 & 1 \\ -1 & 0 \end{bmatrix} \bmod 2 \right\}$$

The above properties of $\theta(z)$ are equivalent to

$$\theta(\gamma z) = \chi_\theta(z) j(\gamma, z)^{\frac{1}{2}} \theta(z) \quad \gamma \in \Gamma_\theta, z \in H$$

with χ_θ a multiplier system of weight $1/2$ for Γ_θ such that

$$\chi_\theta \left(\begin{bmatrix} 0 & 1 \\ -1 & 0 \end{bmatrix} \right) = e^{-\frac{\pi i}{4}}, \ \chi_\theta \left(\begin{bmatrix} 1 & 2 \\ 0 & 1 \end{bmatrix} \right) = 1.$$

In classical analytic number theory, the θ-function appeared in the analysis of the problem of counting $r_h(n)$, the number of ways an integer n can be written as a sum of h squares. Indeed, if $h \in \mathbf{N}$, and if we set $\theta^h(z) = \theta(z)^h$ and expand the product formally then we have

$$\theta^h(z) = 1 + \sum_{n=1}^{\infty} r_h(n) e^{\pi i n z}.$$

There is a remarkable relationship between $\theta(z)$ and another classical automorphic form with the same multiplier. If $h \geq 5$ then we set

$$E_{h,\chi_\theta} = \frac{1}{2} \sum_{(c,d)=1} \chi_\theta(\gamma_{c,d})^{-h} (cz + d)^{-\frac{h}{2}}$$

where $\gamma_{c,d} = \begin{bmatrix} * & * \\ c & d \end{bmatrix} \in \Gamma_\theta$. If $\gamma \in \Gamma_\theta$

$$E_{h,\chi_\theta}(\gamma z) = \chi_\theta(\gamma)^h j(\gamma z)^{\frac{h}{2}} E_h(z).$$

Then $E_{h,\chi_\theta}(z)$ has the q expansion $(q = e^{2\pi i z}$ as usual)

$$1 + \sum_{m=1}^{\infty} \rho_h(m) q^{\frac{m}{2}}$$

with

$$\rho_h(m) = \frac{e^{-\frac{\pi i h}{4}} \pi^{\frac{h}{2}}}{\Gamma(\frac{s}{2})} m^{\frac{h}{2}-1} \sum_{c=1}^{\infty} B_c(m)$$

and

$$B_c(m) = c^{-\frac{h}{2}} \sum_{\substack{0 \leq d < 2c \\ (d,c)=1}} \overline{\chi_\theta(\gamma_{c,d})} e^{\frac{\pi i m d}{c}}.$$

(See [Kn] Chapter 5).

Theorem 1. $\theta^h(z) - E_{h,\chi_\theta}(z)$ *is a cusp form for* Γ_θ, *with multiplier* χ_θ^h.
If $5 \leq h \leq 8$ *then* $\theta^h = E_{h,\chi_\theta}$.

This remarkable result is a special case of the Siegel-Weil theorem. Note that for $h = 5, 6, 7, 8$ it implies a formula for $r_n(h)$. For $h > 8$ it implies an asymptotic formula for $r_n(h)$ (see [Kn; Chapter 5] for details). More recent generalizations that relate θ-functions and Eisenstein series are due to S.Kudla-S.Rallis.

2. Classical automorphic forms.

We will now begin a more systematic study of automorphic forms from the classical viewpoint. For simplicity we will concentrate on the trivial multiplier $\chi = 1$. For the rest of this section, Γ will denote a discrete subgroup of $G = SL(2, \mathbf{R})$ such that $\Gamma \backslash G$ is non-compact and has finite G-invariant covolume (for instance, $\Gamma = SL(2, \mathbf{Z})$ and its subgroups of finite index).

If $x \in \mathbf{R} \cup \{\infty\}$, we set $P_x = \{g \in G \mid g \cdot x = x\}$ and $N_x = [P_x, P_x]$. Then x is said to be a cusp for Γ if $\Gamma \cap N_x \neq \{I\}$. It is not hard to show using that Γ is discrete, that $\Gamma \cap P_x = \Gamma \cap \{\pm I\} N_x$. Write $\Gamma_x = \Gamma \cap P_x$. Let x be a cusp and let $h \in G$ be such that $h \cdot \infty = x$. Then one of the following holds:

(i) If $-I \in \Gamma$, then $h^{-1} \Gamma_x h = \left\{ \pm \begin{bmatrix} 1 & m\lambda(x) \\ 0 & 1 \end{bmatrix} \mid m \in \mathbf{Z} \right\}$, for some $\lambda(x) > 0$.

(ii) If $-I \notin \Gamma$, then $h^{-1} \Gamma_x h$ is cyclic generated by either of $\alpha = \begin{bmatrix} 1 & \lambda(x) \\ 0 & 1 \end{bmatrix}$
or $\beta = \begin{bmatrix} -1 & \lambda(x) \\ 0 & -1 \end{bmatrix}$ for some $\lambda(x) > 0$.

If we replace h by $h\,a(x)$, with $a(x) = \begin{bmatrix} \lambda(x)^{\frac{1}{2}} & 0 \\ 0 & \lambda(x)^{-\frac{1}{2}} \end{bmatrix}$ we may assume that $\lambda(x) = 1$. Also, by conjugating Γ by some $g \in SL(2, \mathbf{R})$, if necessary, we may assume that ∞ is a cusp of Γ. We will use the notation $P = P_\infty$, $N = N_\infty$.

If $\mathcal{C}$ denotes the set of cusps of Γ, then $\mathcal{C}$ is Γ-invariant. Furthermore one proves under the above hypotheses on Γ that $\mathcal{C}$ is non empty and $\Gamma \backslash \mathcal{C}$ is finite ([Le]). We will come back to this critical point in later sections.

If $f : H \to \mathbf{C}$ and $h \in G$ we write (following standard classical notation)

$$f_{|h}(z) = j(h, z)^{-k} f(hz).$$

We note that $f_{|hh'} = (f_{|h})_{|h'}$, $h, h' \in G$.

Definition 1. Fix $k \in \mathbf{Z}$. A classical automorphic form of weight k is a holomorphic function $f : H \to \mathbf{C}$ such that

(i) $f(\gamma z) = j(\gamma, z)^k f(z)$ $\gamma \in \Gamma, z \in H$ (i.e. $f_{|\gamma} = f$, for $\gamma \in \Gamma$).

(ii) f is holomorphic at every cusp of Γ.

Condition (ii) has the following interpretation. Fix $x \in \mathbf{R} \cup \infty$, a cusp of Γ, and let $h \in G$ be such that $h.\infty = x$ chosen so that $\lambda(x) = 1$. Then $f_{|h}$ is an automorphic form of weight k for $\Gamma' = h^{-1}\Gamma h$ and ∞ is a cusp of Γ'. Since

$$f_{|h}(z + 1) = f_{|h}(z)$$

we have a q-expansion

$$f_{|h}(z) = \sum_{j \in \mathbf{Z}} c_j(x) e^{2\pi i z j}$$

with $c_j(x) \in \mathbf{C}$. Now f is said to be holomorphic at x if $c_j(x) = 0$ for any $j < 0$. If $c_j(x) = 0$ for any $j \leq 0$ and any cusp x, f is said to be a cusp form

It is not *a priori* obvious, for a general Γ, that non-zero automorphic forms exist. We now introduce the classical Eisenstein series which settles this question. Let $k \in \mathbf{N}, k \geq 3$. Define

$$E_k(z) = \sum_{\Gamma \cap N \backslash \Gamma} (cz + d)^{-k}. \tag{1}$$

One shows that this series converges uniformly on compacta in H, hence it defines a holomorphic function. Also, if $\tau \in \Gamma$

$$E_k(\tau z) = \sum_\gamma j(\gamma, \tau z)^{-k} = \left(\sum_{\gamma \in \Gamma \cap N \backslash \Gamma} j(\gamma \tau q, z)^{-k} \right) j(\tau, z)^k = E_k(z) j(\tau, z)^k$$

The calculation of the q-expansion at each cusp (with formulae similar to those for the analogous series in section 1) shows that $E_k(z)$ is holomorphic at each cusp. In fact, $c_0(x) \neq 0$ if and only if the cusp x is Γ-equivalent to ∞. In particular, since $c_0(\infty) \neq 0$ it follows that $E_k(z)$ is not identically zero.

Similarly, one can define Eisenstein series associated to any fixed cusp x. If $x = h.\infty$, $h \in G$, then ∞ is a cusp of $\Gamma' = h^{-1}\Gamma h$. Let $E_{k,\Gamma'}$ be defined as in (1) above with Γ' in place of Γ. Set

$$E_{k,x}(z) = (E_{k,\Gamma'})_{|h^{-1}}(z).$$

Then $E_{k,x}(z)$ is a Γ-automorphic form of weight k and one shows that if y is a cusp then $c_0(y) \neq 0$ if and only if y is Γ-equivalent to x. We thus see that by subtracting a linear combination of Eisenstein series one can turn any classical automorphic form into a cusp form.

There is another remarkable family of classical automorphic forms due to Poincaré (which he introduced in order to see that any Riemann surface has a non-zero meromorphic function). If $k \in \mathbf{N}$, $k \geq 3$ and $m \in \mathbf{Z}$ set

$$G_{m,k}(z) = \sum_{\Gamma \cap N \backslash \Gamma} \frac{e^{2\pi i m \gamma z}}{(cz+d)^k}$$

One can prove that if $m > 0$, $G_{m,k}$ defines a cusp form of weight k ([Le]). If $m < 0$, $G_{m,k}$ does not satisfy (ii) in Definition 1. As in the case of Eisenstein series one can define a Poincaré series associated to each cusp x.

Petersson introduced the so-called Petersson inner product to prove that the set $\{G_{m,k}| \ m = 1, 2, 3...\}$ spans the space of holomorphic cusp forms of weight k , $k \geq 3$ (cf. [Le]). If $k = 1, 2$ the above series does not converge. In these cases a generalization of the Poincaré series $G_{m,k}$, is given in [MW, §4]. The holomorphic cusp forms have very similar properties to those of the classical Poincaré series, in particular, they also span the corresponing spaces of cusp forms.

A major breakthrough in the subject, due to H. Maass, was the introduction of a generalization of the Eisenstein series which he called "wave forms". In the modern lexicon the Maass wave forms are now called Eisenstein series. These series give a description of the continuous spectrum of the Laplace-Beltrami operator on $L^2(\Gamma \backslash G / K)$. If $s \in \mathbf{C}$ Re$s > 0$, then the series is given by

$$E(s, z) = (\mathrm{Im}z)^{\frac{s}{2}} \sum_{\Gamma \cap N \backslash \Gamma} |cz + d|^{-s}.$$

It converges uniformly on compacta for Re $s > 2$, and defines a real analytic function on H for s in this region. It is a fundamental result of Selberg that $E(s, z)$ admits a meromorphic continuation to $\mathbf{C}$, holomorphic if Re $s = 0$. We note that $E(s, z)$ is Γ-invariant, but a simple variation of the above series yields a continuous family $E_k(s, z)$, of forms of weight k, for any $k \in \mathbf{Z}$ (which will be discussed in the next section) and the holomorphic Eisenstein series are equal to special values of these series. Finally, by proceeding exactly as above in the case of holomorphic Eisenstein series, one can associate an Eisenstein series to each cusp of Γ. As we shall see in Section 5 these series have natural generalizations to to general reductive groups and are central to the theory of automorphic forms.

3. Automorphic forms on $G = SL(2, \mathbf{R})$.

In this section we will give a group theoretical definition of an automorphic form and we will discuss the relationship with the classical notion. The methods in this section have natural generalizations to arbitrary reductive groups.

We will first introduce some notation and recall some basic facts. As in Section 2, let $G = SL(2, \mathbf{R})$ and let $\Gamma \subset G$ be a discrete, non cocompact, subgroup of finite invariant covolume such that ∞ is a cusp. Let $\mathfrak{g}$ denote the Lie algebra of G. If $x \in G$, $X \in \mathfrak{g}$, $f \in C^\infty(G)$ set

$$L_x f(g) = f(x^{-1}g), \quad R_x f(g) = f(gx),$$

$$L_X f(g) = \frac{d}{dt}\Big|_0 f(exp(-tX)g), \quad R_X f(g) = \frac{d}{dt}\Big|_0 f(gexptX).$$

We will also write $Xf = R_X f$ and sometimes for complicated expressions $R(X)f = R_X f$. The actions L_X, R_X are extended to $U(\mathfrak{g}_{\mathbf{C}})$, the universal enveloping algebra of $\mathfrak{g}_{\mathbf{C}}$, in the usual way. (Here we look at $U(\mathfrak{g}_{\mathbf{C}})$ as the left invariant differential operators on G).

Let $C \in U(\mathfrak{g}_{\mathbf{C}})$ be the Casimir element, that is, $C = \sum_i X_i X^i$ where $\{X_i \mid i = 1, 2, 3\}$ is a basis of $\mathfrak{g}_{\mathbf{C}}$ and $\{X^i\}$ is the dual basis with respect to the invariant form, $(x, y) = \operatorname{tr} xy$, $x, y \in \mathfrak{g}$.

It is easy to check that C is independent of the choice of the basis $\{X_i\}$. In particular, $Ad(x)C = C, x \in G$. Hence C lies in $Z(\mathfrak{g})$, the center of $U(\mathfrak{g})$. Furthermore, $R_C f = L_C f$ if $f \in C^\infty(G)$.

If $g = \begin{bmatrix} a & b \\ c & d \end{bmatrix} \in G$ we set

$$\| g \| = \operatorname{tr}(gg^t)^{\frac{1}{2}} = (a^2 + b^2 + c^2 + d^2)^{\frac{1}{2}}.$$

We note that

$$\| g \| = \| g^{-1} \| \quad g \in G,$$

$$\| g_1 g_2 \| \leq \| g_1 \| \, \| g_2 \| \quad g_1, g_2 \in G,$$

$$\| k_1 g k_2 \| = \| g \| \text{ for } k_1, k_2 \in K = SO(2), \quad g \in G.$$

We now recall the Iwasawa decomposition of G (which, in this case is just the Gram-Schmidt process on the columns). Let

$$K = SO(2), \quad A = \left\{ \begin{bmatrix} y & 0 \\ 0 & y^{-1} \end{bmatrix} \middle| y \in \mathbf{R}, \ y > 0 \right\}, \quad N = \left\{ \begin{bmatrix} 1 & x \\ 0 & 1 \end{bmatrix} \middle| x \in \mathbf{R} \right\}.$$

If $g \in G$ can be written uniquely in the form $g = n(g)a(g)k(g)$ with $n(g) \in N, a(g) \in A, k(g) \in K$. We denote by $\mathfrak{n}$, $\mathfrak{a}$, $\mathfrak{k}$ respectively the Lie algebras of N, A and K. We have $\mathfrak{n} = \mathbf{R}X, \mathfrak{a} = \mathbf{R}H, \mathfrak{k} = \mathbf{R}J$ where

$$X = \begin{bmatrix} 0 & 1 \\ 0 & 0 \end{bmatrix}, H = \begin{bmatrix} 1 & 0 \\ 0 & -1 \end{bmatrix}, J = \begin{bmatrix} 0 & 1 \\ -1 & 0 \end{bmatrix}.$$

Set $Y = \begin{bmatrix} 0 & 0 \\ 1 & 0 \end{bmatrix}$. X, Y, H satisfy the standard commutation relations

$$[H, X] = 2X, \quad [H, Y] = -2Y, \quad [X, Y] = H.$$

Let $\alpha \in \mathfrak{a}^*$ be such that $\alpha(H) = 2$, and set $\rho = \frac{1}{2}\alpha$. If $\nu \in \mathfrak{a}_{\mathbb{C}}^*$ then we set $a^\nu = y^z$ if $a = \begin{bmatrix} y & 0 \\ 0 & y^{-1} \end{bmatrix}$ and $\nu = z\rho$. Set

$$Z = \frac{1}{2}\begin{bmatrix} 1 & i \\ i & -1 \end{bmatrix}, \overline{Z} = \frac{1}{2}\begin{bmatrix} 1 & -i \\ -i & -1 \end{bmatrix}, h = \begin{bmatrix} 0 & -i \\ i & 0 \end{bmatrix}.$$

Then

$$[h, Z] = 2Z, [h, \overline{Z}] = 2\overline{Z}, [Z, \overline{Z}] = h$$

and furthermore

$$\mathrm{tr}\, hZ = \mathrm{tr}\, h\overline{Z} = 0, \ \mathrm{tr}\, h^2 = 2, \ \mathrm{tr}\, Z\overline{Z} = 1$$

This implies that C can be written as

$$C = \frac{1}{2}h^2 + Z\overline{Z} + \overline{Z}Z = \frac{1}{2}h^2 - h + 2Z\overline{Z}. \tag{1}$$

Definition 1. A function $f \in C^\infty(G)$ is called a Γ-automorphic form if it satifies the following four conditions

(A1) $f(\gamma g) = f(g) \quad \gamma \in \Gamma, g \in G$

(A2) dim span $(R_K f) < \infty$

(A3) dim span $\{C^i f | i = 1, 2, ...\} < \infty$

(A4) There exists $r \geq 0$ such that if $X \in U(\mathfrak{g})$ then

$$|Xf(g)| \leq C_X \|g\|^r, \quad \text{for } g \in G.$$

Conditions (A1)-(A3) are natural algebraic conditions. We will see shortly that condition (A4) is a replacement for the condition of holomorphy at the cusps of classical automorphic forms. This condition is also called moderate growth. Without it there is no theory of automorphic forms.

We now explain how to associate to a classical automorphic form f, an automorphic form in this sense. We define $\tilde{f}(g) = j(g, i)^{-k} f(g, i)$. We note that if $\gamma \in \Gamma, g \in G$

$$\tilde{f}(\gamma g) = j(\gamma g, i)^{-k} f(\gamma g \cdot i) = j(\gamma, g \cdot i)^{-k} j(g, i)^{-k} j(\gamma g \cdot i)^k f(g \cdot i) = \tilde{f}(g).$$

Also, if $g = nau(\theta)$ with $n \in N, a \in A, u(\theta) = \begin{bmatrix} \cos\theta & \sin\theta \\ -\sin\theta & \cos\theta \end{bmatrix}$ we have

$$j(g, i) = j(a, i)j(u(\theta), i) = a^{-\rho}e^{-i\theta}.$$

Hence
$$\tilde{f}(gu(\theta)) = j(gu(\theta), i)^{-k}f(g \cdot i) = e^{ik\theta}\tilde{f}(g).$$

Thus $\tilde{f}$ satisfies (A1) and (A2). We will next check (A3). We first note that

$$\frac{d}{dt}\Big|_0 f(g\exp t(X + Y) \cdot i) = \frac{d}{dt}\Big|_0 f(g \cdot (\tanh 2t + \frac{i}{\tanh 2t}))$$

$$= 2\frac{d}{dx}f(g \cdot i)j(g, i)^{-2}$$

$$\frac{d}{dt}\Big|_0 f(g \cdot \exp tHi) = \frac{d}{dt}\Big|_0 f(g \cdot e^{2t}i)$$

$$= 2\frac{d}{dy}f(g \cdot i)j(g, i)^{-2}.$$

Thus if we set $\check{f}(g) = f(g.i)$, then

$$R(\overline{Z})\check{f}(g) = (-2i)j(g, i)^{-2}(\frac{\partial}{\partial x} + i\frac{\partial}{\partial y})f(g.i) = 0.$$

since f is holomorphic. We also note that $R(h)\check{f} = 0$, hence $C\check{f} = 0$ by (1).

Thus if we set $\phi(g) = j(g, i)^{-k}$, then

$$C\tilde{f}(g) =$$
$$R(\frac{1}{2}h^2 - h)\phi(g)\check{f}(g) + 2(R(Z)R(\overline{Z})\phi)(g)\check{f}(g) + 2R(\overline{Z})\phi(g)R(Z)\check{f}(g)$$

We claim that
$$h\phi = k\phi \text{ and } R(\overline{Z})\phi = 0.$$

We will check the first assertion. The second will be left as an exercise for the reader. We have

$$h\phi(g) = (-i)\frac{d}{dt}\Big|_0 j(g\exp t(X - Y), i)^{-k}$$

$$= (-i)j(g, i)^{-k}\frac{d}{dt}\Big|_0 j(\exp t(X - Y), i)^{-k}$$

$$= (-i)j(g,i)^{-k}(-k)\frac{d}{dt}_{|0}(-i\sin t + \cos t) = k\phi(g).$$

In light of the above equations we have

$$C\tilde{f} = (\frac{k^2}{2} - k)\tilde{f} = (\frac{(k-1)^2 - 1}{2})\tilde{f}$$

and (A3) follows.

Before checking (A4) we will recall some results from reduction theory. If $x \in \mathbf{R} \cup \{\infty\}$ is a cusp of Γ, let P_x be as in §2, $N_x = [P_x, P_x]$, $\Gamma_x = \Gamma \cap P_x$. We call P_x a cuspidal parabolic subgroup of G with respect to Γ. In particular, if $x = \infty$ we have $P_\infty = P = \{\pm I\}AN$, $N_\infty = N$, $\Gamma_\infty = \Gamma \cap \{\pm I\}N$. If $x = h.\infty$, $h \in G$, then $P_x = \{\pm I\}A_x N_x$ with $A_x = hAh^{-1}$. We will set

$$(A_x)_t^+ = \{a \in A_x | (h^{-1}ah)^\alpha \geq t\}.$$

A Siegel set for P_x is a set of the form $\mathcal{S}_{P_x,\omega,t} = \omega(A_x)_t^+.K$, where $\omega \subset N_x$ is compact. It is not hard to check that for any $t > 0$ there exists $C = C_t$ such that

$$C^{-1}(h^{-1}ah)^\rho \leq \| a \| \leq C(h^{-1}ah)^\rho$$

for $a \in (A_x)_t^+$.

We fix a complete set of representatives, $\{x_1, ..., x_r\}$ ($x_1 = \infty$), for the Γ-equivalencef classes of cusps of Γ. We pick $h_i \in K$ so that $h_i.\infty = x_i$, $1 \leq i \leq r$, and we set $P_i = P_{x_i} = h_i P h_i^{-1}$. We note that $h_i \mathcal{S}_{P,\omega,t} = \mathcal{S}_{P_i,\omega_i,t}$ with $\omega_i = h_i \omega h_i^{-1}$.

The main result in the reduction theory of (G, Γ) can be stated as follows:

Theorem 1. *Let G, Γ be as above.*

(i) *Let P' be a Γ-cuspidal parabolic subgroup of G. Then there exists $t > 0$ so that if $\gamma \in \Gamma$ satisfies $\gamma \mathcal{S}_{P',t,\omega} \cap \mathcal{S}_{P',t,\omega} \neq \emptyset$ then $\gamma \in \Gamma \cap P'$.*

(ii) *Let P_i be as above $i = 1, 2, ..., r$. Then there exist $t_i > 0, \omega_i \subset N_{x_i}, \omega_i$ compact such that*

$$G = \cup_{i=1}^r \Gamma \mathcal{S}_{P_i,\omega_i,t_i} = \cup_{i=1}^r \Gamma h_i \mathcal{S}_{P,\omega,t_i}.$$

Furthermore for any $t > 0, \Gamma \backslash (G - \cup_1^r \pi(\mathcal{S}_{P_i,\omega_i,t})^o)$ is compact (here $\pi : G \to \Gamma \backslash G$ denotes the canonical projection).

Remark. We note that any subgroup Γ, which has a reduction theory as in Theorem 1, is of finite covolume in G. Indeed, Theorem 1 (i), it is clearly sufficient to show that each Siegel set $\mathcal{S} = \mathcal{S}_{P_x,\omega,t}$ has finite Haar measure in G. Now $G = N_x A_x K$, hence

$$vol(\mathcal{S}) = \int_\omega \int_{(A_x)_t^+} \int_K a^{-2\rho} dk \, da \, dn = C \int_{(A_x)_t^+} a^{-2\rho} da.$$

We have

$$(A_x)_t^+ = \{h \exp s H \, h^{-1} \mid (\exp s H)^\alpha = e^{2s} \geq t\}$$

thus

$$vol(\mathcal{S}) = C \int_{\log(\sqrt{t})} e^{-2s} ds < \infty.$$

In the case when $\Gamma = SL(2, \mathbf{Z})$, we have $r = 1$ in Theorem 1 and if

$$\mathcal{S} = \exp([0,1]X) A_t^+ K, \, t < \sqrt{3}/2$$

then $\Gamma \mathcal{S} = G$ (cf.[Le]). In particular any subgroup Γ of $SL(2,\mathbf{R})$ which is commensurable with $SL(2,\mathbf{Z})$ (that is, $\Gamma \cap SL(2,\mathbf{Z})$ is of finite index in both, Γ and $SL(2,\mathbf{Z})$) is a discrete subgroup of finite covolume, having ∞ as a cusp.

We now prove that $\tilde{f}$ satisfies the estimate in (A4). We first note

$$Im(g \cdot z) = \frac{Im \, z}{|cz + d|^2} \quad g \in G, z \in H \tag{2}$$

hence

$$Im(g \cdot i) = (c^2 + d^2)^{-1} = a(g)^{2\rho}. \tag{3}$$

By Theorem 1 it will be sufficient to show that there exists $C > 0, r > 0$ such that

$$|\tilde{f}(hg)| \leq C \parallel g \parallel^r$$

for $g \in \mathcal{S}_{P,\omega,t}$, if $h = h_i$, $i = 1, 2, ..., r$. Now

$$\tilde{f}(hg) = j(hg, i)^{-k} f(hg, i) = j(g, i)^{-k} f_{|h}(g \cdot i).$$

If $g \in \mathcal{S}_{P,\omega,t}$ then $Im(g.i) \geq t$, hence by holomorphy of f at x

$$|f_{|h}(g \cdot i)| \leq C_h$$

for some $C_h > 0$. Hence by (2) and (3)

$$|\tilde{f}(hg)| \leq C_h a(g)^{k\rho}$$

Finally

$$1 = |ad - bc| \leq (a^2 + b^2)^{\frac{1}{2}} (c^2 + d^2)^{\frac{1}{2}} \leq \parallel g \parallel a(g)^{-\rho}$$

by (3). Thus

$$|\tilde{f}(hg)| \le C_h \parallel g \parallel^k$$

for $g \in S_{p,\omega,t}$, $h = h_i, i = 1, 2, ..., r$. The proof is now complete.

We will now consider the functions $\tilde{E}(s, g)$ associated to the Maass wave forms (cf. Section 2 (2)). We will show they are automorphic forms in the sense of Definition 1. We have, if $Re\, s > 1$,

$$\tilde{E}(s,g) = (Im(g \cdot i))^{\frac{s+1}{2}} \sum_{\Gamma \cap N \backslash \Gamma} |j(\gamma, g \cdot i)|^{-s-1}$$

$$= \sum_{\Gamma \cap N \backslash \Gamma} a(\gamma g)^{(s+1)\rho}$$

in light of (2) and (3). It is clear that $\tilde{E}(s, g)$ verifies (A1) and (A2), wherever it is defined.

More generally, if $m \in \mathbf{Z}$, we consider

$$E_m(s,g) = \sum_{\Gamma \cap N \backslash \Gamma} a(\gamma g)^{(s+1)\rho} \tau_m(k(\gamma g)) \qquad (4)$$

with $\tau_m(u(\theta)) = e^{im\theta}$. Clearly $E_m(s, g)$ is left Γ-invariant and

$$E_m(s, gk) = \tau_m(k) E_m(s, g)$$

if $k \in K$. Also $E_0(s, g) = \tilde{E}(s, g)$. Before showing that $E_m(s, g)$ satisfies (A3) and (A4) we will observe an important fact. Let $m \ge 3$, then

$$E_m(m-1, g) = \sum_{\Gamma \cap N \backslash \Gamma} a(\gamma g)^{m\rho} \tau_m(k(\gamma g))$$

$$= \left(\sum_{\Gamma \cap N \backslash \Gamma} j(\gamma, g \cdot i)^{-m} \right) j(g, i)^{-m} = \tilde{E}_m(g)$$

(see Section 2 (1)). Thus, the holomorphic Eisenstein series are special values of Maass wave forms. This corresponds to the representation theoretic fact that the discrete series embeds in the non-unitary principal series (see below) for special values of the parameter s.

It will be convenient to reinterpret $E_m(s, g)$ via representation theory. If $s \in \mathbf{C}, \varepsilon = 0, 1$ set

$$H^{\varepsilon,s} = \{\, f \in C^\infty(G) | f(na(\pm I)g) = a^{(s+1)\rho}(-1)^\varepsilon f(g), n \in N, a \in A \}$$

$$(\pi_{\varepsilon,s}(x)f)(g) = f(gx), f \in H^{\varepsilon,s}, x, g \in G$$

A direct calculation yields

$$\pi_{\varepsilon,s}(C)f = (\frac{s^2 - 1}{2})f \quad f \in H^{\varepsilon,s} \tag{5}$$

Set $\delta(f) = f(1), f \in H^{\varepsilon,s}$. Then $\delta(\pi_{\varepsilon,s}(n)f) = \delta(f)$ for $n \in N$. If $Re\, s > 1$ then set

$$E_{P,\varepsilon,s}(f) = \sum_{\Gamma \cap N \backslash \Gamma} \delta(\pi_{\varepsilon,s}(\gamma f)).$$

We note that

$$E_{P,\varepsilon,s}(\pi_{\varepsilon,s}(g)f) = \sum_{\Gamma \cap N \backslash \Gamma} a(\gamma g)^{(s+1)\rho} f(k(\gamma g))$$

in particular, if $\varepsilon = (-1)^m$,

$$E_{P,\varepsilon,s}(\pi_{\varepsilon,s}(g)\tau_m) = E_m(s,g)$$

The map $E_{P,\varepsilon,s} : H^{\varepsilon,s} \to C^\infty(\Gamma \backslash G)$ intertwines $\pi_{\varepsilon,s}$ and the right regular action of G on $C^\infty(\Gamma \backslash G)$. Thus (5) implies that $E_m(s,g)$ satisfies (A3).

Also

$$|E_m(s,g)| \le \sum_{\Gamma \cap N \backslash \Gamma} a(\gamma g)^{(Res+1)\rho}$$

$$= \sum_{\Gamma \cap N \backslash \Gamma} a(\gamma)^{(Res+1)\rho} a(k(\gamma)g)^{(Res+1)\rho}$$

$$\le E_0(Res,1) \parallel g \parallel^{Res+1}$$

since $a(g)^\rho \le \parallel g \parallel$. Hence $E_m(s,g)$ is an automorphic form.

Remark. One defines Eisenstein series $E_m(Q,s,g)$ associated to an arbitrary cuspidal parabolic subgroup Q, by the obvious analog of (4). It is shown in exactly the same way as in the case of $E_m(s,g)$ that $E_m(Q,s,g)$ is an automorphic form.

4. Some basic inequalities.

The purpose of this section is to introduce the notion of constant term and to prove some basic estimates which play an important role in the theory. We retain the notation and assumptions of the previous section.

Let x be a cusp for Γ with associated parabolic subgroup

$$P_x = \{\pm I\} A_x N_x.$$

Then $\Gamma \cap N_x \backslash N_x$ is a circle. We normalize the invariant measure on N_x so that $\mathrm{vol}(\Gamma_x \backslash N_x) = 1$. Since ∞ is a cusp, P_x is K-conjugate to P, hence

$A_x \subset \{p \in G | p^t = p, p > 0\}$. We fix $X \in \mathfrak{n}_x$, the Lie algebra of N_x so that $\omega = \exp([0,1]X)$ is a fundamental domain for the action of Γ_x.

Definition 1. Let $f \in C^\infty(\Gamma \backslash G)$. If $\chi \in \hat{N}_x$, $\chi_{|N_x} = 1$, then the (P_x, χ)-Fourier coefficient of f is given by

$$f_{P_x, \chi}(g) = \int_{\Gamma_x \backslash N_x} \chi(n)^{-1} f(ng) dn.$$

If $\chi = 1$, $f_{P_x, \chi} = f_{P_x}$ is called the constant term of f along P_x. If $f_{P_x} = 0$ for any cusp x, f is said to be a cusp form.

If $\Gamma \subset SL(2, \mathbf{Z})$ is a subgroup of finite index such that $h = 1$ then we set $\chi_n(e^{xX}) = e^{2\pi i n x}$. If f is a holomorphic automorphic form then $\tilde{f}_{P_\infty, \chi_n}(1)$ is the coefficient of q^n in the q-expansion of f. We now prove some important estimates due to Langlands.

Theorem 1.

(i) If $f \in C^\infty(\Gamma \backslash G)$ satisfies (A4) and if $f_Q = 0$, for all Q a cuspidal parabolic subgroups, then for any $m \in \mathbf{Z}$ and $t > 0$ there exists $C_{m,t}$ such that

$$|f(g)| \leq C_{m,t} \parallel g \parallel^{-m}$$

for all $g \in S_{Q, \omega, t}$.

(ii) If $f \in C^\infty(\Gamma \backslash G)$ satisfies (A4) and $Cf = \frac{1}{2}(s^2 - 1)f$ then there exists $C_1 > 0$ such that

$$|f(g)| \leq C_1 \parallel g \parallel^{1 + |Re\,s|}, \ Re\,s > 0$$

$$\leq C_1 \|g\| (1 + \log \|g\|), \ Re\,s = 0$$

for any $g \in G$.

Corollary 2. If f is a cusp form, then $Xf \in L^2(\Gamma \backslash G)$ for any $X \in U(\mathfrak{g})$.

Proof. By (i) of the theorem (and reduction theory) Xf is bounded for any $X \in U(\mathfrak{g})$. Since $L^\infty(\Gamma \backslash G) \subset L^2(\Gamma \backslash G)$ the corollary follows.

Proof. (of the theorem) We will estimate $|f(g) - f_Q(g)|$ for $g \in S_{Q, \omega, t}$, a Siegel set. For simplicity of notation we will assume $Q = P$. If $g = nak$, $n \in N$, $n = e^{xX}$, $a \in A$, $k \in K$ then

$$f(nak) - f_P(nak) = f(e^{xX}ak) - \int_0^1 f(e^{(x+t)X}ak) dt$$

$$= -\int_0^1 \int_0^t \frac{d}{ds} f(e^{(x+s)X}ak) \, ds dt$$

$$= \int_0^1 \int_0^t (L_X f)(e^{(x+s)X} ak)\, ds dt$$

$$= -a^{-\alpha} \int_0^1 \int_0^t (R_{Ad(k^{-1})X} f)(e^{(x+s)X} ak)\, ds dt$$

Hence

$$|f(g) - f_P(g)| \leq Ca(g)^{-\alpha} \parallel a(g) \parallel^r \leq C' \parallel g \parallel^{r-2}$$

by (3) in Section 3.

If we replace f by Yf, with $Y \in U(\mathfrak{g})$ then since $f_P = 0$, then the above calculations imply that

$$|Yf(g)| \leq C_Y \parallel g \parallel^{r-2} \tag{1}$$

for $g \in S_{P,\omega,t}$. That is, we have improved the exponent in the estimate in (A.4) by 2. It is now clear that by iteration of the above argument we get the estimate asserted in (i).

To prove (ii) we will first obtain an expression for f_P. Let $H \in \mathfrak{a}$, $X \in \mathfrak{n}$ be as in Section 3. Then $\mathrm{tr}H^2 = 2$, $\mathrm{tr}XY = 1$.

Thus

$$C = \frac{1}{2}H^2 + XY + YX = \frac{1}{2}H^2 + H + 2YX.$$

Since $f_P(ng) = f_P(g)$ for $n \in N$, and

$$L_C f_P = C f_P = \frac{1}{2}(s^2 - 1)f_P$$

it follows that

$$L_{H^2+2H} f_P = (s^2 - 1)f_P.$$

This can be rephrased as

$$((\frac{d}{dt})^2 - 2\frac{d}{dt})f_P(e^{tH}k) = (s^2 - 1)f_P(e^{tH}k).$$

If we solve this simple ordinary differential equation then we have

$$f_P(e^{tH}k) = e^{(s+1)t} f_1(k) + e^{(-s+1)t} f_2(k) \quad (s \neq 0), \tag{2}$$

$$f_P(e^{tH}k) = e^t(t f_1(k) + f_2(k)) \quad (s = 0)$$

where f_1, f_2 are smooth functions on K. This implies that

$$|f_P(g)| \leq C \parallel g \parallel^{1+|Re\, s|}, \text{for all } g \in G.$$

(1) now implies that if $g \in \mathcal{S}_{P,\omega,t}$

$$|f(g)| \leq C' \parallel g \parallel^{1+|Re\,s|}.$$

Since P is arbitrary (ii) follows by reduction theory.

Remark. If $f(gu(\theta)) = e^{il\theta} f(g)$, $l \in \mathbf{Z}$, then in (2) we have

$$f_P(nau(\theta)) = c_+ a^{(s+1)\rho} e^{il\theta} + c_- a^{(-s+1)\rho} e^{il\theta}.$$

for some constants c_+ and c_-.

5. The decomposition of $L^2(\Gamma\backslash G.)$

We now look at methods of analyzing the spectral decomposition of $L^2(\Gamma\backslash G)$. We first explain how to extend the ideas from $SL(2,\mathbf{R})$ to more general groups. The main reference for this section will be [L] (see also [HC]). For the sake of simplicity we confine our attention to arithmetic groups.

If V is a finite dimensional vector space over $\mathbf{C}$, let $\mathcal{P}(V)$ denote the algebra of polynomial functions on V. A subgroup $\mathbf{G}$ of $GL(V)$ is said to be an affine algebraic group if there exist $f_1, ..., f_d \in \mathcal{P}(End(V))$ such that $\mathbf{G} = \{g \in GL(V) \mid f_i(g) = 0\}$.

Examples:

1. $\mathbf{G} = GL(V)$ $(d = 1, f_1 = 0)$.

2. $\mathbf{G} = SL(V)$ $(d = 1, f_1(g) = \det g - 1)$.

3. Let $(,)$ be a symmetric, non-degenerate form on V and let

$$\mathbf{G} = O(V,(,)) = \{g \in GL(V) \mid (gv, gw) = (v, w)\ v, w \in V\}.$$

If $\{v_i \mid i = 1, 2, ..., r\}$ is a basis of V, $O(V)$ is the zero set of the polynomials $f_{i,j}(g) = (gv_i, gv_j) - (v_i, v_j)$ $1 \leq i, j \leq r$.

4. Let ω be a non-degenerate alternating form on V and let

$$Sp(V, \omega) = \{g \in GL(V) \mid \omega(gv, gw) = \omega(v, w),\ v, w \in V\}.$$

If $\mathbf{G} \subset GL(V)$ is an affine algebraic group, we look at $\mathbf{G}$ as an affine algebraic set by embedding $\mathbf{G} \to End(V) \times \mathbf{C}$ via $g \to (g, \det g^{-1})$. The equations defining the image are

$$\tilde{f}_i(g, z) = f_i(g) = 0 \quad 1 \leq i \leq d$$

$$z \det g - 1 = 0$$

Let

$$I(\mathbf{G}) = \{ f \in \mathcal{P}(End(V) \times \mathbf{C}) | f_{|\mathbf{G}} = 0 \}.$$

If $F \subset \mathbf{C}$ is a subfield, an F-vector subspace V_F of V is said to be an F-form of V if any F-basis of V_F is a basis of V over $\mathbf{C}$. Set

$$I_F(\mathbf{G}) = \{ f \in I(\mathbf{G}) | f(End_F(V_F) \times F) \subset F \}.$$

We say that $\mathbf{G}$ is an F-group (or that $\mathbf{G}$ is defined over F if

$$I(\mathbf{G}) = \mathcal{P}(End_{\mathbf{C}}(V) \times \mathbf{C}) I_F(\mathbf{G}).$$

From now on we fix $V_{\mathbf{Q}}$, a $\mathbf{Q}$-structure on V, and we let $\mathbf{G} \subset GL(V)$ be an affine algebraic group defined over $\mathbf{Q}$. We set

$$V_{\mathbf{R}} = V_{\mathbf{Q}} \otimes_{\mathbf{Q}} \mathbf{R} \quad \mathbf{G}_{\mathbf{Q}} = \mathbf{G} \cap GL(V_{\mathbf{Q}}), \quad \mathbf{G}_{\mathbf{R}} = \mathbf{G} \cap GL(V_{\mathbf{R}}).$$

If $V_{\mathbf{Z}} \subset V_{\mathbf{Q}}$ is a $\mathbf{Z}$-form let

$$\mathbf{G}_{\mathbf{Z}} = \mathbf{G} \cap GL(V_{\mathbf{Z}}) = \{ g \in \mathbf{G} \mid g V_{\mathbf{Z}} = V_{\mathbf{Z}} \}.$$

If $\mathbf{H} \subset \mathbf{G}$ is a subgroup which is a $\mathbf{Q}$-group corresponding to the same choice of $V_{\mathbf{Q}}$, $\mathbf{H}$ is called a $\mathbf{Q}$-subgroup of $\mathbf{G}$. An algebraic group $\mathbf{A}$ is said to be a $\mathbf{Q}$-torus if $\mathbf{A}$ is defined over $\mathbf{Q}$ and $\mathbf{A}$ is isomorphic to $GL(1)^r$. If this isomorphism is defined over $\mathbf{Q}$, $\mathbf{A}$ is said to be a $\mathbf{Q}$-split torus.

Subgroups Γ_1, Γ_2 of $\mathbf{G}_{\mathbf{R}}$ are said to be commensurable if $\Gamma_1 \cap \Gamma_2$ is of finite index in both Γ_1 and Γ_2. An arithmetic subgroup of $\mathbf{G}_{\mathbf{R}}$ is a subgroup of $\mathbf{G}_{\mathbf{Q}}$ which is commensurable to $\mathbf{G}_{\mathbf{Z}}$ for some (hence for any) $\mathbf{Z}$-form $V_{\mathbf{Z}}$ of $V_{\mathbf{Q}}$.

We set

$$\mathcal{X}(\mathbf{G}) = \{ \varphi : \mathbf{G} \to GL(1) | \varphi \text{ an algebraic group morphism} \},$$

$$\mathcal{X}_{\mathbf{Q}}(\mathbf{G}) = \{ \varphi \in \mathcal{X}(\mathbf{G}) | \varphi(\mathbf{G}_{\mathbf{Q}}) \subset GL(1, \mathbf{Q}) \}$$

and

$${}^{\circ}\mathbf{G} = \{ g \in \mathbf{G} | \varphi(g)^2 = 1, \text{ for all } \varphi \in \mathcal{X}_{\mathbf{Q}}(\mathbf{G}) \}.$$

An algebraic group $\mathbf{G}$ is said to be reductive if every algebraic representation ρ of $\mathbf{G}$ is completely reducible.

We now recall two fundamental results on discrete subgroups:

Theorem 1. ([B-HC]) *If* $\mathbf{G}$ *is a reductive* $\mathbf{Q}$ *-algebraic group and* $\Gamma \subset \mathbf{G}_{\mathbf{R}}$ *is arithmetic then* $\Gamma \backslash \mathbf{G}_{\mathbf{R}}$ *has finite invariant volume if and only if* $\mathcal{X}_{\mathbf{Q}}(\mathbf{G}^{\circ}) = \{1\}$ *(here* $\mathbf{G}^{\circ}$ *is the identity component of* $\mathbf{G}$*).*

Theorem 2. (Margulis, cf [Z], Ch 6). *Let* $\mathbf{G} \subset GL(V)$ *be a connected, semisimple, center-free algebraic group defined over* $\mathbf{R}$, *with* $\mathbf{R}$-*rank* $\mathbf{G} > 1$. *Let* $V_{\mathbf{R}}$ *be an* $\mathbf{R}$-*form of* V. *If* $\Gamma \subset \mathbf{G_R}$ *is a discrete subgroup of finite co-volume, then* Γ *is arithmetic, that is,* Γ *is commensurable to* $G_{\mathbf{Z}}$ *for some* $\mathbf{Q}$-*form* $V_{\mathbf{Q}}$ *of* $V_{\mathbf{R}}$.

Remark. If $\mathbf{G}$ has $\mathbf{R}$-rank 1, it is known that non-arithmetic discrete subgroups of finite covolume exist: If $\mathbf{G_R} \simeq SO(n,1)$ for $n \geq 2$ (Vinberg, Makarov, Gromov- Piatetski-Shapiro) and if $\mathbf{G_R} \simeq SU(2,1)$ (Mostow).

For the rest of this exposition we will assume that

(1) $\mathbf{G}$ is reductive and defined over $\mathbf{Q}$.

(2) $\mathbf{G} = {}^{o}\mathbf{G}$.

If $\mathbf{G}$ is an affine algebraic group satisfying (1) and (2), let $\mathbf{P}$ be a parabolic subgroup defined over $\mathbf{Q}$ (cf [B],[Z]). Then $\mathbf{P} = \mathbf{MN}$ where $\mathbf{M}$, $\mathbf{N}$ are $\mathbf{Q}$-subgroups of $\mathbf{P}$, $\mathbf{N}$ is the unipotent radical of $\mathbf{P}$ and $\mathbf{M}$ is reductive. Furthermore $\mathbf{M} = {}^{o}\mathbf{M}\mathbf{A}$ where $\mathbf{A}$ is a $\mathbf{Q}$-split torus of $\mathbf{M}$.

We will denote by G, P, M, N, ${}^{o}M$, and A the real points of $\mathbf{G}$, $\mathbf{P}$, $\mathbf{M}$, $\mathbf{N}$, ${}^{o}\mathbf{M}$, and $\mathbf{A}$ respectively. We have that $P = {}^{o}PA$ where ${}^{o}P = {}^{o}MN$. P will be called a cuspidal parabolic subgroup of G. Fix K a maximal compact subgroup of G. Then $G = PK$ and $P \cap K = {}^{o}P \cap K$. If $\pi : P \to P/N \simeq M$ is the canonical projection we set $K_M = \pi(K \cap {}^{o}P)$. Let Γ be an arithmetic subgroup of G. We set $\Gamma_M = \pi(\Gamma \cap {}^{o}P) \simeq \Gamma \cap {}^{o}P/\Gamma \cap N$; then Γ_M is an arithmetic subgroup of M. We also note that $\Gamma \cap N \backslash N$ is compact. If $\mathbf{P}$ is minimal over $\mathbf{Q}$ then $\Gamma \cap {}^{o}P \backslash {}^{o}P$ is compact.

Examples.

1.
$$\mathbf{G} = SL(2, \mathbf{C}), \quad V = \mathbf{C}^2, \; V_{\mathbf{R}} = \mathbf{R}^2, \; V_{\mathbf{Q}} = \mathbf{Q}^2.$$

Then $\mathbf{G_R} = SL(2, \mathbf{R})$, $\mathbf{G_Q} = SL(2, \mathbf{Q})$, and Γ is arithmetic if it is commensurable with $SL(2, \mathbf{Z})$. The Γ-cuspidal parabolic subgroups are exactly the subgroups of the form $P = \mathbf{P_R}$, where $\mathbf{P}$ is a parabolic $\mathbf{Q}$-subgroup of $\mathbf{G}$.

2. Let $F = \mathbf{Q}(\sqrt{2})$, $\mathbf{G} = R_{F|Q}(\mathbf{H})$ where $\mathbf{H} = SL(2, \mathbf{C})$ and $R_{F|Q}$ denotes restriction of scalars (cf. [Z]). Then, in this example, the formalism yields:

$$\mathbf{G} = SL(2, \mathbf{C}) \times SL(2, \mathbf{C}), \; \mathbf{G_R} = SL(2, \mathbf{R}) \times SL(2, \mathbf{R}).$$

Let $\sigma : \mathbf{Q}(\sqrt{2}) \to \mathbf{Q}(\sqrt{2})$ be the automorphism defined by $\sigma(\sqrt{2}) = -\sqrt{2}$. Define

$$\sigma\left(\begin{bmatrix} a & b \\ c & d \end{bmatrix}\right) = \begin{bmatrix} \sigma(a) & \sigma(b) \\ \sigma(c) & \sigma(d) \end{bmatrix} \quad \text{if} \begin{bmatrix} a & b \\ c & d \end{bmatrix} \in SL(2, \mathbf{Q}(\sqrt{2})).$$

Then

$$G_{\mathbf{Q}} = \{(g, \sigma(g)) \mid g \in SL(2, \mathbf{Q}(\sqrt{2}))\}.$$

$$G_{\mathbf{Z}} = \{(g, \sigma(g)) \mid g \in SL(2, \mathbf{Z}(\sqrt{2}))\}.$$

If $\mathbf{P}_1 = \left\{ \begin{bmatrix} a & b \\ 0 & a^{-1} \end{bmatrix} \mid a, b \in \mathbf{C}, a \neq 0 \right\}$ then $\mathbf{P} = R_{F|\mathbf{Q}}(\mathbf{P}_1)$ is a parabolic subgroup of $\mathbf{G}$ defined over $\mathbf{Q}$ and $P = P_1 \times P_1$.

Furthermore $P = {}^{\circ}MA$ with

$$A = \left\{ \left(\begin{bmatrix} \lambda & 0 \\ 0 & \lambda^{-1} \end{bmatrix}, \begin{bmatrix} \lambda & 0 \\ 0 & \lambda^{-1} \end{bmatrix} \right) \mid \lambda > 0 \right\}$$

and

$$ {}^{\circ}M = \left\{ \left(\begin{bmatrix} \pm\lambda & 0 \\ 0 & \pm\lambda^{-1} \end{bmatrix}, \begin{bmatrix} \lambda^{-1} & 0 \\ 0 & \lambda \end{bmatrix} \right) \mid \lambda \in \mathbf{R}^* \right\}$$

Let $P = {}^{\circ}MAN$ be a cuspidal parabolic, and let $\mathfrak{n}$ be the Lie algebra of N. Set $\Phi = \Phi(P, A)$, the set of roots of A on $\mathfrak{n}$. If $\alpha \in \Phi$ then we denote by $\mathfrak{n}_\alpha$ the associated root space. We have $\mathfrak{n} = \oplus \sum_{\alpha \in \Phi} \mathfrak{n}_\alpha$. Set, for $t > 0$,

$$A_t^+ = \{a \in A \mid a^\alpha \geq t \quad \alpha \in \Phi(P, A)\}.$$

We put a symmetric, non-degenerate, bilinear form, B, on $\mathfrak{g}$ such that $B_{|\mathfrak{k}}$ is negative definite (such exist). Then it can be shown that B is positive definite on $\mathfrak{a}$. We set $(\,,\,)$ equal to the dual form on $\mathfrak{a}^*$. A Siegel set for P is a set of the form $\mathcal{S}_{P,\omega,t} = \omega A_t^+ K$ where $\omega \subset N{}^{\circ}M$ is compact, and $t > 0$.

The reduction theory in the present context is completely analogous to the case of $G = SL(2, \mathbf{R})$. As before, there are only a finite number of minimal cuspidal parabolic subgroups of G. If $P_1, ..., P_r$ is a complete set of representatives then there exist compact sets $\omega_i \subset {}^{\circ}P_i$, $t_i > 0$ ($1 \leq i \leq r$) such that $G = \cup_i \Gamma \mathcal{S}_{P_i, \omega_i, t_i}$ (see [B-HC]).

Fix a K-invariant inner product on V and set $\| g \| = \| g \|_{HS}$, for $g \in G$. Then with an appropriate choice of inner product (a so called admissible one) one has

$$\| g \| = \| g^{-1} \|, \quad g \in G,$$

$$\| g_1 g_2 \| \leq \| g_1 \| \, \| g_2 \|, \quad g_1, g_2 \in G,$$

$$\| k_1 g k_2 \| = \| g \| \quad \text{for } g \in G, \, k_1, k_2 \in K.$$

We say that $f \in C^\infty(G)$ is a Γ-automorphic if (as in the case of $SL(2, \mathbf{R})$) the following four conditions are statisfied

(A1) $f(\gamma g) = f(g)$ $\gamma \in \Gamma, g \in G$,

(A2) dim (span $R_K f$) $< \infty$,

(A3) dim span $(Z(\mathfrak{g})f) < \infty$ ($Z(\mathfrak{g})$ the center of $U(\mathfrak{g})$, the space of left invariant differential operators on G).

(A4) there exists $r > 0$ such that, if $X \in U(\mathfrak{g}), g \in G$, then

$$|R_X f(g)| \leq C_X \| g \|^r .$$

Condition (A4) is usually called the condition of uniform moderate growth. The space of Γ-automorphic forms will be denoted $\mathcal{A}(\Gamma \backslash G)$. Then $\mathcal{A}(\Gamma \backslash G)$ has a natural structure of a $\mathfrak{g}$-module, $Xf = R_X f$ for $X \in G$ and a natural structure of a K-module, $kf = R_k f$ for $k \in K$. These two structures are compatible (that is, $kXf = Ad(k)Xkf$, and the action of $\mathfrak{k}$ on f is given by differentiating the action of K on f). Furthermore, condition (A2) implies that if $f \in \mathcal{A}(\Gamma \backslash G)$ then f is K-finite (that is dim(span Kf) $< \infty$). More generally, if V is a $\mathfrak{g}$-module and a K-module such that if $v \in V$ then v is K-finite and the actions of K and $\mathfrak{g}$ are compatible then V is called a $(\mathfrak{g}, K)$-module.

We can now define the notion of Eisenstein series in the present context. We note that if $P = {}^\circ MAN$ is a cuspidal parabolic, then the pair $({}^\circ M, \Gamma_M)$ is in the same class as (G, Γ), that is ${}^\circ M =^\circ \mathbf{M}_R$ where ${}^\circ \mathbf{M}$ is a Q-algebraic group satisfying (1),(2) and Γ_M is arithmetic in ${}^\circ M$. For simplicity we will assume from now on that $K \cap {}^\circ P = K \cap {}^\circ M$. Then $K_M = \pi(K \cap {}^\circ P) \simeq K \cap {}^\circ M$.

Let $\varphi_o : K \to L^2(\Gamma_M \backslash {}^\circ M) \cap \mathcal{A}(\Gamma_M \backslash {}^\circ M)$ be continuous and such that

$$\varphi_o(mk) = R_m \varphi_o(k)$$

for $m \in K \cap M$. Set

$$\varphi(namk) = \varphi_o(k)(m) \tag{1}$$

for $n \in N$, $a \in A$, $m \in {}^\circ M$, $k \in K$. It is easy to check that φ is well defined and lies in $C((\Gamma \cap P)NA \backslash G)$. We define

$$E_{P,\nu}(\varphi_0)(g) = \sum_{\Gamma \cap P \backslash \Gamma} a(\gamma g)^{\nu + \rho} \varphi(\gamma g).$$

This series converges uniformly on compacta in $\{\nu \in \mathfrak{a}_\mathbb{C}^* \mid Re(\nu, \alpha) > (\rho, \alpha), \ \alpha \in \Phi(P, A)\}$ (here $\rho = \frac{1}{2} \sum_{\alpha \in \Phi} \alpha$). Furthermore, if φ_o is K-finite relative to the right regular action then $E_{P,\nu}(\varphi_o) \in \mathcal{A}(\Gamma \backslash G)$ in this region. The following is a fundamental theorem of Langlands generalizing the corresponding theorem of Selberg in the case of and arithmetic subgroup of $SL(2, \mathbf{R})$.

Theorem 1. ([L, Ch 7]) $\nu \mapsto E_{P,\nu}(\varphi)$ *has a meromorphic continuation to* $\mathfrak{a}_{\mathbb{C}}^*$, *which is holomorphic on* $i\mathfrak{a}^*$.

Fix $H \subset L^2(\Gamma_M \backslash {}^\circ M)$ a closed, ${}^\circ M$-invariant irreducible subspace. Then $H_{K \cap M} \subset \mathcal{A}(\Gamma_M \backslash {}^\circ M)$. Set $\sigma(m) = R_{m|H}, m \in {}^\circ M$. Consider for $\nu \in \mathfrak{a}_{\mathbb{C}}^*$ the space of functions of the form

$$\varphi_\nu(namk) = a^{\nu+\rho}\varphi(namk)$$

where φ is defined by **(1)**, with φ_o right K-finite and $\varphi_o(K) \subset H$. This space defines a $(\mathfrak{g}, K)$-module which is equivalent with the generalized principal series $I_{P,\sigma,\nu}$. Furthermore

$$E_{P,\nu} : I_{P,\sigma,\nu} \to \mathcal{A}(\Gamma \backslash G)$$

is a $(\mathfrak{g}, K)$-homomorphism.

If $\alpha \in C_c^\infty(\mathfrak{a}^*)$ then we set

$$E_P(\varphi_o, \alpha)(g) = \int_{\mathfrak{a}^*} \alpha(\nu) E_{P,\nu}(\varphi)(g) d\nu.$$

Then $E_P(\varphi_o, \alpha) \in L^2(\Gamma \backslash G)$ and

$$\| E_P(\varphi_o, \alpha) \|^2 = c_j^{-2} \Big(\int_{K \times \Gamma_M \backslash {}^\circ M} \| \varphi_0(m)(k) \|^2 dm dk \Big) \| \alpha \|_2^2$$

where c_j depends only on the choices of invariant measures ([L]).

Let $L^2(\Gamma \backslash G)_{cont}$ be the closure of the space spanned by functions of the form $E_P(\varphi_o, \alpha)$ with P running over all cuspidal parabolic subgroups . Let $L^2(\Gamma \backslash G)_{disc}$ be the closure of the span of the closed irreducible subspaces of $L^2(\Gamma \backslash G)$. We have

Theorem 2.

$$L^2(\Gamma \backslash G) = L^2(\Gamma \backslash G)_{disc} \oplus L^2(\Gamma \backslash G)_{cont}$$

a Hilbert space direct sum.

Following Langalands we now give a refinement of this decomposition. We say that two cuspidal parabolic subgroups $P = M_P N_P$, $Q = M_Q N_Q$ are associate if there exists $g \in \mathbf{G}_Q$ so that gQg^{-1} and P contain the same minimal parabolic subgroup P_o and such that $M_{gQg^{-1}} = M_P$ relative to some Langlands decomposition of gQg^{-1} and P. Fix $\mathcal{P}$ a complete set of representatives of the association classes of cuspidal parabolic subgroups. If $P \in \mathcal{P}$, let $[P]$ denote the equivalence class of P.

Theorem 3. ([L]) *If P and Q are cuspidal and non associate then*

$$\langle E_P(\varphi_o, \alpha), E_Q(\psi_o, \beta) \rangle = 0$$

for all $\varphi_o, \alpha, \psi_o, \beta$. If $P \in \mathcal{P}$ is fixed, let $L^2_{[P]}(\Gamma\backslash G)$ denote the closure of the space spanned by functions of the form $E_P(\varphi_0, \alpha)$. Then

$$L^2(\Gamma\backslash G) = \oplus_{P \in \mathcal{P}} L^2_{[P]}(\Gamma\backslash G)$$

an orthogonal direct sum.

We note that $[G]$ consists only of G and $L^2_{[G]}(\Gamma\backslash G) = L^2(\Gamma\backslash G)_{disc}$. Langlands' theorem thus reduces the decomposition of $L^2(\Gamma\backslash G)$ to the decomposition of $L^2(\Gamma\backslash G)_{disc}$.

If $P = M_P N_P$ is a cuspidal parabolic subgroup and if $f \in \mathcal{A}(\Gamma\backslash G)$ defines the constant term of f along P by

$$f_P(g) = \int_{\Gamma \cap N_P\backslash N_P} f(ng)dn$$

(dn is normalized so that $\text{vol}(\Gamma \cap N_P\backslash N_P) = 1$).

If $f_P = 0$ for any $P \neq G$, f is called a cusp form. If f is a cusp form, then the argument given for $SL(2, \mathbf{R})$ (cf. Theorem 2 (i)) has a generalization to the case at hand to prove that f is rapidly decreasing on every Siegel set This implies that $f \in L^2(\Gamma\backslash G)$. Let $L^2(\Gamma\backslash G)_{cusp}$ denote the closure of span $\{f \in \mathcal{A}(\Gamma\backslash G) \mid \text{f is a cusp form}\}$. Let

$$L^2(\Gamma\backslash G)_{res} = L^2(\Gamma\backslash G)^{\perp}_{cusp} \cap L^2(\Gamma\backslash G)_{disc}.$$

Langlands has shown that $L^2(\Gamma\backslash G)_{res}$ is the closure of the span of certain residues of Eisenstein series. It is worth pointing out that Langlands' theorem preceded Harish-Chandra's decomposition of $L^2(G)$. Under the influence of Langlands' result Harish-Chandra developed his theory of the constant term for G which was central to his proof of the Plancherel theorem for G. Langlands' theorems are difficult (Chapter 7 in [L] has become legendary). They are, however, a first step in the study of $L^2(\Gamma\backslash G)$. The next step is the development of a tractible trace formula. If $f \in C_c(G)$ then we define an operator, $\pi_{disc,\Gamma}(f) = \pi_\Gamma$, on $L^2(\Gamma\backslash G)_{disc}$ by

$$\pi_\Gamma(f)\varphi(x) = \int_G f(g)\varphi(xg)dg.$$

Recently, Müller ([M]) has shown that if $f \in C^\infty_c(G)$ is K-finite (on the right or the left) then $\pi_\Gamma(f)$ is trace class. For $SL(2, \mathbf{R})$ this result is not terribly difficult since for a given character of K the space of functions in $L^2(\Gamma\backslash G)_{res}$ that transform on the right by that character is finite dimensional. The analogous operator on the cuspidal spectrum was proved to be trace class by relatively elementary methods in [L]. The natural problem is: Calculate the trace of $\pi_\Gamma(f)$. This is extremely difficult and the main progress is due to Arthur (for a survey see [A]), who works in the setting of congruence subgroups (i.e. adelically) and (interestingly) independently of the trace class theorem, and to Osborne and Warner [OW] who work in the general setting but have always taken the trace class theorem as an assumption.

Another important question involves the generalization of the classical constructions of automorphic forms. For the past several years R. Miatello and I have been working on methods of constructing all square integrable forms as residues of meromorphic families of functions on $\Gamma\backslash G$. If G has real rank 1 and $\Gamma\backslash G$ is non-compact we have introduced an analogue of the non-holomorphic Poincaré series which solve the problem. These correspond to Whittaker vectors in the same way that Eisenstein series correspond to conical vectors. More recently we have developed a method using Harish-Chandra's expansion of spherical functions that also applies to the cocompact case.

References

[A] Arthur, J., Eisenstein series and the trace formula, Proc. Symp. in Pure Math. 33 Part I (1979), 253-274.

[B] Borel A., *Introduction aux groupes arithmétiques*, Hermann, Paris, 1969.

[B-HC] Borel A.-Harish Chandra, Arithmetic subgroups of algebraic groups, Annals of Math. (2), 75 (1962), 485-535.

[HC] Harish-Chandra, *Automorphic forms on semisimple Lie groups*, Lecture Notes in Math. 62, Springer Verlag, 1968.

[Kn] Knopp M., *Modular forms and number theory*, Markham Publishing Co., Chicago, 1970.

[L] Langlands R., *On the Functional equations satisfied by Eisenstein series,* Lecture Notes in Math. 544, Springer Verlag, 1976.

[Le] Lehner J., *Discontinuous groups and automorphic functions*, Math. Surveys in Math. V.8, A.M.S, Providence, R.I.

[MW] Miatello R.J, Wallach N.R, Automorphic forms constructed from Whittaker vectors, J. of Func. Anal., 86 (1989), 411-487.

[M] Müller, W., The trace class conjecture in the theory of automorphic forms, Annals of Math. (2), 130 (1989), 473-529.

[OW] Osborne, M.S., Warner, G., *The theory of Eisenstein systems*, Academic Press,1981,

[Z] Zimmer R., *Ergodic theory and semisimple groups*, Birkhäuser, 1984

Dept. of Mathematics
University of California at San Diego,
La Jolla, CA 92093

Analytic and Geometric Realization
of Representations

LECTURES BY WILFRIED SCHMID

NOTES BY JORGE VARGAS

Hyperfunctions. Let $\Delta = \{z \in \mathbb{C} : |z| < 1\}$, $S^1 = \partial\Delta$. For u in $C(S^1)$, let Pu denote the Poisson transform of u:

$$P_u(re^{i\theta}) = \frac{1}{2\pi} \int_0^{2\pi} \frac{1-r^2}{1+r^2 - 2r\cos(\theta - \varphi)} u(e^{i\varphi}) d\varphi.$$

Then Pu is harmonic on Δ, and Pu converges radially to u.

Problem: What is the largest natural domain of P? If $u = \sum_n a_n e^{in\theta}$, then $Pu = \sum_n a_n r^{|n|} e^{in\theta}$. The series Pu must converge for $r < 1$! This is equivalent to $\forall r < 1$, $\exists C = C(r)$ such that $|a_n| \le Cr^{-|n|}$.

What does this condition mean?

Let $C^\omega(S^1)$ be the space of real analytic functions on S^1; $\sum b_n e^{in\theta} \in C^\omega(S^1)$ if and only if $\sum b_n z^n$ converges on some anulus $\{z : r < |z| < r^{-1}\}$ ($r < 1$), or equivalently, if and only if $\exists r < 1$, $D = D(r)$ such that $|b_n| \le Dr^{|n|}$. We use the constants r, D to put an inductive limit topology on $C^\omega(S^1)$. Let $C^{-\omega}(S^1)$ be the strong dual of $C^\omega(S^1)$:

$$C^{-\omega}(S^1) = \{\sum a_n e^{in\theta} : |a_n| \le C(r)r^{-|n|} \quad \forall r < 1\}.$$

The pairing is given by "integration":

$$\langle \sum b_n e^{in\theta}, \sum a_n e^{in\theta} \rangle = \sum_n b_n a_{-n}.$$

We give to the right hand side the topology defined by the seminorms determined by $C(r)$. We have seen that $P : C^{-\omega}(S^1) \cong$ space of harmonic functions on Δ. (In fact, this is a topological isomorphism). We have the inclusions

$$C^\omega(S^1) \subset C^\infty(S^1) \subset L^2(S^1) \subset C^{-\infty}(S^1) \subset C^{-\omega}(S^1).$$

$C^{-\omega}(S^1)$ is called the space of hyperfunctions on S^1. Loosely speaking, hyperfunction is the most general notion of function. To justify the terminology, we must have the notions of restriction and patching of hyperfunction. Recall that there do not exist real analytic partitions of 1!

Supported in part by a John Simon Guggenheim Memorial Fellowship and NSF Grant DMS-87-01578

Let $\mathbb{P}^1 = \mathbb{C} \cup \{\infty\}$. We can re-phrase the definition of $C^{-\omega}(S^1)$ as follows:

$$C^{-\omega}(S^1) = \{ \sum_{n \text{ integer}} a_n e^{in\theta} : |a_n| < \ldots \} =$$

$$= \{\sum_{n \geq 0} a_n e^{in\theta} : |a_n| < \ldots\} \oplus \{\sum_{n \geq 0} a_{-n} e^{-in\theta} : |a_{-n}| < \ldots\}/\mathbb{C} =$$

$$= \mathcal{O}(\Delta) \oplus \mathcal{O}(\mathbb{P}^1 - \bar{\Delta})/\mathbb{C} = \mathcal{O}(\mathbb{P}^1 - S^1)/\mathcal{O}(\mathbb{P}^1) \cong \mathcal{O}(\mathcal{U} - S^1)/\mathcal{O}(\mathcal{U})$$

for any neighborhood $\mathcal{U}$ of S^1 in $\mathbb{C}$.

With this as motivation, we now define hyperfunctions on $\mathbb{R}$. Fix $\mathcal{U} \subset \mathbb{R}$ an open subset. Let $\tilde{\mathcal{U}} \subset \mathbb{C}$ be open, with $\tilde{\mathcal{U}} \cap \mathbb{R} = \mathcal{U}$.

Lemma. Suppose $\tilde{\tilde{\mathcal{U}}} \subset \mathbb{C}$ is open and $\tilde{\tilde{\mathcal{U}}} \cap \mathbb{R} = \mathcal{U}$. Then $\mathcal{O}(\tilde{\mathcal{U}} - \mathcal{U})/\mathcal{O}(\tilde{\mathcal{U}}) = \mathcal{O}(\tilde{\tilde{\mathcal{U}}} - \mathcal{U})/\mathcal{O}(\tilde{\tilde{\mathcal{U}}})$. Furthermore, the isomorphism is natural.

Proof. We may as well assume $\tilde{\tilde{\mathcal{U}}} \supset \tilde{\mathcal{U}}$. The Mayer-Vietoris exact sequence implies

$$0 \to H^\circ(\tilde{\tilde{\mathcal{U}}}, \mathcal{O}) \to H^\circ(\tilde{\tilde{\mathcal{U}}} - \mathcal{U}, \mathcal{O}) \oplus H^\circ(\tilde{\mathcal{U}}, \mathcal{O}) \to$$

$$\to H^\circ(\tilde{\mathcal{U}} - \mathcal{U}, \mathcal{O}) \to H^1(\tilde{\tilde{\mathcal{U}}}, \mathcal{O}) \to \ldots$$

Now, any open set in $\mathbb{C}$ is Stein, hence $H^1(\tilde{\tilde{\mathcal{U}}}, \mathcal{O}) = 0$. The lemma follows.

Definition. $C^{-\omega}(\mathcal{U}) = \mathcal{O}(\tilde{\mathcal{U}} - \mathcal{U})/\mathcal{O}(\tilde{\mathcal{U}})$.

If $\mathcal{U}_1 \subset \mathcal{U}_2 \subset \mathbb{R}$ are open, $\tilde{\mathcal{U}}_1 \subset \tilde{\mathcal{U}}_2 \subset \mathbb{C}$ open, and $\tilde{\mathcal{U}}_i \cap \mathbb{R} = \mathcal{U}_i$, then one has a natural map $C^{-\omega}(\mathcal{U}_2) = \mathcal{O}(\tilde{\mathcal{U}}_2 - \mathcal{U}_2)/\mathcal{O}(\tilde{\mathcal{U}}_2) \longrightarrow \mathcal{O}(\tilde{\mathcal{U}}_1 - \mathcal{U}_1)/\mathcal{O}(\tilde{\mathcal{U}}_1) = C^{-\omega}(\mathcal{U}_1)$, therefore, we have natural "restriction maps".

Lemma. $C^{-\omega}$ is a sheaf. That is, if $\mathcal{U} \subset \mathbb{R}$ is open, $\mathcal{U} = \bigcup_{\alpha \in A} \mathcal{U}_\alpha$ is an open cover and if for each $\alpha \in A$, $f_\alpha \in C^{-\omega}(\mathcal{U}_\alpha)$ is given such that $f_\alpha|_{\mathcal{U}_\alpha \cap \mathcal{U}_\beta} = f_\beta|_{\mathcal{U}_\alpha \cap \mathcal{U}_\beta}$ for $\alpha, \beta \in A$, then there exists a unique $f \in C^{-\omega}(\mathcal{U})$ such that $f|_{\mathcal{U}_\alpha} = f_\alpha$, for each $\alpha \in A$.

Proof. Similar to that of the previous lemma.

Observation. Hyperfunctions can be differentiated:

$$\mathcal{O}(\tilde{\mathcal{U}} - \mathcal{U})/\mathcal{O}(\tilde{\mathcal{U}}) \xrightarrow{d/dz} \mathcal{O}(\tilde{\mathcal{U}} - \mathcal{U})/\mathcal{O}(\tilde{\mathcal{U}});$$

we call this operation "$\frac{d}{dx}$".

Observation. Hyperfunctions can be multiplied by real analytic functions: because any $h \in C^\omega(\mathcal{U})$ extends to $\tilde{h} \in \mathcal{O}(\tilde{\mathcal{U}})$ for some $\tilde{\mathcal{U}} \in \mathbb{C}$ open, such that $\tilde{\mathcal{U}} \cap \mathbb{R} = \mathcal{U}$. Thus multiplication by $\tilde{h}$ yields a map $m(\tilde{h})$

$$\mathcal{O}(\tilde{\mathcal{U}} - \mathcal{U})/\mathcal{O}(\tilde{\mathcal{U}}) \xrightarrow{m(\tilde{h})} \mathcal{O}(\tilde{\mathcal{U}} - \mathcal{U})/\mathcal{O}(\tilde{\mathcal{U}});$$

we call this map "multiplication by h"! So, $C^{-\omega}(\mathcal{U})$ is a module over the ring of linear differential operators with C^ω coefficients.

Observation. If $\mathcal{U}_1 \subset \mathcal{U}_2 \subset \mathbb{R}$ are open, $C^{-\omega}(\mathcal{U}_2) \to C^{-\omega}(\mathcal{U}_1)$ is surjective. In other words, $C^{-\omega}$ is a flabby sheaf!

Proof. Look at the first lemma, set $\tilde{\tilde{\mathcal{U}}} = \mathbb{C} - (\mathbb{R} - \mathcal{U})$; then $\mathbb{C} - \mathbb{R} = \tilde{\tilde{\mathcal{U}}} - \mathcal{U}$, so $C^{-\omega}(\mathbb{R}) \longrightarrow C^{-\omega}(\mathcal{U})$ is surjective.

Suppose $K \subset \mathbb{R}$ is compact, $\tilde{\mathcal{U}}$ a neighborhood of K in $\mathbb{C}$. Let $C^{-\omega}(K)$ denote the space of sections of the sheaf of hyperfunctions over K.

Lemma. $C^{-\omega}(K) \cong \mathcal{O}(\tilde{\mathcal{U}} - K)/\mathcal{O}(\tilde{\mathcal{U}})$.

Proof. Apply Mayer-Vietoris and the flabbiness of $C^{-\omega}$.

Corollary. If $K \subset \mathbb{R}$ is compact, $C^{-\omega}(K) = C^{\omega}(K)'$ (= strong dual of the space of sections of the sheaf of real analytic functions over K).

Corollary. If $\mathcal{U} \subset \mathbb{R}$ is an open, relatively compact set, then $C^{\omega}(\mathcal{U}) = C^{-\omega}(\bar{\mathcal{U}})/ C^{-\omega}(\partial\mathcal{U}) = C^{\omega}(\bar{\mathcal{U}})'/C^{\omega}(\partial\mathcal{U})'$.

Consider an open subset $\mathcal{U} \subset \mathbb{R}$, $\tilde{\mathcal{U}} \subset \mathbb{C}$ as before. We may consider the local cohomology along $\mathcal{U}$, $H^*_{\mathcal{U}}(\tilde{\mathcal{U}}, ...)$. Then we have the long exact sequence
$$0 \longrightarrow H^0_{\mathcal{U}}(\tilde{\mathcal{U}}, \mathcal{O}) \longrightarrow H^0(\tilde{\mathcal{U}}, \mathcal{O}) \longrightarrow H^0(\tilde{\mathcal{U}} - \mathcal{U}, \mathcal{O}) \longrightarrow H^1_{\mathcal{U}}(\tilde{\mathcal{U}}, \mathcal{O}) \longrightarrow$$
$H^1(\tilde{\mathcal{U}}, \mathcal{O}) = 0$. We also have $H^0_{\mathcal{U}}(\tilde{\mathcal{U}}, \mathcal{O}) = 0$. This allows us to re-interpret hyperfunctions in terms of local cohomology: $C^{-\omega}(\mathcal{U}) = H^1_{\mathcal{U}}(\tilde{\mathcal{U}}, \mathcal{O})$.

Now, we are ready to define hyperfunctions on $\mathbb{R}^n$. For this suppose $\mathcal{U} \subset \mathbb{R}^n$ is open and think of $\mathbb{R}^n$ as embedded in $\mathbb{C}^n$; then $H^p_{\mathcal{U}}(\mathbb{C}^n, \mathcal{O}) = 0$ for $p \neq n$ [6].

Definition. $C^{-\omega}(\mathcal{U}) = H^n_{\mathcal{U}}(\mathbb{C}^n, \mathcal{O})$.

Then:

(1) $C^{-\omega}$ is a sheaf; this sheaf is flabby.
(2) For $\mathcal{U} \subset \mathbb{R}^n$ open, relatively compact we have $C^{-\omega}(\mathcal{U}) = C^{-\omega}(\bar{\mathcal{U}}) /C^{-\omega}(\partial\mathcal{U}) \cong C^{\omega}(\bar{\mathcal{U}})'/C^{\omega}(\partial\mathcal{U})'$. The same result is true on any real analytic manifold, modulo the choice of C^{ω} volume form. In particular, if S is a compact C^{ω} manifold with C^{ω} volume form, we have that $C^{-\omega}(S) = C^{\omega}(S)'$ (but there is no good topology on $C^{-\omega}(X)$ if X is a noncompact manifold).
(3) Poincaré, Dolbeault lemmas hold for hyperfunctions!

Why are hyperfunctions useful and important? Because

a) Their duality with the space of real analytic functions.
b) They provide geometric, flabby resolutions of $\mathbb{C}, \mathcal{O}$.
c) Certain differential equations become solvable only in the context of hyperfunctions. For example consider the equation $x^2 \frac{df}{dx} + f = 0$, which has a hyperfunction solution near $x = 0$, but no distribution solution.
d) They come up naturally as boundary values of solutions of linear differential equations.

A final historical remark: the theory of hyperfunctions was developed by Sato, around 1960.

Canonical lifting of Harish-Chandra modules.

We have seen that the Poisson transform $P : C^{-\omega}(S^1) \to H(\Delta) =$ space of harmonic functions on Δ is a topological isomorphism. Now $SU(1,1)$ acts by linear fractional transformations on Δ, S^1, and P is equivariant with respect to the actions on the two function spaces. A slight generalization: let $H_c(\Delta) = c$-eigenspace of the non-Euclidean Laplace operator on $C^\infty(\Delta)$ for any particular $c \in \mathbb{C}$. Then we have an $SU(1,1)$-equivariant topological isomorphism $P_c : C^{-\omega}(S^1) \to H_c(\Delta)$.

Let G be a connected semisimple Lie group, $K \subset G$ a maximal compact subgroup, $X = G/K, D(X) =$ algebra of G-invariant linear differential operators on X. $D(X)$ is a polynomial algebra in r variables, $r = \text{rank}(G/K)$. Fix $\chi : D(X) \to \mathbb{C}$ an algebra homomorphism. $H_\chi(X) = \{f \in C^\infty(X) : Df = \chi(D)f \ \forall D \in D(X)\}$. Then there exists $P_\chi : C^\infty(G/MAN, \mathcal{L}_\chi) \to H_\chi(X)$. Here MAN is a minimal parabolic subgroup and $\mathcal{L}_\chi \to G/MAN$ the line bundle defined by χ. Helgason [5] proved the following facts:

(1) P_χ is an isomorphism on the K-finite level.
(2) P_χ extends to $C^{-\omega}(G/MAN, \mathcal{L}_\chi)$.
(3) $P_\chi : C^{-\omega}(G/MAN, \mathcal{L}_\chi) \cong H_\chi(X)$ is a topological isomorphism if $\text{rank}(X) = 1$.

He also conjectured that 3) is always true without restriction on $\text{rank}(X)$. The conjecture was proved by Kashiwara, Kowata, Minemura, Okamoto, Oshima and Tanaka [7]. As we shall see this conjecture is also a consequence of a fairly general phenomenon on homogeneous spaces of semisimple Lie groups: solutions of invariant systems of linear differential equations often have hyperfunction boundary values.

Fix a semi simple Lie group G in the Harish-Chandra class, $K \subset G$ a maximal compact subgroup, let $\mathfrak{g}, \mathfrak{k}$ be the complexified Lie algebras. By a representation of G we mean a continuous representation, on a complete locally convex Hausdorff topological vector space, with finite length and finite K-multiplicities. We have a functor

$$\{\text{representations}\} \overset{HC}{\to} \{\text{Harish-Chandra modules}\}$$

$$(\pi, V_\pi) \to \text{space of K-finite vectors in } V_\pi$$

This functor is onto (Casselman). However, it is very far from one to one. If a Harish-Chandra module V is the image under HC of a representation (π, V_π), then we call (π, V_π) a *globalization* of V.

Problem. Do there exist functorial globalizations with good properties?
   Yes: Cassellman-Wallach [1,12]: $C^\infty, C^{-\infty}$ globalizations;
   Schmid [8]: minimal $(= C^\omega)$, maximal $(= C^{-\omega})$.

Let V be a Harish-Chandra module. Pick some globalization (π, V_π). Let $V' =$ dual Harish-Chandra module of $V = K$-finite part of the algebraic dual of V. Then $V' \hookrightarrow V'_\pi (=$ continuous dual of $V_\pi)$. For v in V, v' in V' define $f_{v,v'}$ in $C^\infty(G)$ by $f_{v,v'}(g) = \langle v', \pi(g)v \rangle$.

Observations.

(1) $f_{v,v'}$ is real analytic.

(2) The Taylor series at points of K depends only on V, not on V_π, hence the $f_{v,v'}$ are canonically attached to V.

Let $T = \{v'_1, ..., v'_n\}$ be a finite set of $U(\mathfrak{g})$-generators of V'. Define i_T : $V \to C^\infty(G)^n$ by $i_T(v) = (f_{v,v'_1}, f_{v,v'_2}, ..., f_{v,v'_n})$.

Observations.

(1) i_T is $(\mathfrak{g}, K)$-equivariant relative to the right action on $C^\infty(G)$.

(2) i_T is injective (because T is a set of generators).

(3) The induced topology on V is independent of the choice of T.

In fact, choose $\tilde{T} = \{\tilde{v}'_1, ..., \tilde{v}'_m\}$ another set of generators. Choose X_{ij} in $U(\mathfrak{g})$, so that $\tilde{v}'_j = \sum X_{ij} v'_i$. We have the commuting diagram

$$
\begin{array}{ccc}
 & \overset{i_T}{\nearrow} & C^\infty(G)^n \\
V & & \downarrow{\scriptstyle l(X_{ij})} \\
 & \underset{i_{\tilde{T}}}{\searrow} & C^\infty(G)^m
\end{array}
$$

in which all maps are continuous. Now (3) follows, since we can reverse the roles of $\tilde{T}$ and T.

The completion of V in the induced topology (alternatively, but less invariantly, the closure of the image in $C^\infty(G)^n$) is a globalization of V. I shall refer to it as the maximal globalization and denote it by $V_{\max}$. The topology of $V_{\max}$ does not depend on the initial choice of generators. In this sense, $V_{\max}$ is canonically attached to V. By construction, the functor $V \to V_{\max}$ is a right inverse of HC.

Let $V_{\min}$ be the strong topological dual to $(V')_{\max}$. Then $V_{\min}$ is another globablization of V. Just as $V_{\max}$ can be realized as a closed subspace of $C^\infty(G)^n$, $V_{\min}$ can be constructed as a quotient of $C_0^\infty(G)^n$.

Observation. Let (π, V_π) be any globalization of V. Then there exist continuous G-invariant inclusions, such that

$$
\begin{array}{ccccc}
V_{\min} & \overset{\alpha}{\to} & V_\pi & \overset{\beta}{\to} & V_{\max} \\
\cup & & \cup & & \cup \\
V & = & V & = & V
\end{array}
$$

Fix (π, V_π) a Banach globalization; v in V_π is a C^∞ (respectively C^ω)-vector if $g \to \pi(g)v$ is a C^∞ (respectively C^ω) map from G to V_π.

Let V_π^∞ = subspace of C^∞-vectors. Let V_π^ω = subspace of C^ω-vectors. It is clear that G acts on these spaces. We topologize them via the inclusions $V_\pi^\infty \to C^\infty(G, V_\pi)$; $V_\pi^\omega \to C^\omega(G, V_\pi)$, $v \to (g \to \pi(g)v)$. In this way V_π^∞ and V_π^ω become globalizations of V. If V_π is reflexive, define $V_\pi^{-\infty} = ((V_\pi')^\infty)'$, similarly for $V_\pi^{-\omega}$. We have the inclusions $V_\pi^\omega \subset V_\pi^\infty \subset V_\pi \subset V_\pi^{-\infty} \subset V_\pi^{-\omega}$.

Theorem 1. The natural inclusions $V_\pi^{-\omega} \hookrightarrow V_{\max}$, $V_{\min} \hookrightarrow V_\pi^\omega$ are topological isomorphisms.

Theorem (Casselman-Wallach). $V_\pi^\infty, V_\pi^{-\infty}$ depend only on V, not on V_π.

Corollary (of either theorem). $V_\pi^\omega, V_\pi^{-\omega}$ depend only on V, not on V_π.

Corollary (of the proof of Theorem 1). $V \to V_{\min}$, $V \to V_{\max}$ are topologically exact functors.

Let (π, V_π) be a representation of G, where V_π is a space of sections of a homogeneous vector bundle over G/V, V compact, defined by a system of G-invariant linear differential equations; under mild conditions, $V_\pi = (HC(V_\pi))_{\max}$.

Example. $V_\pi = H_\chi(X)$ [5] or $H^s(G/T, \mathcal{O}(L_\lambda))$ [9].

Corollary. Helgason's conjecture, $P_\chi : L^2(G/MAN, \mathcal{L}_\chi) \longrightarrow H_\chi(X)$ is a topological isomorphism. The is an isomorphism on the level of Harish-Chandra modules, hence an isomorphism on the level of maximal globalizations.

Corollary. In many interesting situations, the $\bar{\partial}$ and $\bar{\partial}_b$ operators on homogeneous spaces for G have closed range.

Sketch of a proof for theorem 1.

Proposition (Wallach). The asymptotic coefficients of matrix coefficients are continuous in the topology of $V_{\pi'}$ for any Banach globalization [12].

Proposition (Hecht-Schmid). All embeddings of a Harish-Chandra module V into representations induced from minimal parabolic subgroup MAN can be read off from the asymptotic coefficients [4].

Hence to prove theorem 1 if suffices to prove the theorem for representations induced from a minimal parabolic subgroup. We now consider a particular case. Assume $G = SU(1,1)$, $K = U(1)$, and (π, V_π) an irreducible spherical principal series representation. As K- module $V_\pi \simeq L^2(S^1)_{\text{even}}$, $V = \{\sum a_n e^{in\theta}; a_n = 0$ for all n so that $|n| \gg 0\}$. $V \simeq V', 1$ is a generator of V'. So we get an injection

$$V \hookrightarrow C^\infty(G/K), e^{in\theta} \to f_{e^{in\theta},1} := f_n (n \text{ even})$$

We must show that the induced topology coincides with the hyperfunction topology; continuity in one direction is clear. Thus it suffices to show: For each $r < 1$, there exists a compact set $C \subset \Delta = G/K$ such that $\sup_{g \in C} |f_n(g)| \geq ar^n$ where a does not depend on n. Define $\varphi_n : \mathbb{R} \to \mathbb{C}$, $\varphi_n(x) = f_n(\tanh x)$, φ_n completely determines f_n! φ_n satisfies an ordinary linear differential equation. We want a lower bound for φ_n; it is easy to get an upper bound for any solution. The explicit formula for the determinant of the fundamental matrix (see below), together with the upper bound for any solution, implies the lower bound for the particular solution φ_n. The equation is

$$\frac{d^2\varphi_n}{dx^2} + \frac{1}{\tanh x}\frac{d\varphi_n}{dx} - \frac{n^2}{\sinh^2 x}\varphi_n = c\varphi_n,$$

hence

$$\varphi_n = a_{+,0}e^{sx} + a_{+,1}e^{(s-1)x} + \ldots + a_{-,0}e^{-sx} + a_{-,1}e^{-(s+1)x} + \ldots,$$

where $s^2 = c - 1$, $a_{+,0} = 1$. The determinant of the fundamental matrix is $\exp(-\int \frac{dx}{\tanh x}) = 1/\cosh(x)$.

$\mathcal{D}$-modules and representation theory.

In the construction or classification scheme for Harish-Chandra modules the following general principle holds (encountered also in other context): irreducible modules are obtained as unique irreducible submodules (or quotients) or "standard modules". Standard modules are obtained by a cohomological construction, which is "easy" on the level of Euler characteristic, standard modules occur in setting where there is a vanishing theorem. We have:

(1) Langlands classification: Standard modules are obtained by parabolic induction from "limits of discrete series".
(2) Vogan-Zuckerman: standard modules are obtained by "cohomological induction" from principal series.
(3) Beilinson-Bernstein: attaches standard modules to $K_{\mathbb{C}}$-orbits in the flag variety X.
   1), 2) can be done in one step via derived functors-algebraic construction.
(4) The Hecht-Milicic-Schmid-Wolf construction attaches standard modules to G-orbits on the flag variety.

Fix a connected, linear semi simple Lie group G. Let $K \subset G$ be a maximal compact subgroup, $\mathfrak{g}_0, \mathfrak{k}_0$ the Lie algebras of G and K, let θ be the Cartan involution, and let $\mathfrak{g}, \mathfrak{k}$ be the complexified Lie algebras. Let X be the flag variety of $\mathfrak{g}$. Thus $X = \{\mathfrak{b} \subset \mathfrak{g} : \mathfrak{b}$ is a Borel subalgebra $\}$. Then X is a smooth projective algebraic variety. X is acted on transitively by $G_{\mathbb{C}}$, the complexification of G, $X = G_{\mathbb{C}}/B$. B can be chosen so that $B = H_{\mathbb{C}}N$ where $H_{\mathbb{C}} = $ complexification of a θ-stable Cartan subgroup $H \subset G$ and $\mathrm{Lie}(N)$ is naturally identified with the antiholomorphic tangent space of

X at B. Let $\mathfrak{h}_0 = \mathrm{Lie}(H)$, $\mathfrak{h} = \mathbb{C} \otimes_{\mathbb{R}} \mathfrak{h}_0 = \mathrm{Lie}(H_{\mathbb{C}})$. Fix λ in $\mathfrak{h}^*$; we say that λ is *integral* if $e^\lambda : H_{\mathbb{C}} \to \mathbb{C}^*$ is a well defined character. For such λ, e^λ extends to a character $e^\lambda : B \to \mathbb{C}^*$, trivial on N. Let $\mathcal{L}_\lambda \to X$ be the $G_{\mathbb{C}}$-equivariant holomorphic line bundle on X associated to λ. $\mathcal{L}_\lambda$ has an algebraic structure and the projection is a rational map. Choose a root ordering in $\Phi(g, h)$ such that $\mathfrak{n} = \mathrm{Lie}(N)$ corresponds to the negative roots. Let ρ be equal to one half the sum of the positive roots. Thus $\mathcal{L}_{-2\rho}$ is the canonical line bundle, i.e., $\mathcal{L}_{-2\rho} = \Lambda^{\mathrm{top}} T^{1,0}(M)^*$. Aside: If X is any smooth algebraic variety (for example a projective complex manifold), and $\mathcal{L} \to X$ is an algebraic line bundle, then we can talk of $\mathcal{D}(\mathcal{L})$, the sheaf of algebraic linear differential operators. This is a sheaf of algebras which acts on rational sections of $\mathcal{L} \to X$. $\mathcal{D}(\mathcal{L})$ canonically contains $\mathcal{O}_X$ as the subsheaf of zero degree operators. For λ in $\mathfrak{h}^*$ such that $\lambda - \rho$ is integral, we define $\mathcal{D}_\lambda = \mathcal{D}(\mathcal{L}_{\lambda-\rho})$. In particular $\mathcal{D}_\rho = \mathcal{D}_X = $ sheaf of ordinary scalar operators. $\mathcal{D}_\lambda$ is acted on by $\tilde{G}_{\mathbb{C}}$ because $\mathcal{D}(\mathcal{L}_{\lambda-\rho})$ is contained in $\mathrm{End}_{\mathbb{C}}(\mathcal{L}_{\lambda-\rho})$. For λ in h^*, $\mathcal{L}_{\lambda-\rho}$ makes no sense unless $\lambda - \rho$ is integral, however $\mathcal{D}(\mathcal{L}_{\lambda-\rho})$ does make sense as a $\tilde{G}_{\mathbb{C}}$ invariant, *twisted sheaf of differential operators* (TDO).

Definition. A TDO is a sheaf of algebras, $\mathcal{D}$ together with an inclusion of the structure sheaf $\mathcal{O} : \mathcal{O} \hookrightarrow \mathcal{D}$, such that every point in X has a neighborhood $\mathcal{U}$ together with an isomorphism $i_{\mathcal{U}} : \mathcal{D}|_{\mathcal{U}} \cong \mathcal{D}_X|_{\mathcal{U}'}$ such that

$$
\begin{array}{ccc}
\mathcal{D}|_{\mathcal{U}} & \xrightarrow{\;i_{\mathcal{U}}\;} & \mathcal{D}_X|_{\mathcal{U}} \\
& \diagdown \quad \diagup & \\
& \mathcal{O}|_{\mathcal{U}} &
\end{array}
$$

commutes. Loosely speaking the pair $(\mathcal{O}, \mathcal{D})$ is locally isomorphic to $(\mathcal{O}, \mathcal{D}_X)$. The definition is in either the algebraic or the holomorphic category. Recall that for a projective manifold both categories are equivalent.

Example. $G = SU(1,1)$ $(\cong SL(\mathbb{R}))$, $G_{\mathbb{C}} = SL(2, \mathbb{C})$ acts by linear fractional transformations on $\mathbb{P}^1 = \mathbb{C} \cup \{\infty\} \cong X$ via $z \to$ stabilizer of z in $\mathfrak{g}$, which is a Borel subalgebra of $\mathfrak{g}$. {line bundles on $\mathbb{P}^1$} $\cong \mathbb{Z} \subset \mathbb{C} \cong$ dual space of the complexified Lie algebra of a Cartan subgroup. The isomorphism is: $n \to \mathcal{L}_n$, where $\mathcal{L}_n$ has degree n; for example, $\mathcal{L}_{-2} = $ canonical bundle, $\mathcal{L}_2 = $ holomorphic tangent bundle. All of these bundles are $\tilde{G}_{\mathbb{C}}$-homogeneous! $\rho \leftrightarrow 1$ under the above parametrization; $\mathcal{D}_n = \mathcal{D}(\mathcal{L}_{n-1})$. $\mathbb{P}^1 = \mathcal{U}_0 \cup \mathcal{U}_\infty$ where $\mathcal{U}_0 = \mathbb{C}$, $\mathcal{U}_\infty = \mathbb{P}^1 - \{0\}$. $\mathcal{L}_n$ has meromorphic sections σ_n^0, σ_n^∞ such that σ_n^0 is holomorphic, nowhere zero on $\mathcal{U}_0$, has degree n at ∞. σ_n^∞ is holomorphic, nowhere zero on $\mathcal{U}_\infty$, has degree n at 0. This determines σ_n^0, σ_n^∞, up to a constant multiple. We can renormalize so that $\sigma_n^\infty = z^n \sigma_n^0$ on $\mathcal{U}_0 \cap \mathcal{U}_\infty$. Using $\sigma_n^0, \sigma_n^\infty$, we get explicit trivializations $\mathcal{L}_n|_{\mathcal{U}_0} \simeq \mathcal{O}|_{\mathcal{U}_0}$ via σ_n^0, $\mathcal{L}_n|_{\mathcal{U}_\infty} \simeq \mathcal{O}|_{\mathcal{U}_\infty}$ via σ_n^∞. These two trivializations are related by an automorphism of $\mathcal{O}$ over $\mathcal{U}_0 \cap \mathcal{U}_\infty = \mathbb{C}^*$, namely $\mathcal{O}|_{\mathcal{U}_0 \cap \mathcal{U}_\infty} \simeq \mathcal{O}|_{\mathcal{U}_0 \cap \mathcal{U}_\infty}$, via $f \to z^n f$. These trivializations induce $\mathcal{D}_{n+1}|_{\mathcal{U}_0} = \mathcal{D}(\mathcal{L}_n|_{\mathcal{U}_0}) \simeq \mathcal{D}_X|_{\mathcal{U}_0}$ via

σ_n^0, $\mathcal{D}_{n+1}|_{\mathcal{U}_\infty} = \mathcal{D}(\mathcal{L}_n|_{\mathcal{U}_\infty}) \simeq \mathcal{D}_X|_{\mathcal{U}_\infty}$ via σ_n^∞, which are related over $\mathcal{U}_0 \cap \mathcal{U}_\infty$ by the automorphism $\mathcal{D}_X|_{\mathcal{U}_0 \cap \mathcal{U}_\infty} \simeq \mathcal{D}_X|_{\mathcal{U}_0 \cap \mathcal{U}_\infty}$, $Y \rightarrow z^{-n} \circ Y \circ z^n$ (Y being a local section of $\mathcal{D}_X|_{\mathcal{U}_0 \cap \mathcal{U}_\infty}$). Conjugation of differential operators makes sense even if the function in question is multiple valued, holomorphic and nowhere zero. Hence the above automorphism of $\mathcal{D}_X|_{\mathcal{U}_0 \cap \mathcal{U}_\infty}$ makes perfect sense even if n is in $\mathbb{C}$. This gives $\mathcal{D}_\lambda$, for λ in $\mathbb{C}$.

We now go back to the general case!

For λ in $\mathfrak{h}^*$, we get a $\tilde{G}_{\mathbb{C}}$-equivariant $\mathcal{D}_\lambda$. The infinitesimal action defines an inclusion $\mathfrak{g} \hookrightarrow \Gamma(\mathcal{U}, \mathcal{D}_\lambda)$ for every $\mathcal{U} \subset X$, hence an algebra morphism $U(\mathfrak{g}) \rightarrow \Gamma(\mathcal{U}, \mathcal{D}_\lambda)$. Let us check this for $G = SU(1,1)$; $\mathfrak{g}$ acts on functions by infinitesimal translation. Consider the basis of $\mathfrak{sl}(2, \mathbb{C})$ $H = \begin{pmatrix} 1 & 0 \\ 0 & -1 \end{pmatrix}$, $E_+ = \begin{pmatrix} 0 & 1 \\ 0 & 0 \end{pmatrix}$, $E_- = \begin{pmatrix} 0 & 0 \\ 1 & 0 \end{pmatrix}$. Then $Hz = -2z$, $E_+ z = -1$, $E_- z = z^2$, hence the map $\mathfrak{g} \hookrightarrow \Gamma(\mathcal{D}_X)$ has the following effect:

$$H \rightarrow -2z\frac{\partial}{\partial z}, E_+ \rightarrow -\frac{\partial}{\partial z}; E_- \rightarrow z^2 \frac{\partial}{\partial z}.$$

Recall the definitions of σ_n^0, σ_n^∞; we have $E_+\sigma_n^0 = 0$, $H\sigma_n^0 = n\sigma_n^0$, $E_-\sigma_n^0 = -nz\sigma_n^0$, $E_-\sigma_n^\infty = -nz^{-1}\sigma_n^\infty$, $H\sigma_n^\infty = -n\sigma_n^\infty$, $E_+\sigma_n^\infty = 0$. Thus, the composition $\mathfrak{g} \hookrightarrow \Gamma(\mathcal{D}_{n+1}|_{\mathcal{U}_0}) \simeq \Gamma(\mathcal{D}_X|_{\mathcal{U}_0})$ (via σ_n^0) has the following effect. $H \rightarrow n - 2z\frac{\partial}{\partial z}$; $E_+ \rightarrow -\frac{\partial}{\partial z}$; $E_- \rightarrow -nz + z^2\frac{\partial}{\partial z}$ and $\mathfrak{g} \hookrightarrow \Gamma(\mathcal{D}_{n+1}|_{\mathcal{U}_\infty}) \simeq \Gamma(\mathcal{D}_X|_{\mathcal{U}_\infty})$ (via σ_n^∞) is given by $H \rightarrow -n - 2z\frac{\partial}{\partial z}$, $E_+ \rightarrow -nz^{-1} - \frac{\partial}{\partial z}$, $E_- \rightarrow z^2\frac{\partial}{\partial z}$. These two morphisms are related by $Y \rightarrow z^{-n} \circ Y \circ z^n$ which is the gluing automorphism defining $\mathcal{D}_{n+1}$, hence we get an inclusion $\mathfrak{g} \hookrightarrow \Gamma(\mathcal{D}_n)$. The formulas remain correct if n in $\mathbb{Z}$ is replaced by n in $\mathbb{C}$.

Once more the general case!

Fix λ in $\mathfrak{h}^*$, $\mathcal{D}_\lambda$ the TDO associated to λ. Write $\Gamma(\mathcal{D}_\lambda)$ for the algebra of global sections of $\mathcal{D}_\lambda$. Let $U(\mathfrak{g}) \rightarrow \Gamma(\mathcal{D}_\lambda)$ be as before. Let $\mathcal{Z}(\mathfrak{g})$ be the center of $U(\mathfrak{g})$, and I_λ equal to the annihilator in $\mathcal{Z}(\mathfrak{g})$ of the Verma module of highest weight $\lambda - \rho$.

Theorem (Harish-Chandra).

(1) Every maximal ideal in $\mathcal{Z}(\mathfrak{g})$ is of the form I_λ, λ in $\mathfrak{h}^*$.
(2) $I_\lambda = I_\mu$ if and only if λ, μ are W-conjugate.

Essentially by the construction of $\mathcal{D}_\lambda$, under the map $U(\mathfrak{g}) \rightarrow \Gamma(\mathcal{D}_\lambda)$, I_λ goes to zero. Define $U_\lambda = U(g)/I_\lambda U(g)$; thus U_λ maps into $\Gamma(\mathcal{D}_\lambda)$.

Theorem (Beilinson-Bernstein). This map is an isomorphism.

Caution: $U_\lambda = U_\mu$ if λ, μ are W-conjugate, but $\mathcal{D}_\lambda$ is not isomorphic to $\mathcal{D}_\mu$ if λ is W-conjugate to μ unless $\mu = \lambda$.

Define $\mathcal{M}(\mathcal{D}_\lambda)$ to be the category of quasi-coherent sheaves of $\mathcal{D}_\lambda$-modules (a sheaf is quasi-coherent if it possesses local presentations by free sheaves, not necessarily of finit rank). Define $\mathcal{M}(U_\lambda)$ as the category of left U_λ-modules, that is, $\mathcal{M}(U_\lambda)$ = category of $U(\mathfrak{g})$-modules annihilated by I_λ. Then Γ, the global section functor, maps $\mathcal{M}(\mathcal{D}_\lambda)$ into $\mathcal{M}(U_\lambda)$. Recall the Γ is left exact. Fix V in $\mathcal{M}(U_\lambda)$. For an open set $\mathcal{U}$ in X, $\Gamma(\mathcal{U}, \mathcal{D}_\lambda)$ is a right U_λ-module. Hence, $\Gamma(\mathcal{U}, \mathcal{D}_\lambda) \otimes V$ is a left $\Gamma(\mathcal{U}, \mathcal{D}_\lambda)$-module. Let ΔV denote the sheaf of $\mathcal{D}_\lambda$-modules spanned by the presheaf $\mathcal{U} \longrightarrow \Gamma(\mathcal{U}, \mathcal{D}_\lambda) \otimes V$ (tensor products to be taken over U_λ). ΔV is quasi coherent because any U_λ-module is a quotient of a free module. Δ is a right exact functor and is called the "localization" functor.

Definition. λ in $\mathfrak{h}^*$ is integrally dominant if $\frac{2(\lambda, \alpha)}{(\alpha, \alpha)} \neq -1, -2, -3, \ldots$ for every positive root α. λ is regular if $(\lambda, \alpha) \neq 0$ for every root α.

Theorem (Beilinson-Bernstein).

    a) Suppose λ in $\mathfrak{h}^*$ is integrally dominant, regular. Then for every $\mathcal{F}$ in $\mathcal{M}(\mathcal{D}_\lambda)$, every stalk of $\mathcal{F}$ is generated (over the corresponding stalk of $\mathcal{D}_\lambda$) by $\Gamma(\mathcal{F})$.

    b) If λ in $\mathfrak{h}^*$ is integrally dominant and $\mathcal{F}$ in $\mathcal{M}(\mathcal{D}_\lambda)$ then $H^i(X, \mathcal{F}) = 0$ for $i \neq 0$.

Corrolary. If λ integrally dominant and regular, then Γ is an equivalence of categories with inverse Δ.

More generally, if λ is integrally dominant, then Γ induces an equivalence of categories between a certain quotient of $\mathcal{M}(\mathcal{D}_\lambda)$ and $\mathcal{M}(U_\lambda)$. Note that every λ in $\mathfrak{h}^*$ is W conjugate to at least one λ' in $\mathfrak{h}^*$ which is integrally dominant. Thus $\mathcal{M}(U_\lambda) = \mathcal{M}(U_{\lambda'}) \simeq \mathcal{M}(\mathcal{D}_{\lambda'})$ if λ is also regular. This equivalence has interesting consequences for either Harish-Chandra modules or Verma modules.

Let $\mathcal{M}(U_\lambda, K)$ be the category of U_λ-modules with compatible K-action.

Note. V in $\mathcal{M}(U_\lambda, K)$ is a Harish-Chandra module if it is finitely generated; in particular, the irreducible objects in $\mathcal{M}(U_\lambda, K)$ for all λ in $\mathfrak{h}^*$, module the action of the Weyl group, are exactly the irreducible Harish-Chandra modules.

Let $\mathcal{M}(\mathcal{D}_\lambda, K_\mathbb{C})$ be the category of quasi coherent $\mathcal{D}_\lambda$-modules with compatible $K_\mathbb{C}$-action.

Corollary. If λ is integrally dominant and regular, the functor $\Gamma : \mathcal{M}(\mathcal{D}_\lambda, K_\mathbb{C}) \to \mathcal{M}(U_\lambda, K)$ is an equivalence of categories, with inverse Δ.

Corollary. If λ is integrally dominant but not necessarily regular, $\mathcal{F}$ an irreducible object in $\mathcal{M}(\mathcal{D}_\lambda, K_\mathbb{C})$, then $\Gamma(\mathcal{F})$ is an irreducible Harish-Chandra module or zero. Every irreducible Harish-Chandra module arises in this fashion. The zero option can only happen if λ is singular.

This gives a classification of irreducible Harish-Chandra modules, due to Beilinson-Bernstein, which we will now describe in detail.

Recall that $G_\mathbb{C}$ acts transitively on X.

Lemma. $K_{\mathbb{C}}$ has finitely many orbits in X; these are affinely embedded.

Note.

(1) On generalized flag manifolds $K_{\mathbb{C}}$-orbits are not always affinely embedded.

(2) If $\mathcal{F}$ is an irreducible object in $\mathcal{M}(\mathcal{D}_\lambda, K_{\mathbb{C}})$ then the support of $\mathcal{F}$ is the closure of one $K_{\mathbb{C}}$-orbit.

Fix $Q \subset X$ a $K_{\mathbb{C}}$-orbit.

Lemma (Matsuki). There exists $\mathfrak{b}$ in Q, $\mathfrak{b} = \mathfrak{h} \oplus \mathfrak{n}$ with $\mathfrak{h}$ the complexified Lie algebra of a θ-stable Cartan subgroup $H \subset G$. Such $\mathfrak{b}$ is unique up to K-conjugacy.

Write $\mathfrak{b} = \mathfrak{h} \oplus \mathfrak{n}$ as above. We may and will assume that the normalizer of $\mathfrak{b}$ in G is B. Write $H = TA$, T a compact Lie group, T^0 a torus, A a connected abelian Lie group which acts, via Ad, with positive real eigenvalues. $A \simeq \mathbb{R}^\ell$ for some ℓ. $T_{\mathbb{C}}$ is a Levi component of $K_{\mathbb{C}} \cap B$. Suppose λ in $\mathfrak{h}^*$ is such that $\lambda - \rho$ lifts to a character $\chi : H \to \mathbb{C}^*$. Then we get a $K_{\mathbb{C}}$-homogeneous algebraic line bundle over Q, plus a $\mathfrak{g}$-equivariant holomorphic extension of this line bundle to a Hausdorff neighborhood of Q in X. Denote it by $\mathcal{L}_\chi$.

Important fact. $\mathcal{L}_\chi$ can be described by trivializations and transition functions whose the logarithmic derivatives are algebraic. Hence $\mathcal{L}_\chi$ is an algebraic object over a formal neighborhood of Q in X.

Facts.

(1) $\mathcal{D}_\lambda|_Q$ is a TDO on Q.

(2) $\mathcal{O}_Q(\mathcal{L}_\chi|_Q)$ is a sheaf of $(\mathcal{D}_\lambda|_Q)$- modules.

(3) Let $j : Q \to X$ be the inclusion map. Then $j_+\mathcal{O}_Q(\mathcal{L}_\chi|_Q)$ $(\mathcal{D}_\lambda$-module direct image) is an object in $\mathcal{M}(\mathcal{D}_\lambda, K_{\mathbb{C}})$. An alternative description to follow.

Proposition (Beilinson-Bernstein).

(1) $j_+\mathcal{O}_Q(\mathcal{L}_\chi|_Q)$ has a unique irreducible subsheaf of $\mathcal{D}_\lambda$-modules.

(2) This irreducible subsheaf is supported on the closure of Q.

(3) Every irreducible in $\mathcal{M}(\mathcal{D}_\lambda, K_{\mathbb{C}})$ is of this form, for some Q, χ as above.

$j_+\mathcal{O}_Q(\mathcal{L}_\chi|_Q)$ is called the standard Beilinson-Bernstein sheaf corresponding to the data (Q, λ, χ). If λ is integrally dominant, $H^0(X, j_+\mathcal{O}_Q(\mathcal{L}_\chi|_Q)$ is called the standard Beilinson- Bernstein module corresponding to (Q, λ, χ) (if not equal to zero). Set $J(Q, X) := j_+\mathcal{O}_Q(\mathcal{L}_\chi|_Q)$.

We now give an alternative description of $j_+(\mathcal{O}_Q(\mathcal{L}_\chi|_Q)$. Fix x in Q. Choose local algebraic coordinates $z_1, \ldots, z_d$, $w_1, \ldots, w_{n-d}$ so that Q is given, locally, by $z_1 = \ldots = z_d = 0$. Let s be a locally defined nonzero holomorphic section of $\mathcal{L}_\chi$, which is algebraic in the formal neighborhood. Consider the quotient space

$$\{fs|f \text{algebraic, regular regular near } x \text{ except on } \bigcup_{j=1}^{d}\{z_j = 0\}/$$

$$\sum_{k=1}^{d}\{fs|f \text{ algebraic, near } x \text{ except on } \bigcup_{\substack{j=1 \\ j \neq k}}^{d}\{z_j = 0\}\} \cong$$

$$\cong \{ \sum_{\substack{m_1,\ldots,m_d \geq 1 \\ \text{finite sum}}} \varphi_{m_1,\ldots,m_d}(w_1,\ldots,w_{n-d})z_1^{-m_1}z_2^{-m_2}\ldots z_d^{-m_d}s\}$$

Contrary to appearances, this quotient space has canonical meaning: via residues, the quotient space can be identified with the algebraic analogue of "$\mathcal{L}_\chi$-valued distributions with support on Q". The description of local cohomology in terms of relative open covers shows that this quotient is the stalk at x of the local cohomology sheaf $\mathcal{H}_Q^d(\mathcal{O}(\mathcal{L}_\chi))$ in degree $d = \text{codim}_\mathbb{C} Q$, where cohomology is taken in the algebraic setting; the local cohomology sheaves in degree other than d vanish. The TDO $\mathcal{D}_\lambda$ acts on sections of $\mathcal{L}_\chi$, hence also on the quotient space defined above. Unraveling the definition of $\mathcal{D}$-module direct image, one can see that the quotient is also the stalk at x of the direct image sheaf $j_+\mathcal{O}_Q(\mathcal{L}|_Q)$. Thus:

Fact. $j_+\mathcal{O}_Q(\mathcal{L}_\chi|_Q) = \mathcal{H}_Q^d(\mathcal{O}(\mathcal{L}_\chi))$ as sheaves of $\mathcal{D}_\lambda$-modules.

Corollary. $H^p(X, j_+\mathcal{O}_Q(\mathcal{L}_\chi|_Q)) = H_Q^{p+d}(X, \mathcal{O}(\mathcal{L}_\chi))$.

Now fix a G-orbit S on X.

Fact (Matsuki). There exists $\mathfrak{b} \in S$, $\mathfrak{b} = \mathfrak{h} \oplus \mathfrak{n}$, $\mathfrak{h} = \mathbb{C} \otimes_\mathbb{R} \text{Lie}(H)$ with $H \subset G$, a θ-stable Cartan subgroup.

As before write $H = TA$, $\chi : T \to \mathbb{C}^*$ a character so that $d\chi = \lambda - \rho$. These data determine a G-invariant real analytic line bundle $\mathcal{L}_{\lambda-\rho} \to G/H$ (strictly speaking, the line bundle depends on χ, not only on λ; to simplify the notation, the dependence on χ will not be indicated explicitly). In the language of geometric quantization, the choice of the Borel subalgebra $\mathfrak{b}$ puts a G-invariant polarization on the manifold G/H, which has a natural G-invariant symplectic structure. Indeed, up to the appropriate notion of isomorphism, the G-orbit $S \subset X$ and the polarized quotient manifold G/H determine each other uniquely. A construction of Zuckerman attaches algebraically defined Harish-Chandra modules $A^j(\mathfrak{b}, \chi)$ to the data $(\mathfrak{b}, \chi)$, or equivalently, to the orbit S and line bundle $\mathcal{L}_{\lambda-\rho}$; more recently, these modules were attached more directly to the orbit S by a geometric-analytic process [10,11].

To reacall the definition of the $A^j(\mathfrak{b},\chi)$, use $\lambda - \rho$ to extend χ from T to a character of $H = TA$, and then to a one dimensional $(\mathfrak{b},T)$-module. View $\mathcal{U}(\mathfrak{g})$ as $\mathfrak{b}$-module by left multiplication, as T-module by conujugation, and as right $\mathfrak{g}$-module by right multiplication. The "produced module" $\text{pro}_{\mathfrak{b},T}^{\mathfrak{g},T}(\chi) = \text{Hom}_{\mathfrak{b}}(\mathcal{U}(\mathfrak{g}),\chi)_{[T]}$ (= space of T-finite elements in $\text{Hom}_{\mathfrak{b}}(\ldots)$) then becomes naturally a left $(\mathfrak{g},T)$-module. Since Γ is left exact, and since the domain contains enough injectives, one can define the right derived functors $R^p\Gamma$. The modules $A^j(\mathfrak{b},\chi) = R^j\Gamma(\text{prod}_{\mathfrak{b},T}^{\mathfrak{g},T}(\chi))$ turn out to be Harish-Chandra modules. $A^j(\mathfrak{b},\chi)$ is the j-th Zuckerman module attached to the pair $(\mathfrak{b},\chi)$, or equivalently, to $(S,\mathcal{L}_{\lambda-\rho})$.

Special Cases.

   a) If $\mathfrak{b}$ is defined over $\mathbb{R}$, we get zero except for $j = 0$, for $j = 0$ we obtain a module of the principal series.
   b) If $\mathfrak{b} \cap \overline{\mathfrak{b}} = \mathfrak{h}$, S is open in X, $S \simeq G/H$, and $\mathcal{L}_{\lambda-\rho} \to S$ carries the structure of G-invariant holomorphic line bundle; in this case, $A^j(\mathfrak{b},\chi)$ is the space of K-finite vectors in $H^j(S,\mathcal{O}(\mathcal{L}_{\lambda-\rho}))$.
   c) For certain S, this is cohomological induction, followed by parabolical induction, or vice versa [10,11].

Let $Q \subset X$ be a $K_{\mathbb{C}}$-orbit and let $S \subset X$ be a G-orbit.

Theorem (Matsuki). The following conditions are equivalent:

   i) $Q \cap S : \phi$, and K acts transitively on $Q \cap S$.
   ii) There exists $\mathfrak{b}$ in $Q \cap S$, $\mathfrak{b} = \mathfrak{h} \oplus \mathfrak{n}$, $\mathfrak{h} = \theta(\mathfrak{h})$, $\mathfrak{h} = \mathbb{C} \otimes \text{Lie}(H)$, $H \subset G$ a Cartan subgroup.

If i) and ii) are satisfied, we call Q, S "dual in the sense of Matsuki". Duality establishes an order reversing bijection between $K_{\mathbb{C}}$-orbits and G-orbits.

Suppose Q, S are dual in the sense of Matsuki, and $\mathfrak{b}$ in $Q \cap S$ is as in ii). Let λ in $\mathfrak{h}^*$, $\chi \neq H \to \mathbb{C}^*$, $d\chi = \lambda - \rho$ (no dominance or regularity on λ is assumed). Then we have the standard Beilinson-Bernstein sheaves and their cohomology groups, $H^i(X, j_+\mathcal{O}_Q(\mathcal{L}_\chi|_Q))$. Also, we have the Zuckerman modules $A^p(\mathfrak{b},\chi)$.

Theorem (Hecht-Milicic-Schmid-Wolf). $H^i(X, j_+\mathcal{O}_Q(\mathcal{L}_\chi|_Q))$ and $A^{s-i}(\mathfrak{b},\chi^{-1} \otimes \Lambda^{\text{top}}\mathfrak{n})$ are canonically dual, $s = \dim[\mathfrak{k} \cap \mathfrak{b}, \mathfrak{k} \cap \mathfrak{b}] = \dim_{\mathbb{R}}(Q \cap S) - \dim_{\mathbb{C}}(Q)$.

Special case. $G = K$, then $Q = S = X$ and the statement becomes the Serre duality theorem.

In general, we can use duality to cross back and forth between the two constructions, carrying results which may be obvious on one side to the other side.

For example:

(1) $A^i(\mathfrak{b}, \chi) = 0$ for $i > s$ because on the other side there is no cohomology in degrees below zero.

(2) $H^i(X, j_+\mathcal{O}_Q(\mathcal{L}_\chi|_Q)) = 0$ for $i > s$, because $A^p = 0$ for $p < 0$.

(3) If we make appropriate assumptions about the position of λ we can get more precise results; e.g., if λ is integrally dominant, the Beilinson-Bernstein vanishing theorem translates into the statement that the standard Zuckerman modules $A^p(\mathfrak{b}, \chi)$ vanish for $p \neq s$.

The example of $G = SU(1,1)$, $K = U(1)$ may help to clarify the theorem and its setting. In this case $X = \mathbb{P}^1 = \mathbb{C} \cup \{\infty\}$, $K_\mathbb{C} = \mathbb{C}^*$ with a in $\mathbb{C}^*$ acting by $z \longrightarrow a^2 z$. $K_\mathbb{C}$ has three orbits in X, namely $\{0\}$, $\{\infty\}$ and their common complement $\mathbb{C}^*$. G has three orbits in X. They are Δ, $\mathbb{P}^1 - cl(\Delta)$, $S^1 = \partial\Delta$. The Matsuki correspondence is: $\{0\} \leftrightarrow \Delta$, $\{\infty\} \leftrightarrow \mathbb{P}^1 - cl(\Delta)$, $\mathbb{C}^* \leftrightarrow S^1 = \partial\Delta$. Assume $Q = \{0\}$. The stabilizer of Q is the lower triangular matrix group. Any line bundle $\mathcal{L}_\chi$ corresponding to this orbit extends to a $G_\mathbb{C}$-equivariant line bundle over X. These are of the form $\mathcal{L}_{n-1}$, n in $\mathbb{Z}$. The Beilinson-Bernstein sheaf $j_+\mathcal{O}_{\{0\}}(\mathcal{L}_{n-1}|_{\{0\}})$ is supported at 0 and has no higher cohomology. Its sections are $\mathcal{L}_{n-1}$-valued "algebraic distributions" with support at 0, i.e., principal parts, around 0, of meromorphic sections of $\mathcal{L}_{n-1}$. Hence we have:

$$H^0(\mathbb{P}^1, j_+(\mathcal{O}_{\{0\}}(\mathcal{L}_{n-1}|_{\{0\}}))) = \begin{cases} \text{discrete series for } n \leq 1 \\ \text{limit of discrete series for } n = 0 \\ \text{reducible modules for } n \geq -1 \end{cases}$$

(in the last case the unique irreducible subrepresentation is the finite dimensional representation of highest weight $-(n+1)$). The case $Q = \{\infty\}$ is similar to $Q = \{0\}$. Let ϵ be equal to 0 or 1, $Q = C^*$. At any point of Q, $K_\mathbb{C}$ has isotropy $\{\pm 1\}$. Now $\mathcal{L}_2 \longrightarrow X$ has an essentially unique K-invariant section t, which is nonzero on $\mathbb{C}^*$ ($t = z^2 \sigma_2^\infty$). For λ in $\mathbb{C}^*$, we consider the line bundle over $\mathbb{C}^*$ generated by

(0) $(t^{1/2})^{\lambda-1}$, it is $(K_\mathbb{C}, g)$-equivariant, we denote it by $\mathcal{L}_{\lambda-1,0}$.

(1) $z^{1/2}(t^{1/2})^{\lambda-1}$, it is $(K_\mathbb{C}, g)$-equivariant, we denote it by $\mathcal{L}_{\lambda-1,1}$.

$\mathcal{L}_{\lambda-1,\epsilon}$ are the standard modules attached to Q, λ, ϵ. The holomorphic sections of $\mathcal{L}_{\lambda-1,\epsilon}$ over Q are principal series. Note that in both cases there is no higher cohomology because all orbits happen to be affine. This is very particular, it does not happen for other groups. $\mathcal{O}(\mathcal{L}_{\lambda-1,\epsilon})$ is reducible if and only if $(\lambda - 1 + \epsilon) \in 2\mathbb{Z}$. If so, $\mathcal{O}_{\mathbb{P}^1}(\mathcal{L}_{\lambda-1+\epsilon})$ is the unique irreducible subsheaf.

$$H^i(\mathbb{P}^1, \mathcal{O}_{\mathbb{P}^1}(\mathcal{L}_{\lambda-1+\epsilon})) = \begin{cases} 0 \text{ if either } \lambda = 0, \text{ or } \lambda > 0 \text{ and } i \neq 0, \\ \text{or } \lambda < 0 \text{ and } i \neq 1. \\ \text{Otherwise, is non zero, irreducible and equal} \\ \text{to a finite dimensional representation} \end{cases}$$

We now sketch a proof of the duality theorem. Recall: Q is a $K_{\mathbb{C}}$-orbit, S is the G-orbit dual to Q, in the sense of Matsuki. Fix $\mathfrak{b}$ in $S \cap Q$, $\mathfrak{b} = \mathfrak{h} \oplus \mathfrak{n}$, so that $\mathfrak{h} = \mathbb{C} \otimes \mathrm{Lie}(H)$, $H \subseteq G$ is a θ-stable Cartan subgroup of G. Let λ in $\mathfrak{h}^*$ be such that $\lambda - \rho$ lifts to a character of H. Thus $\lambda - \rho$ gives rise to $\mathcal{L}_\chi$ a line bundle defined near Q; write $M^q(S, \mathcal{L}_{-\lambda-\rho})$ for $A^q(\mathfrak{b}, \chi^{-1} \otimes \Lambda^{\mathrm{top}} \mathfrak{n})$. Since $H^p(X, j_+ \mathcal{O}_Q(\mathcal{L}_\chi|_Q)) = H_Q^{p+d}(X, \mathcal{O}(\mathcal{L}_\chi))$, we must show that the complex $H_Q^p(X, \mathcal{O}(\mathcal{L}_\chi))$ $[d]$ is dual to the complex $M^{s-p}(S, \mathcal{L}_{-\lambda-\rho})$. Let $Y = \{(\mathfrak{b}_1, \mathfrak{b}_2) \in X \times X : \mathfrak{b}_1, \mathfrak{b}_2 \text{ are opposed}\}$. Then Y is an open $G_{\mathbb{C}}$-orbit in $X \times X$ via the diagonal action. Y is isomorphic to the variety of ordered Cartan subgroups of $G_{\mathbb{C}}$. Also $Y = G_{\mathbb{C}}/H_{\mathbb{C}}$. Hence Y is an affine variety. The projection p onto the first factor $p : X \times X \longrightarrow X$, restricted to Y defines an affine map and determines a fibration

$$Y \simeq G_{\mathbb{C}}/H_{\mathbb{C}}.$$
$$\downarrow p$$
$$X \simeq G_{\mathbb{C}}/B$$

Let Q_Y be the $K_{\mathbb{C}}$-orbit in Y defined by $(\mathfrak{h}, \Phi(\mathfrak{b}, \mathfrak{h}))$. Thus $Q_Y \simeq K_{\mathbb{C}}/T_{\mathbb{C}}$. Recall that $H = TA$, $T \subseteq K$, A is a $\mathbb{R}$-split group. Thus Q_Y is an affine variety. We have

$$\begin{array}{ccc} Q_Y & \hookrightarrow & Y \\ \downarrow p & & \downarrow p \\ Q & \hookrightarrow & X \end{array}$$

The vertical arrows are affine fibrations, and $Q_y \hookrightarrow Y$ is an affine embedding. Pull back $\mathcal{L}_\chi$ to a $\mathfrak{g}$-equivariant line bundle defined in a neighborhood of Q_Y in Y. In the following, we shall not pin down the shifts precisely and we shall not be careful when identifying differentials (which will be obvious).

Lemma. $\mathcal{H}_Q^\cdot(\mathcal{O}(\mathcal{L}_\chi)) \simeq p_*(\mathcal{H}_{Q_Y}^\cdot(\mathcal{O}_Y(\mathcal{L}_\chi \otimes \Lambda^\cdot \tau^*_{Y/X})))[\ldots]$ (here $\simeq$ means quasi-isomorphism). The higher direct image vanishes.

Also

$$H_Q^\cdot(X, J(Q, \mathcal{L}_\chi)) = \mathbb{H}^\cdot(X, \mathcal{H}_Q^\cdot(\mathcal{O}(\mathcal{L}_\chi)))[d] =$$

$$= \mathbb{H}^\cdot(Y, \mathcal{H}_{Q_Y}^\cdot(\mathcal{O}_Y(\mathcal{L}_\chi \otimes \Lambda^\cdot \tau^*_{Y/X})))[d] =$$

(by the lemma and Leray spectral sequence)

$$= H^\cdot(\Gamma_Y(\mathcal{H}_{Q_Y}^\cdot(\mathcal{O}_Y(\mathcal{L}_{\lambda-\rho} \otimes \Lambda^\cdot \tau^*_{Y/X})))[d].$$

This last complex computes the cohomology groups of the standard Beilinson-Bernstein sheaves! On the other hand $M^\cdot(S, \mathcal{L}_{\lambda-\rho}) = R^\cdot \Gamma_{\mathfrak{g},T}^{\mathfrak{b},K} (\mathrm{prod}_{\mathfrak{b},T}^{\mathfrak{g},T}(\chi^{-1} \otimes$

$\Lambda^n \mathfrak{n}$)) and $0 \longrightarrow \mathrm{prod}_{\mathfrak{h},T}^{\mathfrak{g},T}(-) \longrightarrow \mathrm{prod}_{\mathfrak{h},T}^{\mathfrak{g},T}(- \otimes \Lambda^{\cdot}\mathfrak{n}^*)$ is an injective resolution in $\mathcal{M}(g,T)$. Therefore $M^{\cdot}(S, \mathcal{L}_{-\lambda-\rho})$ is the cohomology of the complex $\Gamma_{\mathfrak{g},T}^{\mathfrak{g},K}(\mathrm{prod}_{\mathfrak{h},T}^{\mathfrak{g},T}(\chi^{-1} \otimes \Lambda^n \mathfrak{n} \otimes \Lambda^{\cdot}\mathfrak{n}^*))$. This complex is the complex of K-finite forms on G/H at eH, with values in $\mathcal{L}_{-\lambda-\rho}$, and with coefficients being formal power series. The complex is quasi-isomorphic to the complex of $K_{\mathbb{C}}$-finite formal holomorphic forms, with power series coefficients on $G_{\mathbb{C}}/H_{\mathbb{C}}$, at $eH_{\mathbb{C}}$, with values in $\mathcal{L}_{-\lambda-\rho}$. Thus, the complex is quasi-isomorphic to the complex of "sections" of $(\mathcal{L}_{\lambda-\rho})^* \otimes \Lambda^n \tau_X^* \otimes \Lambda^{\cdot}\tau_Y^*$. Here, "section" means: defined and algebraic (i.e. $K_{\mathbb{C}}$-finite) along Q_Y, with formal power series coefficients, normal to Q_Y in Y. Near any given point y in Q_Y, choose local algebraic coordinates $(z_1, \ldots, , w_1, \ldots,)$ z's along Q_Y, w's normal to Q_Y. Also, a "function" can be represented locally as $\sum_J h_J(z)w^J$, $J = (j_1, \ldots,)$, $j_k \geq 0$. The elements of $\mathcal{O}_{Q_y}(\mathcal{O}_y)$ can be represented locally as

$$\sum_{\substack{I \\ \text{finite}}} f_I(z)\omega^I \qquad I = (i_1, \ldots), i_k \geq 1.$$

Also note that $\Lambda^n \tau_X^* \otimes \Lambda^n \tau_{Y/X}^* \simeq \Lambda^{2n}\tau_Y^*$. Hence the complexes pair, by contraction, wedging, projection into $\Gamma_Y(\mathcal{H}_{Q_Y}(\Omega_Y^{2n})$, hence by taking residue into $\Gamma_{Q_Y}(\Omega^{\mathrm{top}}(Q_Y))$, hence into $\mathbb{C}$ by integration over $K/T \simeq Q \cap S$. This is a perfect pairing of complexes of $(\mathfrak{g},K)$-modules. Therefore we have the duality theorem. For details see [2,11].

We now describe some results in [10,11].

Since in a suitable Borel of a given $K_{\mathbb{C}}$-orbit (respectively G-orbit) there is a Cartan subalgebra which is the complexification of the Lie algebra of a Cartan subgroup of G. We get two maps

$$\{K_{\mathbb{C}} - \text{orbits}\} \quad \searrow^{\text{onto}} \quad \left\{ \begin{array}{l} G - \text{conjugacy class of complexification of} \\ \text{Cartan subalgebras of } g \end{array} \right\}$$
$$\{G - \text{orbits}\} \quad \nearrow_{\text{onto}}$$

Dual orbits are mapped into the same image!

Definition. A G-orbit S is said to be maximally real (m.r.), (respectively, maximally imaginary (m.i.)), if among all orbits belonging to the same conjugacy class of Cartan subgroups it has the least possible (respectively, largest possible) dimension. Equivalently, if for $\mathfrak{b}$ in S, $\mathfrak{b} \supset h$ as usual, $\dim(\mathfrak{b} \cap \bar{\mathfrak{b}})$ is maximal, respectively minimal.

By Matsuki duality, we get a notion of type for $K_{\mathbb{C}}$-orbits.

Remark. m.r. orbits fibers over closed G-orbits in a generalized flag variety with fibres which are open orbits of a lower dimensional group M in the flag variety of $\mathfrak{m}_{\mathbb{C}}$. m.i. orbits fiber over open G-orbits in a generalized flag variety with fibers which are closed orbits of a lower dimensional group M in the flag variety of M.

Under appropriate negatively conditions we have that [11] standard modules corresponding to m.r. orbits are parabolically induced from limit of discrete series, i.e. "Langlands standard modules". Standard modules corresponding to m.i. orbits are cohomologically induced from principal series of split groups. i.e. "Vogan-Zuckerman modules".

We now sketch a geometric path for going from one classification into another.

Fix $Q \subset X$ a $K_{\mathbb{C}}$ orbit, $\mathfrak{b}$ in Q, $\mathfrak{b} = \mathfrak{h} \oplus \mathfrak{n}$, $\mathfrak{h}$ defined over $\mathbb{R}$, θ-stable, $\mathfrak{n}$ corresponds to negative roots. Fix α in π (the set of simple roots) and let X_α denote the generalized flag variety of type α. $X \xrightarrow{p_\alpha} X_\alpha$ is a fibration with fiber $\mathbb{P}^1$. Then $Q_\alpha = p_\alpha(Q)$ is a $K_{\mathbb{C}}$-orbits in X_α and $p_\alpha^{-1}(Q_\alpha)$ is a union of $K_{\mathbb{C}}$-orbits in X.

Case I) α is a compact imaginary root; then $p_\alpha^{-1}(Q_\alpha) = Q$.

Case II) α is a complex root, $\theta\alpha > 0$, then $p_\alpha^{-1}(Q_\alpha) = Q \cup Q_1$, $\dim Q_1 = \dim Q + 1$, p_α restricted to Q is an isomorphism onto Q_α and p_α restricted to Q_1 is a fibration with typical fiber $\mathbb{C}$. So p_α restricted to $Q \cup Q_1$ is a fibration over Q_α with fiber $\mathbb{P}^1$.

Case III) α is a complex root, $\theta\alpha < 0$. Then we have the same situation as in II with the roles of Q and Q_1 reversed.

Case IV) α is a real root. Then $p_\alpha^{-1}(Q_\alpha)$ is a union of two (e.g., $SL(n, \mathbb{R})n \geq 3$) or three (e.g., $SL(2, \mathbb{R})$) orbits. p_α restricted to Q has fiber $\mathbb{C}^* = \mathbb{P}^1 - \{0, \infty\}$.

Case V) α is a noncompact imaginary root. It is like in case IV, with the roles of Q and the other orbits reversed.

Remark. In case II, III the conjugacy class of the Cartan corresponding to Q_1 does not change. In Case IV the Cartan subgroup is "made more compact" and case V is "made less compact".

Lemma. If Q is not of type m.r. (respectively of type m.i.), there exists a complex root α in π such such that $\theta\alpha > 0$ (respectively $\theta\alpha < 0$).

Fix Q a $K_{\mathbb{C}}$-orbit in X, α a complex root in π with $\theta\alpha > 0$. We have the following fiber maps

$$
\begin{array}{ccccc}
X & \longleftarrow & Q \cup Q_1 & \longleftarrow & Q \\
\Big\downarrow{\mathbb{P}^1} & & \Big\downarrow{\mathbb{C} \cup \{\infty\}} & \simeq\Big\downarrow{\{\infty\}} & \\
X_\alpha & \longleftarrow & Q_\alpha & = & Q_\alpha
\end{array}
$$

$\mathcal{L}_{\lambda-\rho}$ is defined near Q. The same data defines $\mathcal{L}_{s_\alpha\lambda-\rho+\alpha}$ near Q_1, since as ordered Cartan subgroups $\mathfrak{h} \subset \mathfrak{b}$, $\mathfrak{h} \subset s_\alpha\mathfrak{b}$ are not equal, but related by s_α.

Lemma. Suppose λ is integrally dominant i.e. $2\frac{(\lambda,\alpha)}{(\alpha,\alpha)} \neq -1, -2, \ldots$. Then $H^*(X, j_+(\mathcal{O}_Q(\mathcal{L}_{\lambda-\rho}|_Q))) \simeq H^*(X, j_{1+}(\mathcal{O}_{Q_1}(\mathcal{L}_{s_\alpha\lambda-\rho-\alpha}|_{Q_1})))$.

Proof. The Leray spectral sequence relates this to a computation on the fiber, which are like the case of Verma modules for A_1.

Remark.

(1) It is possible to analyze the case of negative integrality for cases IV and V, on the G- orbit side.

(2) In the Beilinson-Bernstein picture, we get *not only* statements about intertwining operators for standard modules, but operations on the entire category.

(3) We have an equivalence of categories $\mathcal{M}(U_\lambda, K) \simeq \mathcal{M}(\mathcal{D}_\lambda, K)$ for λ integrally dominant and regular, we also have a vanishing theorem for higher cohomology of any $\mathcal{D}_\lambda$-modules if λ is integrally dominant.

(4) We can go from the Beilinson-Bernstein picture to Langlands picture (i.e., m.r. orbits) by a succession of applications of the lemma.

(5) We can go from "dual to Langlands" (on m.r. orbits) by reversing the succession to the Vogan-Zuckerman picture (on m.i. orbits).

(6) A more precise vanishing theorem for m.r. orbits is obtained by the duality theorem, in fact, we have the vanishing of cohomology if $(\lambda, \alpha) \geq 0$ for all imaginary positive roots. We can use the lemma to get a more general vanishing theorem on all orbits.

Example. Fix $G = SL(2, \mathbb{C})$. Thus $\mathfrak{g} = \mathfrak{sl}(2, \mathbb{C}) \times \mathfrak{sl}(2, \mathbb{C})$ $X = \mathbb{P}^1 \times \mathbb{P}^1$ ($\mathbb{P}^1 = \mathbb{C} \cup \{\infty\}$), $K_\mathbb{C}$ is the diagonal image of $SL(2, \mathbb{C})$ in $G_\mathbb{C} = SL(2, \mathbb{C}) \times SL(2, \mathbb{C})$, $K_\mathbb{C}$ acts diagonally on X; $G \simeq \{(g, \bar{g}) \in SL(2, \mathbb{C}) \times SL(2, \mathbb{C})\}$, G acts on X in the obvious manner. The $K_\mathbb{C}$-orbits are: $Q =$ diagonal on $\mathbb{P}^1 \times \mathbb{P}^1$, $Q_1 =$ complement. Q is of type m.r.; Q_1 is of type m.i. The G orbits are: $S_1 = \{(z, \bar{z}) \in \mathbb{P}^1 \times \mathbb{P}^1, z \in \mathbb{P}^1\}$, $S =$ complement. There is one Cartan subgroup, up to conjugacy $H = TA$, T is a connected one dimensional real torus, A is isomorphic to $\mathbb{R}$. If λ is in h^*, λ lifts to a character of H if and only if λ is T-integral.

Observation.

(1) Any T-integral λ is either integral with respect to all roots, or with respect to none.

(2) An irreducible Harish-Chandra module for $SL(2, \mathbb{C})$ is either a finite dimensional irreducible representation or an irreducible principal series.

Case 0) $\lambda \notin \Lambda$, then λ is regular and integrally dominant, hence all standard modules are irreducible and we have no higher cohomology. Any standard module can be realized either on a closed or on an open orbit.

—Langlands modules: on the closed orbit.

—Vogan-Zuckerman modules: on the closed orbit.

—Beilinson-Bernstein, can pick out orbit depending on λ.

Case 1) $\lambda \in \Lambda$, then $\mathcal{L}_\lambda \longrightarrow X$ is globally defined, we have the exact sequence of sheaves

$$0 \longrightarrow \mathcal{O}(\mathcal{L}_{\lambda-\rho}) \longrightarrow j_+\mathcal{O}_Q(\mathcal{L}_{\lambda-\rho}|_Q) \longrightarrow J(Q, \mathcal{L}_{\lambda-\rho}) \longrightarrow 0.$$

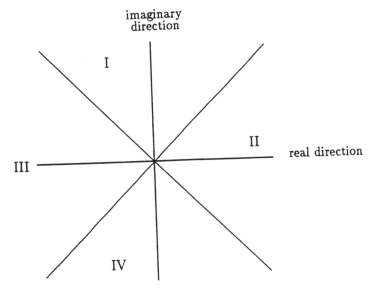

Since Q_1 is affine, we have that

$$H^p(X, j_+(\mathcal{O}_{Q_1}(\mathcal{L}_{Q_1}(\mathcal{L}_{\lambda-\rho}|_{Q_1}))) = 0 \text{ for all } p > 0.$$

We also have

$$H^p(X, \mathcal{O}(\mathcal{L}_{\lambda-\rho})) = \begin{cases} 0 \text{ for } p \neq 0, \quad \text{chamber } I \\ 0 \text{ for } p \neq 1, \quad \text{chamber } II, III \\ 0 \text{ for } p \neq 2, \text{ chamber } IV \\ \text{non zero finite dimensional irreducible in the} \\ \quad \text{remaining cases if } \lambda \text{ is regular} \\ 0 \text{ for all } p \text{ if } \lambda \text{ is singular.} \end{cases}$$

Therefore $H^p(X, j_+(\mathcal{O}_Q(\mathcal{L}_{\lambda-\rho}|_Q))) = 0$ for $p > 0$, in the chambers I, II, III and not for the IV chamber.

(I) $\qquad Q_1 \leftrightarrow$ finite dimensional submodule.
$\qquad\qquad Q \leftrightarrow$ irreducible quotient of the above module.

(II,III) $\qquad Q_1 \leftrightarrow$ irreducible
$\qquad\qquad Q \leftrightarrow$ finite dimensional quotient, the above
$\qquad\qquad\qquad$ representation is its unique irreducible
$\qquad\qquad\qquad$ submodule

(IV) $Q_1 \hookleftarrow$ has a finite dim quotient
              $Q \hookleftarrow$ has a finite dim quotient in H^0

Langlands:

finite dimensional$\longrightarrow Q_1, I$

irreducible principal series$\longrightarrow Q_1$ half of I, half of II

Vogan-Zuckerman
finite dimensional$\longrightarrow Q_1$ half of I, half of II

irreducible principal
series$\longrightarrow Q, I$

Beilinson-Bernstein
finite dimensional$\longrightarrow Q_1, I$

irreducible principal
series$\longrightarrow Q, I$

REFERENCES

1. W. Casselman, *Canonical Extensions of Harish-Chandra modules to representations of G*, Canadian Journal of Mathematics (1989), 385-438.
2. H. Hecht, D. Milicic, W. Schmid and J. Wolf, *Localization and standard modules for real semisimple Lie groups I: The duality theorem*, Inv. Math. **90**, 297-332.
3. H. Hecht, D. Milicic, W. Schmid and J. Wolf, *Localization and standard modules for real semisimple Lie groups II* (to appear).
4. H. Hecht and W. Schmid, *On the asymptotics of Harish-Chandra modules*, Crelle, Band **343**, 169-183.
5. S. Helgason, *A duality for symmetric spaces with applications to group representations I, II*, Advances in Math., 154; II **t.22**, 1187-219.
6. M. Kashiwara, R. Kawai and A. Kimura, *Foundations of Algebraic Analysis*, Princeton Mathematical Series **37**, Princeton University Press.
7. M. Kashiwara, A. Kowata, K. Minemura, K. Okamoto, T. Oshima and M. Tanaka, *Eigenfunctions of invariant differential operators on a symmetric space*, Ann. of Math. **t. 107**, 1-39.
8. W. Schmid, *Boundary value problems for group invariant differential equations*, Société Mathématique de France, Astérisque, hors série, 311-321.
9. W. Schmid, *Homogeneous Complex Manifolds and Representation of semisimple Lie Groups*, in Representation Theory and Harmonic Analysis on Semisimple Lie Groups, Sally-Vogan editors, Math Surveys and monograph AMS (v. 31), p. 223-286..
10. W. Schmid, J. Wolf, *Globalizations of Harish-Chandra Modules*, Bull. AMS (New Series) **17, number 1** (Lul. 1987), 117-120.
11. W. Schmid, J. Wolf, *Goemetric Quantization and Derived Functor Modules for Semisimple Lie Groups* (to appear).

12. N. Wallach, *Asymptotic expansions of generalized matrix entries of real reductive groups*, Lie Group representations I, Lecture Notes in Math. 1024, Springer Verlag, 1983, pp. 287-369.

Department of Mathematics
Harvard University
Cambridge, MA 02138

INTRODUCTION TO QUANTIZED
ENVELOPING ALGEBRAS

GEORGE LUSZTIG

1. INTRODUCTION

1.1 Let $\mathbf{g}$ be a split semisimple Lie algebra over $\mathbf{Q}$. Among the $\mathbf{Q}$–linear maps $f : \mathbf{g} \to R$ (where R is an associative algebra over $\mathbf{Q}$) satisfying

$$f([x, x']) = f(x)f(x') - f(x')f(x) \quad \forall x, x' \in \mathbf{g},$$

one can find a "universal" one, i.e. one through which all other such maps factorize uniquely; the corresponding algebra is denoted $\bar{\mathbf{U}}$ and is called the *(universal) enveloping algebra* of $\mathbf{g}$. The importance of $\bar{\mathbf{U}}$ is that the problem of studying the representations of the Lie algebra $\mathbf{g}$ is seen to be equivalent to that of studying the modules over the algebra $\bar{\mathbf{U}}$ hence it can be approached using techniques of non-commutative ring theory.

1.2 The enveloping algebra $\bar{\mathbf{U}}$ of $\mathbf{g}$ is in a natural way a *Hopf algebra*; in particular, it is equipped with an algebra homomorphism

$$\Delta : \bar{\mathbf{U}} \to \bar{\mathbf{U}} \otimes \bar{\mathbf{U}}$$

(called *comultiplication*). Note that Δ is given by the universal property of $f : \mathbf{g} \to \bar{\mathbf{U}}$ applied to the map

$$f' : \mathbf{g} \to \bar{\mathbf{U}} \otimes \bar{\mathbf{U}} \quad (f'(P) = f(P) \otimes 1 + 1 \otimes f(P)).$$

It is *cocommutative* in the sense that the composition of Δ with the involution $\bar{\mathbf{U}} \otimes \bar{\mathbf{U}} \to \bar{\mathbf{U}} \otimes \bar{\mathbf{U}}$ (switching factors) is equal to Δ.

1.3 While the Lie algebra structure of $\mathbf{g}$ is completely rigid, there is no a priori reason why the same should be true for the Hopf algebra structure of $\bar{\mathbf{U}}$. In 1985, Drinfeld [D] and Jimbo [J] independently discovered a remarkable such deformation called a *quantized enveloping algebra* or a *quantum group*; one of the main features of this new Hopf algebra is that its comultiplication is no longer cocommutative. (In the case of the Lie algebra of SL_2 this deformation has been found earlier by Kulish, Reshetikhin and Sklyanin.)

In Drinfeld's definition, this deformation is infinitesimal, i.e. it is a Hopf algebra over the field of formal power series, while in Jimbo's definition it is a Hopf algebra over the field of rational functions in one variable.

Supported in part by National Science Foundation Grant DMS 8702842 .

As shown in [L1], one can naturally define this deformation over the ring $\mathbf{Z}[v, v^{-1}]$ where v is an indeterminate. We can therefore specialize v to be any complex number, for example a root of 1.

1.4 The quantized enveloping algebras seem to be very interesting objects for several reasons. First, these algebras are important for statistical mechanics (from which their discovery came), conformal field theory and knot theory. In addition, a number of things about algebraic groups, Lie algebras and their representations are now seen to be shadows from the corresponding quantized objects, and this gives new insight about them. For example, the finite dimensional irreducible representations of $\mathbf{g}$ can be deformed into representations of the quantized enveloping algebra; this fact has led to a construction of a canonical basis of these representations with very favourable properties [L6].

The Frobenius map on a semisimple group over an algebraically closed field of characteristic p is seen to be the shadow of a "Frobenius map" in characteristic zero, from a quantum group with parameter $exp(2\pi i/p)$ to an ordinary semisimple group over $\mathbf{C}$ (see [L2],[L5]).

This fact will probably have applications to the study of phenomena in characteristic p. For example, it is likely (cf. [L2],[L4]) that to a large extent, the theory of modular representations of a semisimple group over an algebraically closed field of characteristic p is the same as the theory of finite dimensional representations of the corresponding quantum group at a p−th root of 1.

Another reason which makes the quantized enveloping algebras interesting (at least to me) is that the finite dimensional representation theory of these algebras (in the case when the parameter is a root of 1) seems to be very closely related with a part of the representation theory at a negative central charge for an affine algebra. (See [L3] for a conjecture in this direction).

1.5 This paper is only an introduction to the subject and will not try to cover all the topics mentioned above. In preparing this paper I have benefited greatly from notes taken at my lectures by Nicolas Andruskiewitsch; I wish to thank him here.

2. DEFINITION OF U

2.1 Assume given an $n \times n$ matrix $(a_{ij})_{1 \leq i,j \leq n}$ of integers such that

$$a_{ii} = 2, \quad a_{ij} \leq 0 \text{ for } i \neq j.$$

We also assume that there exist integers $d_1, \ldots, d_n \in \{1, 2, \ldots\}$ such that the matrix $(d_i a_{ij})$ is symmetric, positive definite. (Thus, (a_{ij}) is a Cartan matrix.) If we require that $d_1 + \cdots + d_n$ is as small as possible, then the integers $d_1, \ldots, d_n$ are uniquely determined. We then have $d_i \in \{1, 2, 3\}$.

To these data corresponds in the usual way a semisimple Lie algebra over $\mathbf{Q}$; let $\bar{\mathbf{U}}$ be the corresponding enveloping algebra. This is an associative algebra over $\mathbf{Q}$ with generators $e_i, f_i, h_i (1 \leq i \leq n)$ and relations

$$h_i h_j = h_j h_i$$

$$h_i e_j - e_j h_i = a_{ij} e_j, \quad h_i f_j - f_j h_i = -a_{ij} f_j$$
$$e_i f_j - f_j e_i = \delta_{ij} h_i$$
$$\sum_{r+s=1-a_{ij}} (-1)^s \binom{1-a_{ij}}{s} e_i^r e_j e_i^s = 0, \text{ if } i \neq j$$
$$\sum_{r+s=1-a_{ij}} (-1)^s \binom{1-a_{ij}}{s} f_i^r f_j f_i^s = 0, \text{ if } i \neq j.$$

2.2 We shall use the following notation. For any integers M, N, d with $N \geq 0$ and $d > 0$ we define

$$\begin{bmatrix} M \\ N \end{bmatrix}_d = \prod_{s=1}^{N} \frac{v^{d(M+1-s)} - v^{d(-M-1+s)}}{v^{ds} - v^{-ds}} \in \mathbf{Q}(v),$$

$$[N]_d^! = \prod_{h=1}^{N} \frac{v^{dh} - v^{-dh}}{v^d - v^{-d}} \in \mathbf{Q}(v).$$

(Here, v is an indeterminate.)

Following Drinfeld [D] and Jimbo [J], we shall attach to the data in 2.1 the "quantized enveloping algebra" $\mathbf{U}$. This is an associative algebra over $\mathbf{Q}(v)$, with generators E_i, F_i, K_i, K_i^{-1} $(1 \leq i \leq n)$ and relations:

$$K_i K_j = K_j K_i, \quad K_i K_i^{-1} = K_i^{-1} K_i = 1$$
$$K_i E_j = v^{d_i a_{ij}} E_j K_i, \quad K_i F_j = v^{-d_i a_{ij}} F_j K_i$$
$$E_i F_j - F_j E_i = \delta_{ij} \frac{K_i - K_i^{-1}}{v^{d_i} - v^{-d_i}}$$
$$\sum_{r+s=1-a_{ij}} (-1)^s \begin{bmatrix} 1-a_{ij} \\ s \end{bmatrix}_{d_i} E_i^r E_j E_i^s = 0, \text{ if } i \neq j$$
$$\sum_{r+s=1-a_{ij}} (-1)^s \begin{bmatrix} 1-a_{ij} \\ s \end{bmatrix}_{d_i} F_i^r F_j F_i^s = 0, \text{ if } i \neq j.$$

2.3 $\mathbf{U}$ is a Hopf algebra with comultiplication Δ defined by
$$\Delta E_i = E_i \otimes 1 + K_i \otimes E_i, \quad \Delta F_i = F_i \otimes K_i^{-1} + 1 \otimes F_i, \quad \Delta(K_i) = K_i \otimes K_i.$$

2.4 Let $\Omega, \Psi : \mathbf{U} \to \mathbf{U}^{opp}$ be the $\mathbf{Q}$-algebra homomorphisms defined by
$$\Omega E_i = F_i, \quad \Omega F_i = E_i, \quad \Omega K_i = K_i^{-1} \quad \Omega v = v^{-1}$$
$$\Psi E_i = E_i, \quad \Psi F_i = F_i, \quad \Psi K_i = K_i^{-1}, \quad \Psi v = v.$$

2.5 Let $\mathbf{U}^+$ (resp. $\mathbf{U}^-$) be the subalgebra of $\mathbf{U}$ generated by the elements E_i (resp. F_i) for all i. Let $\mathbf{U}^0$ be the subalgebra of $\mathbf{U}$ generated by the elements K_i, K_i^{-1} for all i.

It is known (see [Ro]) that multiplication defines an isomorphism of $\mathbf{Q}(v)$−vector spaces
$$\mathbf{U}^- \otimes \mathbf{U}^0 \otimes \mathbf{U}^+ \cong \mathbf{U}.$$

2.6 Let $^- : \mathbf{U} \to \mathbf{U}$ be the $\mathbf{Q}$-algebra involution defined by
$$E_i \to E_i, \quad F_i \to F_i, \quad K_i \to K_i^{-1}$$
for all i and $v \to v^{-1}$. This maps $\mathbf{U}^+$ into itself.

3. AN EXAMPLE

3.1 In this section we shall give, following [BLM], a geometric interpretation for the quantized enveloping algebra $\mathbf{U}$ corresponding to the Lie algebra of SL_m $(m \geq 2)$, which corresponds to a Cartan matrix with $a_{ij} = -1$ for $|i - j| = 1$, $a_{ij} = 0$ for $|i - j| > 1$ and $n = m - 1$. (We take $d_i = 1$.)

The geometric interpretation given in [BLM] is in terms of flags in a vector space over a finite field.

3.2 The first step is to construct geometrically some finite dimensional algebras $\mathbf{K}_d$ (one for each integer $d \geq 0$).

As a $\mathbf{Q}(v)$–vector space, $\mathbf{K}_d$ has basis elements e_A where A runs over the (finite) set Θ_d of all $m \times m$ matrices with integer, ≥ 0 entries with sum of entries equal to d. The multiplication is given by

$$e_A e_{A'} = \sum_{A'' \in \Theta_d} g_{A,A',A''} e_{A''}$$

for certain

$$g_{A,A',A''} = c_0 + c_1 v^2 + \cdots + c_N v^{2N} \in \mathbf{Z}[v];$$

to specify $g_{A,A',A''}$ it is enough to specify the integer $g_{A,A',A'';q} = c_0 + c_1 q + \cdots + c_N q^N$ for any q, a power of a prime number.

Let V be a vector space of dimension d over a finite field F_q with q elements. Let $\mathcal{F}$ be the set of all m–step filtrations $V_1 \subset V_2 \subset \cdots \subset V_m = V$. The group $GL(V)$ acts naturally on $\mathcal{F}$. There is a natural bijection

$$GL(V) \backslash \mathcal{F} \times \mathcal{F} \cong \Theta_d$$

from the set of orbits of $GL(V)$ on $\mathcal{F} \times \mathcal{F}$ (diagonal action) to the set Θ_d. It is defined by $f, f' \rightarrow A = (A_{ij})$ where $f = (V_1 \subset V_2 \subset \cdots \subset V_m)$, $f' = (V_1' \subset V_2' \subset \cdots \subset V_m')$, $A_{ij} = \dim(X_{ij}/X_{i,j-1})$. (We set $V_0 = V_0' = 0$, $X_{ij} = V_{i-1} + (V_i \cap V_j')$ $(1 \leq i \leq m, 0 \leq j \leq m)$.)

Let $\mathcal{O}_A \subset \mathcal{F} \times \mathcal{F}$ be the $GL(V)$–orbit corresponding to the matrix $A \in \Theta_d$. For any $A, A', A'' \in \Theta_d$, the integer $g_{A,A',A'';q}$ is defined as follows: we choose $(f_1, f_2) \in \mathcal{O}_{A''}$ and we let $g_{A,A',A'';q}$ be the number of $f \in \mathcal{F}$ such that $(f_1, f) \in \mathcal{O}_A$, $(f, f_2) \in \mathcal{O}_{A'}$. This completes the definition of the $\mathbf{Q}(v)$–algebra $\mathbf{K}_d$.

For each $A = (A_{ij}) \in \Theta_d$, let

$$[A] = v^{-s} e_A \text{ where } s = \sum_{i,j,k,l;i \geq k;j < l} A_{ij} A_{kl}.$$

The elements $[A]$ form a $\mathbf{Q}(v)$–basis of $\mathbf{K}_d$. We have

$$[A][A'] = \sum_{A'' \in \Theta_d} \tilde{g}_{A,A',A''} [A'']$$

with $\tilde{g}_{A,A',A''} \in \mathbf{Q}(v)$.

3.3 Next, we shall define a $\mathbf{Q}(v)$−algebra $\mathbf{K}$ (without unit) whose structure constants are obtained from those of $\mathbf{K}_d$ by a limit procedure (for $d \to \infty$). As a $\mathbf{Q}(v)$−vector space, $\mathbf{K}$ has a basis $[A]$ indexed by the set $\tilde{\Theta}$ consisting of all $m \times m$ matrices A with integer entries such the entries off diagonal are ≥ 0. The multiplication is given by

$$[A][A'] = \sum_{A'' \in \tilde{\Theta}} G_{A,A',A''}[A'']$$

for certain elements $G_{A,A',A''} \in \mathbf{Q}(v)$. Assume that the sum of entries of A is the same as that of A' and that of A''. (If this is not satisfied, we take $G_{A,A',A''} = 0$.)

For $A \in \tilde{\Theta}$ and $p \in \mathbf{Z}$ we set

$$_pA = A + pI \in \tilde{\Theta}$$

where $I = (\delta_{i,j})$ is the identity $m \times m$ matrix. Clearly, if p is large enough, we have $_pA \in \Theta_d$, $_pA' \in \Theta_d$, $_pA'' \in \Theta_d$ for some $d \geq 0$. Let v' be an indeterminate (independent of v). One shows that there is a unique element $\tilde{G}_{A,A',A''}(v,v') \in \mathbf{Q}(v)[v']$ such that

$$\tilde{g}_{_pA,_pA',_pA''} = \tilde{G}_{A,A',A''}(v,v^{-p})$$

for all sufficiently large p. We then set

$$G_{A,A',A''} = \tilde{G}_{A,A',A''}(v,1).$$

This completes the definition of the $\mathbf{Q}(v)$−algebra $\mathbf{K}$.

3.4 Let Θ^0 be the set of all $m \times m$ matrices with integer, ≥ 0 entries with all diagonal entries equal to zero. Given $A \in \Theta^0$ and $\mathbf{j} = (j_1, \ldots, j_m) \in \mathbf{Z}^n$ such that $j_1 + \cdots + j_m = 0$, we define

$$A(\mathbf{j}) = \sum_Z v^{z_1 j_1 + \cdots + z_m j_m}[A + Z]$$

where Z runs over the set of all diagonal matrices with diagonal entries $(z_1, \ldots, z_m) \in \mathbf{Z}^m$. The elements $A(\mathbf{j})$ as above span (and form a basis of) a $\mathbf{Q}(v)$−vector subspace $\mathbf{K}_\infty$ of the $\mathbf{Q}(v)$−vector space of (possibly) infinite linear combinations of the basis elements of $\mathbf{K}$. It turns out that the product of two elements of $\mathbf{K}_\infty$ (computed term by term using the product in $\mathbf{K}$) is well defined and is again an element of $\mathbf{K}_\infty$. Thus, $\mathbf{K}_\infty$ is a $\mathbf{Q}(v)$−algebra. Moreover, there is a unique algebra isomorphism

$$\mathbf{U} \cong \mathbf{K}_\infty$$

such that

$E_i \to \sum_A [A]$ (A runs over all matrices in $\tilde{\Theta}$ with entry 1 in the $(i, i+1)$ position and entry zero in all other off-diagonal positions);

$F_i \to \sum_A [A]$ (A runs over all matrices in $\tilde{\Theta}$ with entry 1 in the $(i+1, i)$ position and entry zero in all other off-diagonal positions);

$K_i \to \sum_Z v^{z_i - z_{i+1}}[Z]$ (Z runs over all diagonal matrices in $\tilde{\Theta}$ aand z_i are the diagonal entries of Z).

This completes our description of $\mathbf{U}$, following [BLM].

4. THE ALGEBRA U

4.1 We return to the setup in §2. In [C], Chevalley has stated that any finite dimensional $\bar{\mathbf{U}}$−module admits a $\mathbf{Z}$−lattice stable under the subring $\bar{U}$ of $\bar{\mathbf{U}}$ generated by the elements

$$e_i^{(N)} = e_i^N/N!, \quad f_i^{(N)} = f_i^N/N!;$$

in [K], Kostant constructed a nice $\mathbf{Z}$−basis of $\bar{U}$. We will construct a quantum analogue of $\bar{U}$.

We shall use the notation

$$E_i^{(N)} = ([N]_{d_i}^!)^{-1}E_i^N, \quad F_i^{(N)} = ([N]_{d_i}^!)^{-1}F_i^N \quad (N \geq 0).$$

4.2 Let $\mathcal{A}$ be the subring $\mathbf{Z}[v, v^{-1}]$ of $\mathbf{Q}(v)$. Following [L1], we define U to be the $\mathcal{A}$-subalgebra of $\mathbf{U}$ generated by the elements

$$E_i^{(N)}, F_i^{(N)}, K_i, K_i^{-1} \quad (1 \leq i \leq n, N \geq 0).$$

Let U^+ (resp. U^-) be the $\mathcal{A}$-subalgebra of U generated by the elements $E_i^{(N)}$ (resp. $F_i^{(N)}$) for $N \geq 0$ and all i.

4.3 For any $i \in [1, n]$, and any integer $N \geq 0$ we set

$$\begin{bmatrix} K_i \\ N \end{bmatrix} = \prod_{s=1}^{N} \frac{v^{d_i(-s+1)}K_i - v^{d_i(s-1)}K_i^{-1}}{v^{d_is} - v^{-d_is}} \in \mathbf{U}^0.$$

We have $\begin{bmatrix} K_i \\ N \end{bmatrix} \in U$ for all $N \geq 0$. Let U^0 be the $\mathcal{A}$-subalgebra of U generated by the elements $\begin{bmatrix} K_i \\ N \end{bmatrix}$ and K_i, K_i^{-1} ($i \in [1, n]$, $N \geq 0$).

5. HIGHEST WEIGHT $\mathbf{U}$−MODULES

5.1 This section is essentially an exposition of the results of [L1] relating certain representations of $\mathbf{U}$ with representations of $\bar{\mathbf{U}}$.

Let M be a $\mathbf{U}$−module. For any $\mathbf{k} = (k_1, \ldots, k_n) \in \mathbf{Z}^n$, define $M^{\mathbf{k}} = \{\xi \in M | K_i\xi = v^{d_ik_i}\xi \ \forall i\}$; this is the *weight space* corresponding to $\mathbf{k}$. A vector $\xi \in M$ is said to be *primitive* if $\xi \neq 0$, $\xi \in M^{\mathbf{k}}$ for some $\mathbf{k} \in \mathbf{Z}^n$ and $E_i\xi = 0$ for all i. We say that M is a *highest weight module* for $\mathbf{U}$ if there exists a primitive vector $\xi \in M$ such that $\mathbf{U}\xi = M$.

Then ξ is uniquely determined up to a non-zero scalar; we have $\xi \in M^{\mathbf{k}}$ where $\mathbf{k}$ is also uniquely determined by M. (We say that $\mathbf{k}$ is the *highest weight* of M.) Moreover, the map $\mathbf{U}^- \to M$ given by $f \to f\xi$ is surjective. If, in addition, the last map is an isomorphism of $\mathbf{Q}(v)$−vector spaces, we say that M is a *Verma module*.

For any $\mathbf{k} \in \mathbf{Z}^n$ there is a Verma module with highest weight $\mathbf{k}$; it is unique, up to isomorphism. We denote it $M_{\mathbf{k}}$. For any $\mathbf{k} \in \mathbf{Z}^n$ there is a highest weight $\mathbf{U}$−module which is simple as a $\mathbf{U}$−module and which has highest weight $\mathbf{k}$; it is unique, up to isomorphism. We denote it $L_{\mathbf{k}}$. Now

$M_{\mathbf{k}}$ has a unique simple quotient module which is necessarily isomorphic to $L_{\mathbf{k}}$.

5.2 Given $\mathbf{k} \in \mathbf{Z}^n$, we have that $L_{\mathbf{k}}$ is finite dimensional if and only if $\mathbf{k} \in \mathbf{N}^n$. In this case, we consider a primitive vector ξ of $L_{\mathbf{k}}$. Let $L_{\mathbf{k},\mathcal{A}}$ be the U–submodule of $L_{\mathbf{k}}$ generated by ξ.

Let $\bar{L}_{\mathbf{k}} = \mathbf{Q}L_{\mathbf{k},\mathcal{A}}/\mathbf{Q}(v-1)L_{\mathbf{k},\mathcal{A}}$. The endomorphisms $E_i, F_i, \begin{bmatrix} K_i \\ 1 \end{bmatrix}, K_i$ of $L_{\mathbf{k},\mathcal{A}}$ define by extension af scalars endomorphisms of $\bar{L}_{\mathbf{k}}$ denoted $e_i, f_i,$ $h_i, \bar{K}_i$. We have $\bar{K}_i = 1$ and e_i, f_i, h_i satisfy the relations of $\bar{U}$ hence define a $\bar{U}$–module structure on $\bar{L}_{\mathbf{k}}$. One can show that this is a simple (finite dimensional) $\bar{U}$–module with highest weight $\mathbf{k}$ and that the dimensions of its weight spaces (which are given explicitly by Weyl's character formula) are the same as the dimensions of the corresponding weight spaces of $L_{\mathbf{k}}$.

Hence the finite dimensional U–modules $L_{\mathbf{k}}$ have essentially the same structure as in the classical case (of $\bar{U}$). It also follows that the finite dimensional simple $\bar{U}$–modules admit "quantum deformations".

6. Braid group action

6.1 For the material in this section, see [L4],[L5].

Let X (resp.Y) be the free abelian group with basis ϖ_i (resp.$\check{\alpha}_i$), $1 \le i \le n$. Let $<,>: Y \times X \to \mathbf{Z}$ be the bilinear pairing such that $< \check{\alpha}_i, \varpi_j > = \delta_{ij}$ and let $\alpha_j \in X$ be defined by $< \check{\alpha}_i, \alpha_j > = a_{ij}$. Define $s_i : X \to X, s_i :$ $Y \to Y$ by $s_i(x) = x- < \check{\alpha}_i, x > \alpha_i, s_i(y) = y- < y, \alpha_i > \check{\alpha}_i$. We identify $GL(X) = GL(Y)$ using the pairing above and we let W be its (finite) subgroup generated by $s_1, \ldots, s_n$. Let Π be the set consisting of $\alpha_1, \ldots, \alpha_n$, let $R = W\Pi \subset X$ and let $R^+ = R \cap (\mathbf{N}\alpha_1 + \cdots + \mathbf{N}\alpha_n)$.

It is well known that the following are a complete set of relations of W:

$$s_i^2 = 1$$

$$s_i s_j s_i \cdots = s_j s_i s_j \ldots \quad \text{if } i \ne j.$$

In the last equation both products involve 2 (resp. 3,4,6) factors according to whether $a_{ij}a_{ji}$ is equal to 0 (resp. 1,2,3).

The braid group B is defined by generators $T_1, \ldots, T_n$ and relations

$$T_i T_j T_i \cdots = T_j T_i T_j \ldots \quad \text{if } i \ne j$$

like the second set of relations defining W; but T_i^2 is not assumed to be 1.

6.2 In the classical case, the Weyl group, or rather a small finite covering of it acts naturally by automorphisms on the Lie algebra, hence also on the enveloping algebra. This has a quantum analogue; however, we now have only a braid group action and it preserves only the algebra structure of U, not the comultiplication.

For any $i \in [1, n]$, there is a unique algebra automorphism T_i of U such that

$$T_i E_i = -F_i K_i, \qquad T_i E_j = \sum_{r+s=-a_{ij}} (-1)^r v^{-d_i s} E_i^{(r)} E_j E_i^{(s)} \quad \text{if } i \ne j$$

$$T_i F_i = -K_i^{-1} E_i, \qquad T_i F_j = \sum_{r+s=-a_{ij}} (-1)^r v^{d_i s} F_i^{(s)} F_j F_i^{(r)} \quad \text{if } i \neq j$$

$$T_i K_j = K_j K_i^{-a_{ij}}.$$

Its inverse T_i^{-1} is given by

$$T_i^{-1} E_i = -K_i^{-1} F_i, \qquad T_i^{-1} E_j = \sum_{r+s=-a_{ij}} (-1)^r v^{-d_i s} E_i^{(s)} E_j E_i^{(r)} \quad \text{if } i \neq j$$

$$T_i^{-1} F_i = -E_i K_i, \qquad T_i^{-1} F_j = \sum_{r+s=-a_{ij}} (-1)^r v^{d_i s} F_i^{(r)} F_j F_i^{(s)} \quad \text{if } i \neq j$$

$$T_i^{-1} K_j = K_j K_i^{-a_{ij}}.$$

We have $T_i^{-1} = \Psi T_i \Psi$ and $T_i \Omega = \Omega T_i$.

6.3 The automorphisms T_i satisfy the relations of B; hence they define an action of the braid group B on the algebra $\mathbf{U}$. It is convenient to slightly modify these automorphisms. Let $r_i : \mathbf{U} \to \mathbf{U}$ be the algebra automorphism defined by

$$r_i E_j = (-1)^{a_{ij}} E_j, \quad r_i F_j = (-1)^{a_{ij}} F_j, \quad r_i K_j = K_j.$$

Consider the composition $\tilde{T}_i = T_i^{-1} r_i$. The automorphisms $\tilde{T}_i$ still satisfy the relations of B; hence they define an action of the braid group B on the algebra $\mathbf{U}$.

7. A BASIS OF U

7.1 The $\mathbf{Z}$ basis of $\bar{U}$ described in [K] can been generalized to the quantum case. We shall explain the result.

Let w_0 be the longest element of W and let ν be the number of elements of R^+. Let $\mathcal{X}$ be the set of all sequences $\mathbf{i} = (i_1, \ldots, i_\nu)$ in $[1, n]$ such that $s_{i_1} \ldots s_{i_\nu} = w_0$. This set is known to be non-empty.

Given $\mathbf{i} = (i_1, \ldots, i_\nu) \in \mathcal{X}$ and $\mathbf{c} = (c_1, \ldots, c_\nu) \in \mathbf{N}^\nu$ we define

$$E_{\mathbf{i}}^{\mathbf{c}} = E_{i_1}^{(c_1)} \tilde{T}_{i_1}(E_{i_2}^{(c_2)}) \tilde{T}_{i_1} \tilde{T}_{i_2}(E_{i_3}^{(c_3)}) \ldots \tilde{T}_{i_1} \tilde{T}_{i_2} \ldots \tilde{T}_{i_{\nu-1}}(E_{i_\nu}^{(c_\nu)}).$$

The following holds.

For any $\mathbf{i} \in \mathcal{X}$, the elements

$$E_{\mathbf{i}}^{\mathbf{c}} \quad (\mathbf{c} \in \mathbf{N}^\nu)$$

are contained in U^+; they form an $\mathcal{A}$-basis of U^+ and a $\mathbf{Q}(v)$−basis of $\mathbf{U}^+$.

(This is proved in [L4], [L5] for one particular choice of $\mathbf{i}$ in each case. The general case can be reduced to this, see [DL].) We shall denote this basis by $B_{\mathbf{i}}$.

Applying Ω to this basis, we get an $\mathcal{A}$-basis of U^- and a $\mathbf{Q}(v)$−basis of $\mathbf{U}^-$.

Next, one shows that the elements

$$\prod_{i=1}^{n}\left(K_i^{\delta_i}\begin{bmatrix}K_i\\t_i\end{bmatrix}\right),(t_i\geq 0,\delta_i=0\text{ or }1)$$

form an $\mathcal{A}$-basis of U^0 and a $\mathbf{Q}(v)$-basis of $\mathbf{U}^0$.

Finally, one shows that multiplication defines an isomorphism of $\mathcal{A}$-modules

$$U^-\otimes_\mathcal{A} U^0\otimes_\mathcal{A} U^+\cong U.$$

Combining these statements we obtain a basis of U as an $\mathcal{A}$-module which is also a basis of $\mathbf{U}$ as a $\mathbf{Q}(v)$-vector space.

7.2 U inherits from $\mathbf{U}$ a Hopf algebra structure; we shall write the formula for comultiplication:

$$\Delta(E_i^{(N)})=\sum_{b=0}^{N}v^{d_ib(N-b)}E_i^{(N-b)}K_i^b\otimes E_i^{(b)}$$

$$\Delta(F_i^{(N)})=\sum_{a=0}^{N}v^{-d_ia(N-a)}F_i^{(a)}\otimes K_i^{-a}F_i^{(N-a)}.$$

8. A FROBENIUS MAP IN CHARACTERISTIC ZERO

The reference for this section is [L3],[L5].

8.1 If p is a prime number, we may form the algebra $\bar{U}_{F_p}=\bar{U}\otimes F_p$ over the finite field F_p. The algebra $\bar{U}$ is generated by the elements $e_i^{(N)},f_i^{(N)}$ and we denote their images in $\bar{U}_{F_p}$ by the same letters.

Now there is a natural homomorphism of F_p-algebras (*Frobenius map*) $Fr:\bar{U}_{F_p}\to\bar{U}_{F_p}$ defined on the generators by

$$Fr(e_i^{(N)})=e_i^{(N/p)}\text{ if }p\text{ divides }N\text{ and }Fr(e_i^{(N)})=0,\text{ otherwise };$$

$$Fr(f_i^{(N)})=f_i^{(N/p)}\text{ if }p\text{ divides }N\text{ and }Fr(f_i^{(N)})=0,\text{ otherwise }.$$

(This is the transpose of the map induced by Frobenius on the coordinate algebra of a semisimple group over F_q.) It turns out that Fr can be obtain by specialization from a certain "Frobenius map" $\widetilde{Fr}$ in characteristic zero.

8.2 We fix an integer $l\geq 1$ which is relatively prime to any entry of the Cartan matrix; in particular, l is odd. Let $\mathcal{B}$ (resp. $\mathbf{B}$) be the quotient of $\mathcal{A}$ (resp. $\mathbf{Q}[v,v^{-1}]$) by the ideal generated by the l-th cyclotomic polynomial. We have $\mathcal{B}\subset\mathbf{B}$. We denote the image of v in $\mathcal{B}$ again by v; then $v^l=1$ in $\mathbf{B}$. We regard $\mathcal{B}$ and $\mathbf{B}$ as $\mathcal{A}$-algebras via the ring homomorphism $\mathcal{A}\to\mathcal{B}$ which takes v to v, and we form the $\mathcal{B}$-algebra $U_\mathcal{B}=U\otimes_\mathcal{A}\mathcal{B}$ (resp. the $\mathbf{B}$-algebra $U_\mathbf{B}=U\otimes_\mathcal{A}\mathbf{B}$). We shall use the same notation for an element of U and for its image under the canonical homomorphism $U\to U_\mathcal{B}$ or $U\to U_\mathbf{B}$.

One can show that there is a unique **B**−algebra homomorphism $\widetilde{Fr}$: $U_{\mathbf{B}} \to \bar{U} \otimes_{\mathbf{Q}} \mathbf{B}$ such that

$$\widetilde{Fr}(E_i^{(N)}) = e_i^{(N/l)} \text{ if } l \text{ divides } N \text{ and } \widetilde{Fr}(E_i^{(N)}) = 0, \text{ otherwise ;}$$

$$\widetilde{Fr}(F_i^{(N)}) = f_i^{(N/l)} \text{ if } l \text{ divides } N \text{ and } \widetilde{Fr}(F_i^{(N)}) = 0, \text{ otherwise ,}$$

$$\widetilde{Fr}(K_i) = 1.$$

We call it the *Frobenius map*. It restricts to a homomorphism of $\mathcal{B}$−algebras $\widetilde{Fr} : U_{\mathcal{B}} \to \bar{U} \otimes_{\mathbf{Z}} \mathcal{B}$. Now note that $K_i^{2l} = 1$ in $U_{\mathcal{B}}$ and in $U_{\mathbf{B}}$ and that K_i^l is central in $U_{\mathcal{B}}$ and in $U_{\mathbf{B}}$. Let $U'_{\mathcal{B}}$ (resp. $U'_{\mathbf{B}}$) be the quotient of $U_{\mathcal{B}}$ (resp. $U_{\mathbf{B}}$) by the ideal generated by the central elements $K_1^l, \ldots, K_n^l$. Clearly, $\widetilde{Fr}$ is zero on this ideal hence it factorizes through the quotient $U_{\mathcal{B}}$ (resp. $U_{\mathbf{B}}$).

From the resulting homomorphism $\widetilde{Fr}' : U'_{\mathcal{B}} \to \bar{U} \otimes_{\mathbf{Z}} \mathcal{B}$ one can recover the Frobenius map of 8.1 as follows.

Assume for a moment that $l = p$, a prime number. We may identify $\mathcal{B}/(v-1)\mathcal{B} = F_p$ as rings. Applying the functor $\otimes_{\mathcal{B}} F_p$, both $U'_{\mathcal{B}}$ and $\bar{U} \otimes_{\mathbf{Z}} \mathcal{B}$ become $\bar{U}_{F_p}$ and the homomorphism $\widetilde{Fr}' : U'_{\mathcal{B}} \to \bar{U} \otimes_{\mathbf{Z}} \mathcal{B}$ becomes the homomorphism Fr of 8.1. In this sense, one can regard $\widetilde{Fr}$ as a quantization of Fr or as a lifting of Fr to characteristic zero.

8.3 There is one significant difference between the classical Frobenius map Fr and its non-classical analogue $\widetilde{Fr}$. In the classical case one can consider powers Fr^s ($s = 1, 2, \ldots$) while in the non-classical case this does not make sense.

This is reflected in the following property of binomial coefficients. Let $M \geq N \geq 1$ be integers. If p is a prime number, we can write uniquely $M = M_0 + M_1 p + M_2 p^2 + \ldots$, $N = N_0 + N_1 p + N_2 p^2 + \ldots$, with $M_i, N_i \in [0, p-1]$, and we have the well known congruence (with many factors in the right hand side):

$$\binom{M}{N} \equiv \binom{M_0}{N_0}\binom{M_1}{N_1}\binom{M_2}{N_2} \cdots \pmod{p}.$$

On the other hand, if we write $M = m_0 + m_1 l$, $N = n_0 + n_1 l$, we have the equality (with only two factors in the right hand side):

$$\begin{bmatrix} M \\ N \end{bmatrix}_{d_i} = \begin{bmatrix} m_0 \\ n_0 \end{bmatrix}_{d_i} \binom{m_1}{n_1} \in \mathbf{B}.$$

Note also that, while the F_p−algebra $\bar{U}_{F_p}$ has infinitely many generators

$$e_i, e_i^{(p)}, e_i^{(p^2)}, \ldots, f_i, f_i^{(p)}, f_i^{(p^2)}, \ldots$$

and it cannot be generated by finitely many elements, the **B**−algebra $U_{\mathbf{B}}$ is generated by finitely many elements:

$$E_i, E_i^{(l)}, F_i, F_i^{(l)}, K_i \quad (1 \leq i \leq n).$$

9. THE ALGEBRA u

The reference for this section is [L3],[L4],[L5].

9.1 The classical Frobenius map $Fr : \bar{U}_{F_p} \to \bar{U}_{F_p}$ is a homomorphism of Hopf algebras. It has a kernel (in the sense of Hopf algebras); it is the subalgebra of $\bar{U}_{F_p}$ generated by the elements e_i, f_i. It is an F_p–vector space of finite dimension $p^{2\nu+n}$; it can be interpreted as the Hopf algebra dual to the coordinate algebra of the (infinitesimal) kernel of the Frobenius map on the corresponding semisimple group in characteristic p or as a restricted enveloping algebra.

9.2 Similarly, the non-classical Frobenius map $\widetilde{Fr} : U_{\mathbf{B}} \to \bar{U} \otimes_{\mathbf{Q}} \mathbf{B}$ is a Hopf algebra homomorphism; it has a kernel in the sense of Hopf algebras. This is a Hopf subalgebra $\mathbf{u}$ of $U_{\mathbf{B}}$ of finite dimension $l^{2\nu+2n}$ over $\mathbf{B}$. It is the subalgebra of $U_{\mathbf{B}}$ generated by the elements E_i, F_i, K_i. (We have $E_i^l = F_i^l = 0, K_i^{2l} = 1$ in $U_{\mathbf{B}}$.) The quotient of $\mathbf{u}$ by the ideal generated by the central elements $K_i^l - 1$ is again a Hopf algebra; it has dimension $l^{2\nu+n}$. In the case where $l = p$, this can be regarded as a quantization of the finite dimensional Hopf algebra over F_p considered above. One expects (cf. [L5]) that they have the same representation theory. (This would have very interesting consequences for the representation theory of groups in characteristic p.)

9.3 One of the ingredients of the proof of existence of $\widetilde{Fr}$ given in [L4] was the construction of certain algebra derivations of $\mathbf{u}$. More precisely, it was shown that, for any $i \in [1, n]$, the maps $x \to \delta_i(x) = E_i^{(l)} x - x E_i^{(l)}$ and $x \to \delta_i'(x) = F_i^{(l)} x - x F_i^{(l)}$ are well defined derivations of the algebra $\mathbf{u}$. Their exponentials generate an interesting algebraic group of automorphisms of the algebra $\mathbf{u}$.

10. REPRESENTATIONS OF $U_{\mathbf{B}}$

10.1 We want to understand the category $\mathcal{C}$ of finite dimensional $U_{\mathbf{B}}$–modules on which the central elements K_i^l act as identity. In the case where $l = 1$ this is just the classical situation of finite dimensional representations of $\bar{U}$. However, if $l > 1$, this category is more complicated since its objects are not in general semisimple.

The simple objects of $\mathcal{C}$ are classified, as in the classical case, by elements $\mathbf{k} \in \mathbf{N}^n$ as follows. Let $\mathbf{k} \in \mathbf{N}^n$; as in 5.2, consider the simple $\mathbf{U}$–module $L_{\mathbf{k}}$ (of finite dimension over $\mathbf{Q}(v)$) and a primitive vector ξ in it; let also $L_{\mathbf{k},A}$ be the U–submodule of $L_{\mathbf{k}}$ generated by ξ. Let $\mathcal{M}_{\mathbf{k}} = L_{\mathbf{k},A} \otimes_A \mathbf{B}$. This is then a $U_{\mathbf{B}}$–module in a natural way. One can show that it has a unique simple quotient; we denote it $\mathcal{L}_{\mathbf{k}}$. The resulting map $\mathbf{k} \to \mathcal{L}_{\mathbf{k}}$ is a bijection between $\mathbf{N}^n$ and the set of isomorphism classes of simple objects in $\mathcal{C}$.

10.2 If M is a $U_{\mathbf{B}}$–module in $\mathcal{C}$, we can define the $\mathbf{k}$–weight space $M^{\mathbf{k}}$ for any $\mathbf{k} = (k_1, \ldots, k_n) \in \mathbf{Z}^n$ by

$$M^{\mathbf{k}} = \{\xi \in M | K_i \xi = v^{d_i k_i} \xi, \begin{bmatrix} K_i \\ l \end{bmatrix} \xi = \begin{bmatrix} k_i \\ l \end{bmatrix}_{d_i} \xi \ \forall i\}.$$

Then M is the direct sum of its weight spaces. If $M = \mathcal{M}_\mathbf{k}$ with $\mathbf{k} \in \mathbf{N}^n$, then the weight spaces of M have the same dimensions as the corresponding weight spaces of $L_\mathbf{k}$; in particular, their dimension is given by Weyl's character formula.

However, the dimensions of the weight spaces of $\mathcal{L}_\mathbf{k}$ are unknown. There is only a conjectural formula for them (see [L3]); it is in terms of certain polynomials introduced by Kazhdan and the author for the affine Weyl group.

10.3 In the study of representations of semisimple groups in characteristic p, there is a reduction due to Steinberg of the general case to the case of highest weights which are " restricted" (i.e. belong to $[0, p-1]$); this is done by the means of Steinberg's tensor product theorem.

This has a quantum analogue ([L3]) which we now explain. Given $\mathbf{k} \in \mathbf{N}^n$, we can write uniquely $\mathbf{k} = \mathbf{k}' + l\mathbf{k}''$ where $\mathbf{k}' \in [0, l-1]^n$ and $\mathbf{k}'' \in \mathbf{N}^n$. Then one can show that $\mathcal{L}_\mathbf{k} \cong \mathcal{L}_{\mathbf{k}'} \otimes \mathcal{L}_{l\mathbf{k}''}$ as $U_\mathbf{B}$−modules (the tensor product of two $U_\mathbf{B}$−modules is defined in terms of the comultiplication of $U_\mathbf{B}$). Moreover, $\mathcal{L}_{l\mathbf{k}''}$ is obtained by regarding the simple $\bar{\mathbf{U}}$−module with highest weight module $\mathbf{k}''$ as a $U_\mathbf{B}$−module, via the homomorphism $\widetilde{Fr} : U_\mathbf{B} \to \bar{\mathbf{U}} \otimes_\mathbf{Q} \mathbf{B}$. Hence, the weight structure of $\mathcal{L}_{l\mathbf{k}''}$ is known from the classical case. There is one significant difference between Steinberg's tensor product theorem and the tensor product theorem in the quantum case: in the first case, the tensor product involves many factors, while in the second case it involves only two factors. This is due to the phenomenon described in 8.3.

One can show that the quantum tensor product theorem described above remains true in the case when the Cartan matrix is replaced by an extended Cartan matrix (of affine type).

10.4 If $\mathbf{k} \in [0, l-1]^n$, the $U_\mathbf{B}$−module $\mathcal{L}_\mathbf{k}$ remains irreducible upon restriction to $\mathbf{u}$. Moreover, any irreducible $\mathbf{u}$−module on which the K_i^l act as 1 is obtained exactly once as such a restriction. (see [L3],[L4]). This is analogous to a result of Curtis in the representation theory of groups in characteristic p.

For further results on the representations of $U_\mathbf{B}$ (in the case where l is a prime), see [APW].

11. THE CASE $n = 1$

11.1 We shall illustrate the results of §10 in the case where the $n = 1$; this corresponds to the Lie algebra of SL_2. We shall take $d_1 = 1$. We follow [L3].

If $k \in \mathbf{N}$, the $U_\mathbf{B}$−module $\mathcal{M}_k$ has a $\mathbf{B}$−basis $x_k, x_{k-2}, x_{k-4}, \ldots, x_{-k}$ where x_{k-2i} is in the $(k-2i)$−weight space and $x_{k-2i} = F_1^{(i)} x_k$ $(0 \le i \le k-1)$.

We have

$$F_1 x_{k-2i} = \frac{v^{i+1} - v^{-i-1}}{v - v^{-1}} x_{k-2i-2},$$

$$F_1^{(l)} x_{k-2i} = \begin{bmatrix} i+l \\ l \end{bmatrix}_1 x_{k-2i-2l},$$

$$E_1 x_{k-2i} = \frac{v^{k-i+1} - v^{-k+i-1}}{v - v^{-1}} x_{k-2i+2},$$

$$E_1^{(l)} x_{k-2i} = \begin{bmatrix} k-i+l \\ l \end{bmatrix}_1 x_{k-2i+2l},$$

with the convention that $x_{k-2i} = 0$ for $i < 0$ or $i > k$.

11.2 Write $k = k' + lk''$ with $0 \le k' \le l - 1$ and $k'' \ge 0$. If $k'' = 0$ or if $k' = l - 1$, then the $U_{\mathbf{B}}$−module $\mathcal{M}_k$ is simple, hence it is isomorphic to $\mathcal{L}_k$. Assume now that $k' \ne l-1$ and $k'' \ge 1$. Let $\tilde{k} = (l-2-k') + l(k''-1)$. Then the kernel of the canonical (surjective) homomorphism $\mathcal{M}_k \to \mathcal{L}_k$ is isomorphic as a $U_{\mathbf{B}}$−module to $\mathcal{L}_{\tilde{k}}$.

12. A CANONICAL BASIS OF $\mathbf{U}^+$

12.1 In [L6] a canonical basis of $\mathbf{U}^+$ has been constructed under the assumption that $a_{ij} = a_{ji}$ (simply laced case); the general case was only briefly sketched there. I have arrived at that construction when trying to understand Ringel's approach [Ri] to $\mathbf{U}^+$ in terms of representations of quivers. One of the main points of [L6] was a study of how representations of quivers degenerate; this has allowed the introduction of methods from intersection cohomology and has lead to the canonical basis.

As observed in *loc. cit.*, from the point of view of geometry of quiver representations, the non-simply laced case appears as a twisted version of a simply laced case of higher rank. This allows one to transfer many results of *loc. cit.* to the non-simply laced case.

Here we shall review the purely algebraic part of the construction of the canonical basis.

For any $\mathbf{i} \in \mathcal{X}$ we denote by $\mathcal{L}_{\mathbf{i}}$ the $\mathbf{Z}[v^{-1}]$−submodule of $\mathbf{U}^+$ generated by $B_{\mathbf{i}}$ (see 7.1).

First one shows that $\mathcal{L}_{\mathbf{i}}$ is independent of the choice of $\mathbf{i} \in \mathcal{X}$; we denote it $\mathcal{L}$.

Second, one shows that the image of $B_{\mathbf{i}}$ under the canonical projection $\pi : \mathcal{L} \to \mathcal{L}/v^{-1}\mathcal{L}$ is independent of the choice of $\mathbf{i} \in \mathcal{X}$; we denote it β.

Finally, one shows that for any element $b \in \beta$ there is a unique element $\tilde{b} \in \mathcal{L}$ which is fixed by $^- : \mathbf{U}^+ \to \mathbf{U}^+$ and satisfies $\pi(\tilde{b}) = b$. The elements $\{\tilde{b} | b \in \beta\}$ form a $\mathbf{Z}[v^{-1}]$−basis of $\mathcal{L}$, an $\mathcal{A}$−basis of U^+ and a $\mathbf{Q}(v)$−basis of $\mathbf{U}^+$.

We call it the *canonical basis*.

12.2 From the results in 12.1 it follows that, given $\mathbf{i}, \mathbf{i}' \in \mathcal{X}$, there is a unique bijection $R_{\mathbf{i}}^{\mathbf{i}'} : \mathbf{N}^{\nu} \cong \mathbf{N}^{\nu}$ such that $R_{\mathbf{i}}^{\mathbf{i}'}(\mathbf{c}) = \mathbf{c}'$ if and only if $\pi(E_{\mathbf{i}}^{\mathbf{c}}) = \pi(E_{\mathbf{i}'}^{\mathbf{c}'})$.

The bijections $R_{\mathbf{i}}^{\mathbf{i}'}$ are of interest, since they are the main ingredient in the purely combinatorial formula for the dimension of finite dimensional simple $\bar{\mathbf{U}}$−modules (see *loc. cit*) or of their weight spaces.

These bijections are piecewise linear. We will illustrate this in a number of examples.

12.3 Assume first that $n = 2$ and $a_{12} = a_{21} = 0$. Let $\mathbf{i} = (1,2)$, $\mathbf{i}' = (2,1)$. Then $R_{\mathbf{i}}^{\mathbf{i}'}(a,b) = (a',b')$ where $(a',b') = (b,a)$.

Assume next that $n = 2$ and $a_{12} = a_{21} = -1$. Let $\mathbf{i} = (1,2,1)$, $\mathbf{i}' = (2,1,2)$. Then $R_{\mathbf{i}}^{\mathbf{i}'}(a,b,c) = (a',b',c')$ where

$$(a',b',c') = (b+c-a,a,b)$$

if $a \le c$ and

$$(a',b',c') = (b,c,a+b-c)$$

if $a \ge c$. This is proved in *loc. cit.*.

12.4 Assume next that $n = 3$ and $a_{12} = a_{21} = a_{23} = a_{32} = -1$, $a_{13} = a_{31} = 0$. Let $\mathbf{i} = (1,3,2,1,3,2)$, $\mathbf{i}' = (2,1,3,2,1,3)$. Then $R_{\mathbf{i}}^{\mathbf{i}'}(a,b,c,d,e,f) = (a',b',c',d',e',f')$ is given as follows:

If $b+c \ge f+e$, $a+c \le d+f$, and $a \ge d$, then

$$(a',b',c',d',e',f') = (c,d,d+e+f-a-c,a+c-d,b+c+d-e-f,e).$$

If $b \ge e$, $a \le d$, $c \ge f$, then

$$(a',b',c',d',e',f') = (c+d-a,a,e,f,b+c+d-e-f,c+e-f).$$

If $b+c \ge f+e$, $a \le d$, $c \le f$, then

$$(a',b',c',d',e',f') = (c+d-a,a,e+f-c,c,b+c+d-e-f,e).$$

If $b \le e$, $a \ge d$, $c \ge f$, then

$$(a',b',c',d',e',f') = (c+e-b,d,b,f,c+d-f,a+c+e-d-f).$$

If $b \le e$, $a \le d$, $c \ge f$, then

$$(a',b',c',d',e',f') = (c+d+e-a-b,a,b,f,c+d-f,c+e-f).$$

If $b+c \le f+e$, $a+c \ge d+f$, $b \ge e$, then

$$(a',b',c',d',e',f') = (c,d+e+f-b-c,e,b+c-e,d,a+c+e-d-f).$$

If $b+c \le f+e$, $a+c \le d+f$, and $a+b+c \ge d+e+f$, then

$$(a',b',c',d',e',f') = (c,d+e+f-b-c,d+e+f-a-c,a+b+2c-e-f-d,d,e).$$

If $a+c \ge d+f$, $b \le e$, $c \le f$, then

$$(a',b',c',d',e',f') = (c+e-b,d+f-c,b,c,d,a+c+e-d-f).$$

If $b + c \leq f + e$, $a + c \leq d + f$, $c \leq f$ and $a + b + c \leq d + e + f$, then

$$(a', b', c', d', e', f') = (e + f + d - a - b, a, b, c, d, e).$$

If $b + c \geq f + e$, $a + c \geq d + f$, $b \geq e$ and $a \geq d$, then

$$(a', b', c', d', e', f') = (c, d, e, f, b + c + d - e - f, a + c + e - d - f).$$

This can be proved by repeated application of 12.3.

12.5 Assume next that $n = 2$ and $a_{12} = -2$, $a_{21} = -1$. Let $\mathbf{i} = (2, 1, 2, 1)$, $\mathbf{i'} = (1, 2, 1, 2)$. Then $R_{\mathbf{i}}^{\mathbf{i'}}(a, c, d, f) = (a', c', d', f')$ is given as follows.
If $a \geq d$ and $a + c \geq d + f$, then

$$(a', c', d', f') = (c, d, f, a + c - f).$$

If $a \geq d$, $a + c \leq d + f$ and $2a + c \geq 2d + f$, then

$$(a', c', d', f') = (c, 2d + f - a - c, 2a + 2c - 2d - f, d).$$

If $c \leq f$, $a + c \leq d + f$ and $2a + c \leq 2d + f$, then

$$(a', c', d', f') = (2d + f - 2a, a, c, d).$$

If $a \leq d$ and $c \geq f$, then

$$(a', c', d', f') = (c + 2d - 2a, a, f, a + d - f).$$

This can be deduced from 12.4 regarding the quiver geometry for type B_2 as a twisted version of the geometry for type A_3.

12.6 In general, the bijections $R_{\mathbf{i}}^{\mathbf{i'}}$ are compositions of bijections which correspond essentially to rank 2, hence they can be explicitly determined from 12.3, 12.5, except when there are factors G_2, when a computation like the one above remains to be done.

12.7 We will describe explicitly (part of) the canonical basis in low rank. We shall use the notation

$$\begin{bmatrix} M \\ N \end{bmatrix}' = v^{-N(M-N)} \begin{bmatrix} M \\ N \end{bmatrix}_1 \in 1 + v^{-1}\mathbf{Z}[v^{-1}]$$

$$\begin{bmatrix} M \\ N \end{bmatrix}'' = v^{-2N(M-N)} \begin{bmatrix} M \\ N \end{bmatrix}_2 \in 1 + v^{-2}\mathbf{Z}[v^{-2}]$$

for $M \geq N \geq 0$.

Assume first that $n = 1$. Then the canonical basis is $\{E^{(a)} | a \geq 0\}$.
Assume next that $n = 2$ and $a_{12} = a_{21} = -1$. We have the identity

$$E_1^{(a)} E_2^{(b)} E_1^{(c)} = \sum_p v^{-p^2 + p(a+c-b)} \begin{bmatrix} a + p \\ a \end{bmatrix}' E_{(1,2,1)}^{(a+p, c-p, b-c+p)}$$

where p is subject to $0 \leq p \leq c$, $p \geq c - b$. It follows that, in the case where $b \geq a + c$, we have

$$E_1^{(a)} E_2^{(b)} E_1^{(c)} \equiv E_{(1,2,1)}^{(a,c,b-c)} \text{ modulo } v^{-1}\mathcal{L};$$

hence, $E_1^{(a)} E_2^{(b)} E_1^{(c)}$ belongs to the canonical basis.

Similarly, $E_2^{(a)} E_1^{(b)} E_2^{(c)}$, where $b \geq a + c$, belongs to the canonical basis. Each element of the canonical basis is obtained in one of these two ways, and in fact uniquely, except for the fact that $E_1^{(a)} E_2^{(b)} E_1^{(c)} = E_2^{(c)} E_1^{(b)} E_2^{(a)}$ in the case where $b = a + c$.

12.8 Assume now that $n = 2$ and $a_{12} = -2, a_{21} = -1$. We have the identity

$$E_1^{(a)} E_2^{(b)} E_1^{(c)} E_2^{(d)} = \sum_{u,r} v^{\theta} \begin{bmatrix} d+r \\ d \end{bmatrix}'' \begin{bmatrix} a+u \\ a \end{bmatrix}' E_{(1,2,1,2)}^{(a+u,c+r-b-u,2b+u-2r-c,d+r)}$$

where u, r are subject to $u \geq 0$, $r \geq 0$, $c + r - b - u \geq 0$, $2b + u - 2r - c \geq 0$ and

$$\theta = -2r^2 - u^2 + 2ru + 2r(d+b-c) + u(a+c-2b).$$

It follows that, in the case where $c \geq b + d$ and $2b \geq a + c$, we have

$$E_1^{(a)} E_2^{(b)} E_1^{(c)} E_2^{(d)} \equiv E_{(1,2,1,2)}^{(a,c-b,2b-c,d)} \text{ modulo } v^{-1}\mathcal{L};$$

hence, $E_1^{(a)} E_2^{(b)} E_1^{(c)} E_2^{(d)}$ belongs to the canonical basis.

Similarly we see that $E_2^{(a)} E_1^{(b)} E_2^{(c)} E_1^{(d)}$ belongs to the canonical basis provided that $b \geq a + c$ and $2c \geq b + d$.

However, the monomials just described do not exhaust the canonical basis. For example, the element $E_1^{(2)} E_2 E_1 - E_1^{(3)} E_2$ belongs to the canonical basis and is not a monomial as above.

12.9 Assume now that $n = 3$ and $a_{12} = a_{21} = a_{23} = a_{32} = -1$, $a_{13} = a_{31} = 0$.

If $c \geq a + d$, $c \geq b + e$ and $d + e \geq c + f$, then $E_1^{(a)} E_3^{(b)} E_2^{(c)} E_1^{(d)} E_3^{(e)} E_2^{(f)}$ belongs to the canonical basis.

REFERENCES

[APW] H. H. Andersen, P. Polo and Wen K., *Representations of quantum algebras*, Inv. Math. (1991).

[BML] A. A. Beilinson, R. MacPherson and G. Lusztig, *A geometric setting for the quantum deformation of GL_n*, Duke Math. J. **61** (1990).

[C] C. Chevalley, *Certains schémas de groupes semisimples*, Séminaire Bourbaki (1961/62).

[D] V. G. Drinfeld, *Hopf algebras and the Yang-Baxter equation*, Soviet Math. Dokl. **32** (1985), 254-258.

[DL] M. Dyer and G. Lusztig, *Appendix*, Geom. Dedicata (1990).

[J] M. Jimbo, *A q-difference analogue of $U(\mathbf{g})$ and the Yang-Baxter equation*, Lett. Math. Phys. **10** (1985), 63-69.

[K] B. Kostant, *Groups over $\mathbf{Z}$*, Proc. Symp. Pure Math. **9** (1966), 90-98, Amer. Math. Soc.

[L1] G. Lusztig, *Quantum deformations of certain simple modules over enveloping algebras*, Adv. in Math. **70** (1988), 237-249.

[L2] ———, *Modular representations and quantum groups*, Contemp. Math. **82** (1989), 59-77, Amer. Math. Soc..

[L3] ———, *On quantum groups*, J. Algebra **131** (1990), 466-475.

[L4] ———, *Finite dimensional Hopf algebras arising from quantized universal enveloping algebras*, Jour. Amer. Math. Soc. **3** (1990), 257-296.

[L5] ———, *Quantum groups at roots of 1*, Geom. Dedicata (1990).

[L6] ———, *Canonical bases arising from quantized enveloping algebras*, J. Amer. Math. Soc. **3** (1990), 447-498; *Common trends in mathematics and quantum filed theoreis*, ed, J. Eguchi et al., Progress of Theor. Physics **102**, 175-201.

[Ri] C. M. Ringel, *Hall algebras and quantum groups*, Inv. Math..

[Ro] M. Rosso, *Finite dimensional representations of the quantum analog of the enveloping algebra of a complex simple Lie algebra*, Comm. Math. Phys. **117** (1988), 581-593.

Massachusetts Institute of Technology
Department of Mathematics
Cambridge, MA 02139

The Vanishing of Scalar Curvature on 6 Manifolds, Einstein's Equation, and Representation Theory

Bertram Kostant[*]

1. A Distinguished Unitary Representation of $SO(4,4)$

Let G be a real semisimple Lie Group, which we may assume to be connected or even simply connected, and let $\mathfrak{g}_0 = \text{Lie}(G)$. The dual vector space of $\mathfrak{g}_0$ will be denoted by $\mathfrak{g}_0^*$. Also, $\mathfrak{g}$ and $\mathfrak{g}^*$ will denote the complexifications of $\mathfrak{g}_0$ and $\mathfrak{g}_0^*$, respectively. We recall that the group G and the Lie algebra $\mathfrak{g}$ operate on $\mathfrak{g}$ by the adjoint representation and on $\mathfrak{g}^*$ by the so-called coadjoint representation. Moreover, set $G_{\mathbb{C}} = Ad(\mathfrak{g})$. Since the bilinear form $(x,y) = tr(ad\,x\,ad\,y)$ on $\mathfrak{g} \times \mathfrak{g}$ is nonsingular, we can identify $\mathfrak{g}$ and $\mathfrak{g}^*$, which we shall do whenever it is convenient.

An element $x \in \mathfrak{g}$ is called *nilpotent* if $ad\,x$ is nilpotent. Under the identification of $\mathfrak{g}$ with $\mathfrak{g}^*$ we can also talk about nilpotent elements in $\mathfrak{g}^*$ and of a nilpotent $G_{\mathbb{C}}$-orbit $O \subset \mathfrak{g}^*$. For example, if $G = SL(n,\mathbb{R})$ the number of nilpotent $G_{\mathbb{C}}$-orbits in $\mathfrak{g}$ is $p(n)$, the number of partitions of n into positive integers: In fact, the conjugacy classes of nilpotent matrices are classified just by the sizes of the Jordan-boxes in their normal Jordan form.

The universal enveloping algebra $U(\mathfrak{g})$ carries a natural ascending filtration by vector-subspaces $\{U_n(\mathfrak{g})\}$ such that the associated graded algebra

$$gr\,U(\mathfrak{g}) = \oplus_{n \geq 0} U_n(\mathfrak{g})/U_{n-1}(\mathfrak{g})$$

is isomorphic to the symmetric algebra $S(\mathfrak{g})$ by the Poincaré-Birkhoff-Witt theorem.

Given an irreducible representation π of G on a Hilbert space H, there is an associated and irreducible representation $\pi : U(\mathfrak{g}) \to \text{End}(H^{\infty})$ of $U(\mathfrak{g})$ on the space H^{∞} of C^{∞}-vectors in H. Let J denote the annihilator of H^{∞} in $U(\mathfrak{g})$. Then the associated graded ideal

$$gr(J) = \oplus_{n \geq 0}(J \cap U_n(\mathfrak{g}))/(J \cap U_{n-1}(\mathfrak{g}))$$

in $S(\mathfrak{g})$ defines a set of common zeros in $\mathfrak{g}^*$ which is called the *associated variety* $V(gr(J))$ of J or π.

---

[*]We wish to thank Juan Tirao for his contribution to these notes.

Since J is a two-sided ideal, J and hence $gr(J)$, are stable under the adjoint action of $G_{\mathbb{C}}$, so the associated variety is a union of $G_{\mathbb{C}}$-orbits. Moreover, the following is true:

Theorem 1 *(Borho, J-L Brylinski, Joseph, A; Vogan, D., Barbasch, D; see [B]). The associated variety of the primitive ideal J is always the closure of a single nilpotent orbit O.*

This theorem sets up a map $J \to O$ from the set $\mathcal{J}$ of primitive ideals of $U(\mathfrak{g})$ which arise from Hilbert space irreducible representations of G to the set $\mathcal{O}$ of nilpotent $G_{\mathbb{C}}$-orbits in $\mathfrak{g}^*$. Furthermore, one has

$$\dim U(\mathfrak{g})/\mathfrak{g} = \dim 0$$

where the left side is the Gelfand-Kirillov dimension of $U(\mathfrak{g})/\mathfrak{g}$. When J arises from an irreducible representation π, one-half this number is referred to here as the Gelfand-Kirillov dimension of π. It may be computed directly from the dimension of the K-type in H^{∞}.

If $G_{\mathbb{C}}$ is simple, from many points of view a particularly interesting case arises when J is the Joseph ideal and accordingly O is the orbit of the highest root vector — having then minimal positive dimension.

For example, if $G = M(2n, \mathbb{R})$ the metaplectic representation π' : $M(2n, \mathbb{R}) \to U(L_2(\mathbb{R}^n))$ has as its associated variety the minimal coadjoint orbit $O \simeq (\mathbb{C}^{2n} - \{0\})/\mathbb{Z}_2$ of $\mathfrak{sp}(2n, \mathbb{R})$.

We shall be interested in a construction of a unitary irreducible representation π, the analogue of π', of the universal covering group G of $SO(4,4)_e$ of minimal positive Gelfand-Kirillov dimension.

Let $\mathbb{R}^{p+q+2} = \mathbb{R}^{p+1} \oplus \mathbb{R}^{q+1}$ be an orthogonal direct sum with respect to a bilinear form B which is positive definite on $\mathbb{R}^{p+1}$ and negative definite on $\mathbb{R}^{q+1}$. Then $SO(p+1, q+1)_e$ is the identity component of the special orthogonal group of $\mathbb{R}^{p+q+2}$ with respect to B. Also, $SO(p+1) \times SO(q+1)$ is a maximal compact subgroup of $SO(p+1, q+1)_e$.

If π were an induced representation then G would operate on a manifold of dimension $p + q - 1$, which is one half the dimension of the orbit of the highest root vector. However G does not operate on a manifold of such low dimension. It does however operate on a manifold of dimension $p + q$. In fact, let $C \subseteq \mathbb{R}^{p+q+2}$ be the $(p + q + 1)$-dimensional cone defined by putting $C = \{v \in \mathbb{R}^{p+q+2} : B(v, v) = 0\}$ and let C^* be the set of all nonzero elements in C. Clearly, G and $\mathbb{R}^* = \mathbb{R} - \{0\}$ operate on C^* and these actions commute. Let $X^{p+q} = C^*/\mathbb{R}^*$. Then G operates on X^{p+q}, and this action is transitive because it is so on C^*. Let S^p and S^q be, respectively, the unit spheres in $\mathbb{R}^{p+1}$ and $\mathbb{R}^{q+1}$ (i. e. $|B(x, x)| = 1$). One notes that $S^p \times S^q \subset C^*$ and the projection $C^* \to X^{p+q}$ gives rise by restriction to a $SO(p+1) \times SO(q+1)$-equivariant two to one map from $S^p \times S^q$ onto X^{p+q}.

From this we get a whole family of representations of G. Since C^* is a submanifold of $\mathbb{R}^{p+q+2}$, for any $s \in \mathbb{C}$, we may consider the space $\Gamma^s(X)$

of all $\mathbb{C}$-valued C^∞-functions on C^* which admits the homogeneity

$$f(t\,v) = |t|^s\, f(v)$$

for any $v \in C^*$ and $t \in \mathbb{R}^*$. Clearly we may identify $\Gamma^s(X)$ as the space of all C^∞-sections of a certain G-homogeneous line bundle L^s over X. The G-module structure of $\Gamma^s(X)$ is given by $(g.f)(v) = f(g^{-1}v)$ where $g \in G$, $f \in \Gamma^s(X)$ and $v \in C^*$. Generically these G-modules are irreducible. Obviously one has a pairing

$$\Gamma^r(X) \otimes \Gamma^s(X) \longrightarrow \Gamma^{r+s}(X)$$

for any $r, s \in \mathbb{C}$ and in addition this is a map of G-modules. Also one has the obvious identification

$$\Gamma^0(X) = C^\infty(X)$$

Our next concern is with $\Gamma^{-2p}(X)$ when $p = q$. We shall see that $\Gamma^{-2p}(X)$ as a G-module can be identified with the space of all smooth volume elements on $S^p \times S^p$.

Let x_i, $i = 1, \ldots, 2p+2$, be the canonical coordinate functions on $\mathbb{R}^{2p+2}$. The Euler operator $E = \Sigma\, x_i \frac{\partial}{\partial x_i}$, as a vector field is tangent to C^* because it is tangent to rays. One notes that

$$Ef = sf \quad \text{for any} \quad f \in \Gamma^s(X).$$

Also, if $\theta(E)$ denotes Lie differentiation, one clearly has

$$\theta(E)(dx_1 \wedge \ldots \wedge dx_{2p+2}) = (2p+2)dx_1 \wedge \ldots \wedge dx_{2p+2}$$

Now let $\epsilon_i = 1$, for $1 \le i \le p+1$, and $\epsilon_i = -1$ for $p+2 \le i \le 2p+2$, and set $Q = \Sigma\, \epsilon_i x_i^2$. Then $EQ = 2Q$. Also $(dQ)_v$ is non-zero for any $v \in C^*$. Hence there exists a $(2p+1)$-form $\tilde{\omega}$ of $\mathbb{R}^{2p+2}$ defined along C^* such that

$$dQ \wedge \tilde{\omega} = dx_1 \wedge \ldots \wedge dx_{2p+2}$$

on C^*. Furthermore, since C is defined by the equation $Q = 0$, one knows that the volume element ω on C^* defined by pulling back $\tilde{\omega}$, is independent of $\tilde{\omega}$. From this one gets that ω is invariant under G and that

$$\theta(E)(\omega) = 2p\,\omega$$

Let $\iota(E)$ denote the operator of the interior product by E on differential forms. Then $\iota(E)\omega$ is a G-invariant $2p$-form on C^* such that

$$\theta(E)(\iota(E)\omega) = 2p\,\iota(E)\omega \quad \text{and} \quad \iota(E)(\iota(E)\omega) = 0$$

since $\theta(E)$ and $\iota(E)$ commute and $\iota(E)\,\iota(E) = 0$. Now consider $f\,\iota(E)\omega$ for $f \in \Gamma^{-2p}(X)$. Clearly $f\,\iota(E)\omega$ is annihilated by $\iota(E)$ and $\theta(E)$. Thus

$f\iota(E)\omega$ descends to a volume element on X. Moreover, $f \to f\iota(E)\omega$ is clearly a G-isomorphism of $\Gamma^{-2p}(X)$ with the space of all smooth volume elements on X.

As we observed, we have a pairing of $\Gamma^s(X)$ and $\Gamma^{-s-2p}(X)$ into $\Gamma^{-2p}(X)$. But then one has a scalar-valued bilinear pairing of $\Gamma^s(X)$ and $\Gamma^{-s-2p}(X)$ defined by putting

$$< f, h > = \int_X f h \, \iota(E)\omega$$

Furthermore this pairing is G-invariant. In particular $\Gamma^{-p}(X)$ is a pre-Hilbert G-module, but $s = -p$ does not lead to a representation of Gelfand-Kirillov dimension $1(p+q-1) = (\text{with } p = q)1(2p-1)$. The correct value of s turns out to be $s = 1 - p$, as we shall see.

Let $\Delta = \sum_{i=1}^{2p+2} \epsilon_i \left(\frac{\partial}{\partial x_i}\right)^2$ be the Laplace-Beltrami operator on $\mathbb{R}^{2p+2}$ defined with respect to the pseudo-Riemannian structure corresponding to Q. If $f \in \Gamma^s(X)$ then f clearly admits a C^∞ extension $\tilde{f}$, homogeneous of degree s, to an open neighborhood of C^* in $\mathbb{R}^{2p+2}$, stable under multiplication by $\mathbb{R}^*$. But since dQ does not vanish on C^* it follows that $\tilde{f}$ vanishes on C^* if and only if $\tilde{f} = Qh$ for some C^∞ function h, homogeneous of degree $s - 2$. If we also denote by Q the operator of multiplication by the function Q, we have

$$\Delta \tilde{f} = Q\Delta h + [\Delta, Q]h$$

and

$$[\Delta, Q] = \sum_{i,j=0}^{2p+2} \epsilon_j \epsilon_i \left[\left(\frac{\partial}{\partial x_j}\right)^2, x_i^2\right] = \sum_{j=1}^{2p+2} \left[\left(\frac{\partial}{\partial x_j}\right)^2, x_j^2\right]$$

$$= \sum_{j=1}^{2p+2} \left(2 + 4x_j\frac{\partial}{\partial x_j}\right) = 2(2p+2) + 4E$$

But then

$$[\Delta, Q]h = (2(2p+2) + 4(s-2))h = 0$$

if $s = 1 - p$. Thus $\Delta \tilde{f} = Q\Delta h$, which implies that $\Delta \tilde{f}$ vanishes on C^*. Consequently, the Laplacian Δ induces a linear map from $\Gamma^{1-p}(X)$ into $\Gamma^{-1-p}(X)$, by taking $\Delta f = \Delta \tilde{f}|C^*$; this clearly intertwines the actions of G on $\Gamma^{1-p}(X)$ and $\Gamma^{-1-p}(X)$.

We now consider the harmonic component (i.e. the kernel of Δ on $\Gamma^{1-p}(X)$) H^∞ of $\Gamma^{1-p}(X)$ and the cokernel $(H^\infty)'$ of $\Delta : \Gamma^{1-p}(X) \to \Gamma^{-1-p}(X)$. As one easily shows this leads to the following exact sequence of G-maps

$$0 \longrightarrow H^\infty \longrightarrow \Gamma^{1-p}(X) \xrightarrow{\Delta} \Gamma^{-1-p}(X) \longrightarrow (H^\infty)' \longrightarrow 0$$

Since Δ is symmetric with respect to the pairing $\Gamma^{1-p}(X) \otimes \Gamma^{-1-p}(X) \to \mathbb{C}$ introduced above, H^∞ and $(H^\infty)'$ turn out to be the dual G-modules. Moreover H^∞ has the right Gelfand-Kirillov dimension.

If M is a manifold with a pseudo-Riemannian metric g, we say that a diffeomorphism $T : M \to M$ is a *conformal transformation* if $T^* g = fg$ for some positive C^∞ function f.

Now we shall see that we can identify the G-invariant cone $\mathcal{M} = \{f > 0 : f \in \Gamma^{-2}(X)\}$ with the space of all metrics on X in a conformal class. For any $v \in C^*$ the differential $d\pi_v$ of the projection map $\pi : C^* \to X$ induces an isomorphism, denoted also by $d\pi_v$, of $v^\perp / \mathbb{R}v$ onto $T_{\pi(v)}(X)$. Since $d\pi_{sv} = s^{-1} d\pi_v$, $s \in \mathbb{R}^*$, the map from $\mathrm{Hom}\,(\mathbb{R}v, v^\perp / \mathbb{R}v)$ into $T_{\pi(v)}(X)$ given by $\alpha \to d\pi_v(\alpha(v))$ is an isomorphism. Hence $T_{\pi(v)}(X) \simeq \mathrm{Hom}(\mathbb{R}v, v^\perp / \mathbb{R}v)$. Now, since $v \in v^\perp$ if $v \in C^*$, B induces a nonsingular bilinear form $(\,,\,)$ on $v^\perp / \mathbb{R}v$ of signature (p, q). Hence, given $\alpha, \beta \in T_{\pi v}(X)$ and $f \in M$ we can define

$$g(\alpha, \beta) = f(v)(\alpha(v), \beta(v))$$

yielding the announced identification.

Hence when $p = q$, we have $G = \widetilde{SO(p+1, p+1)}_e$ and the exact sequence of G-modules

$$O \longrightarrow H^\infty \longrightarrow \Gamma^{1-p}(X) \xrightarrow{\Delta} \Gamma^{-1-p}(X) \longrightarrow (H^\infty)' \longrightarrow 0.$$

We have also seen that when $s = -2p$, $\Gamma^{-2p}(X)$ may be identified with the space of C^∞ volume elements on x. Thus $\Gamma^{1-p}(X)$ and $\Gamma^{-1-p}(X)$ are dual G-modules. When $s = -2$, the G-invariant cone $\mathcal{M} \subset \Gamma^{-2}(X)$ may be identified with the space of all metrics on X in a conformal class. Because of these facts we shall focus our attention on the case where $1 - p = -2$, i.e. $p = 3$. Thus we shall be considering $G = \widetilde{SO(4, 4)}_e$ and the exact sequence

$$0 \longrightarrow H^\infty \longrightarrow \Gamma^{-2}(X) \xrightarrow{\Delta} \Gamma^{-4}(X) \longrightarrow (H^\infty)' \longrightarrow 0.$$

Let $\eta : G \to SO(4, 4)_e$ be the covering homomorphism and let $K = \eta^{-1}(SO(4) \times SO(4))$. Then K is a maximal compact subgroup of G, hence simply connected, and therefore $K \simeq SU(2) \times SU(2) \times SU(2) \times SU(2)$. Therefore the unitary dual $\hat{K}$ of K may be parameterized by the set of all 4-tuples (k, l, m, n) where k, l, m, n are nonnegative integers. The corresponding presentation is just $V_k \otimes V_l \otimes V_m \otimes V_n$, where V_n denotes the irreducible representation of $SU(2)$ of dimension $n + 1$.

As we pointed out before, we have a K-invariant 2-to-1 covering map of $S^3 \times S^3$ onto X which allows us to identify $\Gamma^{-2}(X)$ with the antipode map invariant functions $C_o^\infty(S^3 \times S^3)$ in $C^\infty(S^3 \times S^3)$. Making the identification $SU(2) \simeq S^3$ the action of K on $S^3 \times S^3$ corresponds to the action of

$$SU(2) \times SU(2) \times SU(2) \times SU(2)$$

on $SU(2) \times SU(2)$ given by left and right multiplication as follows

$$(\alpha, \beta, \gamma, \delta) \cdot (u, v) = (\alpha\, u \beta^{-1}, \gamma\, v \delta^{-1})$$

Therefore if $\Gamma^{-2}(X)_K$ denotes the space of K-finite elements in $\Gamma^{-2}(X)$, by the Peter-Weyl theorem, we get

$$\Gamma^{-2}(X)_K = \oplus_{m,n}\, (m, m, n, n)$$

as K-modules.

Now the Laplace-Beltrami operator Δ on $\mathbb{R}^4 \times \mathbb{R}^4$ is given by

$$\Delta = \Delta_1 - \Delta_2$$

where $\Delta_1 = \sum_{i=1}^{4} \left(\dfrac{\partial}{\partial x_i}\right)^2$ and $\Delta_2 = \sum_{i=5}^{8} \left(\dfrac{\partial}{\partial x_i}\right)^2$. Thus the action of Δ on $\Gamma^{-2}(X) = C^{\infty}(S^3 \times S^3)$ is given by

$$\Delta = \tilde{\Delta}_1 - \tilde{\Delta}_2$$

where $\tilde{\Delta}_1, \tilde{\Delta}_2$ are the differential operators on $C^{\infty}(S^3 \times S^3)$ defined by the action of the usual Laplace-Beltrami operator of S^3 on the first and on the second variable, respectively. By considering the spectral values of the Laplace-Beltrami operator on the 3-sphere one sees that if H_K^{∞} denotes the space of K-finite elements in the harmonic component H^{∞} of $\Gamma^{-2}(X)$ we get

$$H_K^{\infty} = \oplus_m\, (m, m, m, m).$$

Let $\xi(x) = 1/\sum_{i=1}^{4} x_i^2$ where $x = (x_1, \ldots, x_8) \in C^*$. Then $\xi \in H^{\infty} \subset \Gamma^{-2}(X)$ is fixed by K. Moreover it can be proved that H_K^{∞} is a cyclic $U(\mathfrak{g})$-module with the spherical vector ξ as cyclic generator, and from this it follows that H_K^{∞} is $U(\mathfrak{g})$-irreducible (See [K]). Also, from the explicit K-module structure of H_K^{∞} one easily obtains its Gelfand-Kirillov dimension. In fact, it is known that if

$$\phi(t) = \sum_{\|\gamma\| \leq t} m_\gamma \dim(\gamma)$$

where m_γ is the multiplicity of the K-type γ in H_K^{∞}, then $\dim(H_K^{\infty}) = s$ (See [V]) in case $\phi(t)$ is asymptotic to t^s where $t \to \infty$. In our case,

$$\phi(t) = \sum_{m \leq t} (m+1)^4 \simeq t^5.$$

Therefore $\dim(H_K^{\infty}) = 5$.

2. The Principle of Triality and the Unitary Structure on H

The Lie algebra $\mathfrak{g}_0$ of G, which we identify with the Lie algebra of $SO(4,4)_e$ via the differential of the covering homomorphism $\eta : G \to SO(4,4)_e$, is the split real form of the complex simple Lie algebra D_4. Let $\mathfrak{a}_0 \subset \mathfrak{g}_0$ be the split Cartan subalgebra. Thus $\dim \mathfrak{g}_0 = 28$ and $\dim \mathfrak{a}_0 = 4$. Let $\mathfrak{a}_0'$ be the real dual space of $\mathfrak{a}_0$. Looking at the representation of $\mathfrak{g}_0$ on $\mathbb{R}^8$ defined by η, we see that we can choose a weight basis $v_1, \ldots, v_4, u_1, \ldots, u_4$ of $\mathbb{R}^8$ with respect to $\mathfrak{a}_0$ such that, if $\lambda_i \in \mathfrak{a}_0'$ is the weight corresponding to v_i then $-\lambda_i$ is the weight corresponding to u_i, $i = 1, \ldots, 4$. Moreover, one knows that the λ_i are the basis of $\mathfrak{a}_0'$ and the set of roots $\Delta = \Delta(\mathfrak{g}_0, \mathfrak{a}_0)$ is given by

$$\Delta = \{\pm\lambda_i \pm \lambda_j\}, \qquad 1 \le i < j \le 4.$$

The corresponding set $\pi = \{\alpha_1, \ldots, \alpha_4\}$ of simple positive roots may then be labelled so that $\alpha_1 = \lambda_1 - \lambda_2$, $\alpha_2 = \lambda_3 + \lambda_4$, $\alpha_3 = \lambda_3 - \lambda_4$ and $\alpha_4 = \lambda_2 - \lambda_3$. Then the corresponding Dinkin diagram is

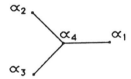

Now, for any $\varphi \in \Delta$, let $n_i(\varphi) \in \mathbb{Z}$, $i = 1, \ldots 4$, be defined so that

$$\varphi = \sum_{i=1}^{4} n_i(\varphi)\alpha_i.$$

We wish to consider three reductive subalgebras $\mathfrak{g}_i \subset \mathfrak{g}_0$, $i = 1, 2, 3$, all isomorphic to $\mathfrak{gl}(4, \mathbb{R})$. These subalgebras are defined by putting

$$\mathfrak{g}_i = \mathfrak{a}_0 + \sum_{n_i(\varphi)=0} \mathbb{R}\, e_\varphi$$

where the sum is over all $\varphi \in \Delta$ such that $n_i(\varphi) = 0$, and $e_\varphi \in \mathfrak{g}_0$ is a nonzero root vector corresponding to ϕ. Let $\mathfrak{n} = \sum_{\varphi \ge 0} \mathbb{R}e_\varphi$ so that

$$\mathfrak{p}_i = \mathfrak{g}_i + \mathfrak{n},$$

$i = 1, 2, 3$, is a maximal parabolic subsalgebra of $\mathfrak{g}_0$. Let $P_i \subset G$ be the normalizer under Ad of $\mathfrak{p}_i$. As one knows $\mathfrak{p}_i$ is the subalgebra of $\mathfrak{g}_0$ corresponding to P_i.

Let Σ_3 be the permutation group of the set $\{1, 2, 3\}$. From the automorphism theory of Lie algebras, one knows that for each $\sigma \in \Sigma_3$ there exists a unique automorphism σ of $\mathfrak{g}_0$ which stabilizes $\mathfrak{a}_0$, $\sigma e_{\alpha_4} = e_{\alpha_4}$ and

$$\sigma e_{\alpha_i} = e_{\alpha_{\sigma(i)}}, \qquad i = 1, 2, 3.$$

Since G is simply connected, σ induces a corresponding automorphism of G which we continue to denote by σ. Clearly $\sigma \mathfrak{p}_i = \mathfrak{p}_{\sigma(i)}$ and hence $\sigma P_i = P_{\sigma(i)}$. Thus, if for any $g \in G$ we let g_i be its left coset representative in G/P_i, one has a diffeomorphism

$$\sigma : G/P_i \longrightarrow G/P_{\sigma(i)} \tag{1}$$

given by $\sigma(g_i) = (\sigma(g))_{\sigma(i)}$, which is such that for any $a \in G$

$$\sigma(a \cdot g_i) = \sigma(a) \cdot \sigma(g_i).$$

Here the dots denote the action of G on G/P_i, $i = 1, 2, 3$.

Each of the three parabolic subgroups P_i defines a real projective variety on which G-operates. We have already been considering the case when $i = 1$. In fact the map $G \to X$ given by $g \to g \cdot \bar{v}_1$ defines an isomorphism

$$G/P_1 \longrightarrow X. \tag{2}$$

Since B has an index 4 one knows that an isotropic subspace W is maximal if and only if $\dim W = 4$. Let J be the set of all maximally isotropic subspaces of $\mathbb{R}^8$. If U and V are respectively the span of the u_i and v_i, then clearly, $U, V \in J$. The action of G on $\mathbb{R}^8$ induces the action of G on J. One knows (see [C]): There are exactly two orbits for the action of G on J. In fact, if $W_1, W_2 \in J$ the condition

$$\dim(W_1 \cap W_2) \quad \text{is even}$$

is an equivalence relation on J. There are two equivalence classes and these are the two orbits of G on J. In particular, the two orbits are

$$Y = \{W \in J : \dim(W \cap V) \quad \text{is even}\}$$

and

$$Y' = \{W \in J : \dim(W \cap V) \quad \text{is odd}\}$$

Now, the correspondence $g \to g \cdot V$ induces a G-isomorphism

$$G/P_2 \longrightarrow Y. \tag{3}$$

Similarly G/P_3 can be identified with Y'.

The automorphism τ' of order 3 of G defined by the cyclic permutation $(2,3)(1,2)$ is unique to the case of 8 dimensions and is referred to as triality in [C]. But in fact the automorphism τ of order 2 defined by the transposition $(1,2)$ is also unique to 8 dimensions and readily defines triality. We find it convenient to deal with τ instead of τ', and even though it has order

2 we will refer to it as a triality. Recalling (1), (2) and (3) we recover the following classical result: There exists a unique diffeomorphism

$$\tau_X : X \longrightarrow Y$$

such that $\tau_X(\overline{v}_1) = V$ and for any $w \in C^*$ and $g \in G$

$$\tau_X(g \cdot \overline{w}) = \tau(g) \cdot \tau_X(\overline{w})$$

Now let $O(\mathbb{R}^4, \mathbb{R}^4)$ be the space of all linear isomorphisms $\gamma : (\mathbb{R}^4, O) \to (O, \mathbb{R}^4)$ such that $Q(\gamma(v)) = -Q(v)$. If $W \in J$, given the components of any $w \in W$ with respect to the decomposition $\mathbb{R}^8 = \mathbb{R}^4 \oplus \mathbb{R}^4$, it is clear that there is a unique element $\gamma_W \in O(\mathbb{R}^4, \mathbb{R}^4)$ such that if $\mathbb{R}^4 = (\mathbb{R}^4, O)$, then

$$W = \{(x, \gamma_W x) : x \in \mathbb{R}^4\}.$$

Furthermore, one clearly has that the map $J \to O(\mathbb{R}^4, \mathbb{R}^4)$, $W \to \gamma_W$ is a bijection.

Let $O(\mathbb{R}^8)$ be the orthogonal group of $\mathbb{R}^8$ with respect to B, and let $O(\mathbb{R}^4)$ (resp. $SO(\mathbb{R}^4)$) be the orthogonal group (resp. special orthogonal group) of the first summand in the decomposition $\mathbb{R}^8 = \mathbb{R}^4 \oplus \mathbb{R}^4$. Now, $O(\mathbb{R}^8)$ operates on J. In particular $O(\mathbb{R}^4) \subset O(\mathbb{R}^8)$ operates on J. The map

$$O(\mathbb{R}^4) \to O(\mathbb{R}^4, \mathbb{R}^4), \quad h \to \gamma_V \circ h^{-1}$$

is bijective. Hence the map

$$O(\mathbb{R}^4) \to J, \quad h \to hV$$

is a bijection. Moreover, since $SO(4,4)_e \cap O(\mathbb{R}^4) = SO(\mathbb{R}^4)$ it follows that Y is stable under the action of $SO(\mathbb{R}^4)$ and that the map

$$SO(\mathbb{R}^4) \to Y, \quad h \to hV$$

is a bijection, so that Y is a principal $SO(\mathbb{R}^4)$-homogeneous space.

Now, given $W \in Y$ let dW be a 4-form on W defining a Lebesgue measure on W, i.e. $dW \in \Lambda^4(\mathbb{R}^8)$ is defined up to an element in $\mathbb{R}^*$. The vector field $E = \sum_{i=1}^{8} x_i \dfrac{\partial}{\partial x_i}$ is tangent to W and $\theta(E)dW = 4\,dW$. Then $\beta = \iota(E)dW$ is a 3-form on W such that

$$\theta(E)(\beta) = 4\beta \quad \text{and} \quad \iota(E)\beta = 0$$

Since $W^* \subset C^*$, given $f \in \Gamma^{-4}(X)$ we may consider $f\beta$ on W^*. Clearly $f\beta$ is annihilated by $\iota(E)$ and $\theta(E)$. Thus $f\beta$ descends to a volume element on $P(W) = W/\mathbb{R}^*$ and we can define

$$\varphi_f(W, dW) = \int_{P(W)} f\beta.$$

The map $dW \to \varphi_f(W, dW)$ is clearly linear. Let L denote the line bundle over Y whose fiber L_W over $W \in Y$ is the line $\Lambda^4(W) \subset \Lambda^4(\mathbb{R}^8)$ and let $\Gamma^{-2}(Y)$ denote the space of C^∞ sections of L^*. Then, $\varphi_f(W) = \varphi_f(W, \cdot) \in L_W^*$ and we may consider the map

$$\Gamma^{-4}(X) \to \Gamma^{-2}(Y), \quad f \to \varphi_f.$$

This map is G-equivariant, and we call it a Radon Transform (RT).

Let us go back to the exact sequence of $(\mathfrak{g}, K)$-modules

$$0 \to H_K^\infty \to \Gamma^{-2}(X)_K \xrightarrow{\Delta} \Gamma^{-4}(X)_K \to (H^\infty)'_K \to 0.$$

We assert that it may be extended as follows:

$$
\begin{array}{ccccccccc}
0 & \to & H_K^\infty & \to & \Gamma^{-2}(X)_K & \xrightarrow{\Delta} & \Gamma^{-4}(X)_K & \to & (H^\infty)'_K & \to & 0 \\
& & & & & & \downarrow{\scriptstyle RT} & & \swarrow & & \\
& & & & \Gamma^{-2}(X)_K & \xleftarrow{\tau} & \Gamma^{-2}(Y)_K & & & &
\end{array}
\tag{4}
$$

That is first of all RT factors through $(H^\infty)'_K$. By looking at the K-type decompositions of $\Gamma^{-2}(X)_K$ and $\Gamma^{-4}(X)_K$ we see that

$$\Delta(\Gamma^{-2}(X)_K) = \oplus_{m \neq n} (m, m, n, n).$$

Given matrix coefficients f_m and f_n, corresponding respectively to the irreducible representations of $SU(2)$ of dimensions $m + 1$ and $n + 1$, it is enough to see that

$$\int_{P(W)} (f_m \otimes \overline{f}_n)\beta = 0$$

for any choice of Lebesgue measure dW on W and $W \in Y$. This is in fact the case when $m \neq n$ by the Schur orthogonality relations.

The map $\tau : \Gamma^{-2}(Y)_K \to \Gamma^{-2}(X)_K$ that appears in (4) is the restriction of the map

$$\tau : \Gamma^{-2}(Y) \to \Gamma^{-2}(X) \tag{5}$$

induced by the triality map $\tau_X : X \to Y$, so that (5) is a τ-twisted G-isomorphism. However the image of the composition of RT with (5) lies in H_K^∞. On the other hand the action of G on H^∞ is easily seen to be stable under the automorphism of G induced by triality. Hence we can untwist and consequently get a $(\mathfrak{g}, K)$-isomorphism from $(H^\infty)'_K$ onto H_K^∞ which allows us to define a $(\mathfrak{g}, K)$-invariant Hermitian bilinear form $< \ | \ >$ on H_K^∞. Now, if $f_1, f_2 \in H_K^\infty$ belong to the K-type component (m, m, m, m), then one computes

$$< f_1 \,|\, f_2 > = (m + 1)(f_1, f_2)$$

where $(\,,\,)$ is the Peter-Weyl inner product defined by

$$(A_1 \otimes B_1, A_2 \otimes B_2) = tr(A_1\, A_2^*)\, tr(B_1\, B_2^*)\,.$$

Hence the Hermitian bilinear form $<\,,\,>$ on H_K^∞ is positive definite. Let $\mathcal{H}$ be the Hilbert space completion of H_K^∞ relative to $<\,,\,>$. By Harish-Chandra's well-known result the (g, K)-module H_K^∞ defines an irreducible unitary representation

$$\pi : G \to U(\mathcal{H})\,.$$

Since $\mathcal{H}$ has a spherical vector and the center of G lies in K the representation π annihilates the center of G. In particular, π defines a distinguished irreducible unitary representation of $SO(4,4)_e$.

Remark. If one considers the real and imaginary part of the Hermitian form in C^4 with signature $++--$ one sees immediately that

$$U(2,2) = SO(4,4)_4 \cap Sp(8, \mathbb{R})\,.$$

As in the case of the metaplectic representation, it is interesting to consider the restriction $\pi|U(2,2)$.

We point out that triality gives

$$\tau(U(2,2)) = SO(4,2)_e \times SO(2)$$

and indeed the subgroups $SO(4,2)_e$ and $SO(2)$ behave again as Howe pairs. Moreover, π restricted to $SO(4,2)_e$ is isomorphic to the discrete part of

$$L^2(SO(4,2)_e/SO(3,2)_e)\,.$$

The overriding motivation for us is that we see $SO(4,4)$ as the point of departure from classical groups to the exceptional Lie groups. The outer automorphism τ' of D_4 corresponding to the cyclic permutation $(1,\,2,\,3)$ has the exceptional Lie algebra $G_2 \subset D_4$ as its fixed subalgebra. Also τ' defines 3 inequivalent 8-dimensional representations $\mathbb{C}_i^8$, $i = 1,2,3$ of D_4 and induces a Lie algebra structure on

$$D_4 \oplus \mathbb{C}_1^8 \oplus \mathbb{C}_2^8 \oplus \mathbb{C}_3^8 = F_4$$

which is the exceptional Lie algebra F_4. Replacing $\mathbb{C}_i^8$ by $\text{End}\ \mathbb{C}_i^8$ it induces a Lie algebra structure on

$$(D_4 \oplus D_4) \oplus \text{End}\ \mathbb{C}_1^8 \oplus \text{End}\ \mathbb{C}_2^8 \oplus \text{End}\ \mathbb{C}_3^8 = E_8$$

which is the exceptional Lie algebra E_8.

3. Riemannian Geometry

A pseudo-Riemannian metric of signature (p, q) on a smooth manifold M of dimension $n = p + q$ is a smooth symmetric 2-tensor g on M such that, at each point x of M, g_x is non-degenerate on $T_x M$, with signature (p, q). We call (M, g) a pseudo-Riemannian manifold. The fundamental Theorem of (pseudo)-Riemannian geometry is: given a pseudo-Riemannian manifold (M, g), there exists a unique linear connection ∇ on M, called the *Levi-Civita connection* of g, such that $\nabla g = 0$ and ∇ is torsion free. This amounts to proving the existence and uniqueness of a rule ∇, which assigns to each smooth vector field X on M, a linear mapping ∇_X of the vector space $X(M)$, of all smooth vector fields on M, into itself satisfying the following conditions:

$$\begin{aligned}
\nabla_{fX+hY} &= f\nabla_X + h\nabla_Y \\
\nabla_X(fY) &= f\nabla_X(Y) + (Xf)Y \\
\nabla_X(Y) &- \nabla_Y(X) = [X, Y] \\
\nabla_X g &= 0
\end{aligned}$$

for $f, h \in C^\infty(M)$, $X, Y \in X(M)$, where for the 4th relation, the operator ∇_X is extended to a unique derivation of the mixed tensor field algebra on M, preserving the type of tensors, commuting with all contractions and such that $\nabla_X(f) = X(f)$, $f \in C^\infty(M)$.

The curvature tensor field R of the Levi-Civita connection is called the *Riemannian curvature tensor* of (M, g) and it is defined by

$$R(X, Y) = [\nabla_X, \nabla_Y] - \nabla_{[X,Y]}$$

for all $X, Y \in X(M)$. Thus R measures how much ∇ fails to be a Lie algebra representation. For any $p \in M$, $R(X, Y)_p$ is a skew-symmetric operator on $T_p(M)$ with respect to g_p. The operator depends only on X_p and Y_p.

If $(U; u^1, \ldots, u^n)$ is a local coordinate system of M, the differential 1-forms du^i define a $C^\infty(U)$ basis of $\Gamma(T^*U)$, and we denote by $\partial_i = \frac{\partial}{\partial u_i}$ the dual basis of $\Gamma(TU)$. Also, the tensor fields $du^i \otimes du^j$ are a basis of $T^{(2,0)}U$ at each point of U, so we may write the restriction of g to U as

$$g = \sum_{i,j=1}^{n} g_{ij} du^i \otimes du^j,$$

where the g_{ij}'s are C^∞ functions on U satisfying $g_{ij} = g_{ji}$. Now, the Levi-Civita connection ∇ is characterized by its values on the basis $\{\partial_i\}$. We get

$$\nabla_{\partial_i} \partial_j = \sum_{k=1}^{n} \Gamma_{ij}^k \partial_k,$$

where the "Christoffel symbols" Γ_{ij}^K are given by

$$\Gamma_{ij}^k = \frac{1}{2} \sum_{\ell=1}^n g^{k\ell}(\partial_i g_{\ell j} + \partial_j g_{\ell i} - \partial_\ell g_{ij}),$$

(at each point of U, $(g^{k\ell})$ is the inverse matrix of (g_{ij})). Finally, the curvature tensor R has components R_{ijk}^ℓ given by

$$R(\partial_i, \partial_j)\partial_k = [\nabla_{\partial_i}, \nabla_{\partial_j}]\partial_k = \sum_{\ell=1}^n R_{ijk}^\ell \partial_\ell,$$

where

$$R_{ijk}^\ell = \partial_i(\Gamma_{jk}^\ell) - \partial_j(\Gamma_{ik}^\ell) + \sum_{m=1}^n (\Gamma_{jk}^m r_{im}^\ell - \Gamma_{ik}^m \Gamma_{jm}^\ell).$$

The *Ricci curvature tensor (or Ricci tensor)* Ric R is the 2-tensor

$$\text{Ric}\, R(X, Y) = tr(Z \to R(X, Z)Y)$$

where tr denotes the trace of the linear map $Z \to R(X, Z)Y$. The Ricci tensor has components R_{ij} given by

$$R_{ij} = \sum_{k=1}^n R_{ikj}^k.$$

Note that the Ricci tensor is symmetric. This follows from the fact that the Levi-Civita connection has no torsion.

The *scalar curvature* of a pseudo-Riemannian manifold (M, g) is the function Scal $g = tr$ Ric where Ric : $TM \to TM$ at each point of M is defined by

$$r(X, Y) = g(\text{Ric}(X), Y).$$

In local coordinates the expression of Scal g is

$$\text{Scal}\, g = \sum_{i=1}^n R_i^i = \sum_{i,j=1}^n R_{ij}g^{ij}.$$

A pseudo-Riemannian manifold (M, g) always possesses a differential operator of particular interest, the so-called *Laplace-Beltrami operator* which is locally defined by

$$\Delta_g f = -\frac{1}{\gamma} \sum_{i,j} \partial_i(g^{ij}\gamma\partial_j f) \tag{6}$$

where $\gamma = |\det g_{ij}|^{1/2}$ and $f \in C^\infty(M)$. It is not difficult to see that the expression on the right-hand side of (6) is independent of the particular coordinate system used.

Now we are interested in seeing what happens to these invariants under conformal change of the metric. Hence, let $g' = e^{2\sigma}g$ where $\sigma \in C^{\infty}(M)$. The basic fact that we shall use is that the bundle in which the curvature tensor naturally lives, according to its symmetry properties, is not irreducible under the action of the orthogonal group, and consequently has a natural decomposition into irreducible components.

Let $V = T_p(M)$ and let $\mathcal{A}$ (respectively $\mathcal{S}$) denote the space of all skew symmetric (respectively symmetric) operators on V with respect to g. For example, the linear map Ric belongs to $\mathcal{S}$ and $R(X, Y)$ lies in $\mathcal{A}$ for any $X, Y \in T_p(M)$. In fact, the Riemannian curvature tensor of (M, g) satisfies the following properties:
- (a) $R(X, Y) = -R(Y, X)$
- (b) $g(R(X, Y)Z, U) = -g(R(X, Y)U, Z)$
- (c) $g(R(X, Y)Z, U) = g(R(Z, U)X, Y)$
- (d) Bianchi's identity: $R(X, Y)Z + R(Y, Z)X + R(Z, X)Y = 0$

Property (c) follows from (a), (b) and (d).

An *algebraic curvature tensor* at p is an element of $(\otimes^3 V^*) \otimes V$ which satisfies the symmetry properties (a), (b), (c), and (d). We shall denote by $\mathcal{C}$ the space of all algebraic curvature tensors at p. The space $\mathcal{A}$ has a non singular symmetric bilinear form defined by $(A, B) = \frac{1}{2}tr(AB)$. Using the metric g we may identify $\mathcal{A}$ with the space of alternating 2-tensors in $V \otimes V$. Given $R \in \mathcal{C}$ and $A = \Sigma X_i \otimes Y_i \in \mathcal{A}$ we let

$$R(A) = \Sigma R(X_i, Y_i) \in \mathcal{A}$$

Then $(R(A), B) = (A, R(B))$, $A, B \in \mathcal{A}$. In this way we can identify $\mathcal{C}$ with a subspace of the space $\mathcal{S}(\mathcal{A})$ of all symmetric operators on $\mathcal{A}$. In fact, this subspace of $\mathcal{S}(\mathcal{A})$ is defined by the linear equation which corresponds to Bianchi's identity. Moreover one has a natural identification

$$\mathcal{S}(\mathcal{A}) = \mathcal{C} \oplus \Lambda^4 V$$

Notice that when dim $M = 4$ we have:

$$
\begin{aligned}
\dim \mathcal{A} &= \dim \Lambda^2 V = 6 & \dim \mathcal{S} &= \dim S^2 V = 10 \\
\dim \mathcal{S}(\mathcal{A}) &= \dim S^2 \mathcal{A} = 21 & \dim \mathcal{C} &= 20
\end{aligned}
$$

Given $S \in \mathcal{S}$ let $\tau(S) : \mathcal{A} \to \mathcal{A}$ be the linear map defined by

$$\tau(S)A = AS + SA$$

Then it is not difficult to see that $\tau(S) \in \mathcal{C}$. Moreover, it follows that the linear map

$$\tau : \mathcal{S} \to \mathcal{C}$$

is injective.

The Ricci tensor associated to an algebraic curvature tensor defines an element of S; hence we also have a linear map

$$\text{Ric} : C \to S$$

Let $W = \ker \text{Ric}$. It can be proved that the following decomposition holds

$$C = W \oplus \tau(S)$$

Now $S = S_0 \oplus \mathbb{R} I$, where S_0 denotes the space of all trace zero elements in S. Thus we come to the fundamental result.

Theorem 2 *If $n \geq 4$, the $O(g)$-module C has the following decomposition into the unique irreducible subspaces*

$$C = W \oplus \tau(S_0) \oplus \tau(\mathbb{R} I)$$

The *Weyl tensor* W of an n-dimensional pseudo-Riemannian manifold (M, g) $(n \geq 4)$ is the component in W of its curvature tensor.

Let us go back to consider a conformal change $g' = e^{2\sigma} g$ of the metric. We want to construct an element $D_\sigma \in S$ associated to $\sigma \in C^\infty(M)$. Given $X \in X(M)$ let

$$A_X Y = -\nabla_Y X, \quad Y \in V$$

The linear map $A_X : V \to V$ is symmetric for all $p \in M$ if and only if X is a gradient field, i.e. $X = X_f$ for some $f \in C^\infty(M)$, where X_f is defined by

$$df(Z) = g(X_f, Z), \quad Z \in X(M)$$

Therefore, if we set

$$D_\sigma = A_{X_\sigma} + X_\sigma \otimes X_\sigma - \frac{1}{2} g(X_\sigma, X_\sigma) I$$

we get an element in S. Now we can express the various invariants of g' in terms of those of g and the derivatives of σ. We shall only consider the curvature tensor and the scalar curvature. We have

(a) $\qquad R' \;=\; R + \tau(D_\sigma)$

(b) Scal $g' \;=\; e^{-2\sigma}(\text{Scal } g + 2(n-1)\Delta_g \sigma - (n-2)(n-1)g(X_\sigma, X_\sigma))$

Notice that from (a) it follows that $W' = W$, i.e. the Weyl tensor is a conformal invariant of the manifold.

The following proposition is of special interest for us

Proposition 1 *Let (M, g) be a 6-dimensional manifold of zero scalar curvature. Let $g' = e^{2\sigma} g$. Then Scal $R' = 0$ if and only if $\Delta_g(e^{2\sigma}) = 0$.*

Proof. Given $p \in M$ we may choose a local coordinate system such that $g(\partial_i, \partial_j)(p) = \epsilon_i \delta_{ij}$ where $\epsilon_i = \pm 1$. Then

$$
\begin{aligned}
\Delta_g(e^{2\sigma})(p) &= -\sum_{ij} \partial_i(g^{ij}\gamma\partial_j(e^{2\sigma})) \\
&= -2\sum_{ij} \partial_i(g^{ij}\gamma(\partial_j\sigma)e^{2\sigma}) \\
&= (2\Delta_g(\sigma)(p) - 4\Sigma\epsilon_i(\partial_i\sigma)^2(p))e^{2\sigma}(p) \\
&= (2\Delta_g(\sigma)(p) - 4g(X_\sigma, X_\sigma)(p))e^{2\sigma}(p)
\end{aligned}
$$

Now

$$
\mathrm{Scal}\, R' = e^{-2\sigma}(10\Delta_g\sigma - 20g(X_\sigma, X_\sigma)) = 5e^{-4\sigma}\Delta_g(e^{2\sigma})
$$

which completes the proof of our proposition. Let us go back to our

representation of $SO(4,4)_e$ defined on the space $H^\infty \subset \Gamma^{-2}(X)$, where $X = S^3 \times S^3$. We had

$$
0 \to H^\infty \to \Gamma^{-2}(X) \overset{\Delta}{\to} \Gamma^{-4}(X) \to (H^\infty)' \to 0.
$$

In the first section of this exposition we also identified a G-invariant cone $\mathcal{M} \subset \Gamma^{-2}(X)$ with a whole conformal class of metrics on X. Now we are interested in characterizing the G-invariant cone $\mathcal{M}_0 = \mathcal{M} \cap H^\infty$. We shall establish the following.

Theorem 3 *The cone $\mathcal{M}_0$ is in fact the set of all metrics g on X in the conformal class of a flat metric with the additional property that $\mathrm{Scal}\, g = 0$. In particular, for this special case the vanishing of scalar curvature is a conformal invariant*

Proof. Let us recall that $X = C^*/\mathbb{R}^*$, where C^* is the set of all nonzero isotropic vectors in $\mathbb{R}^8$. If $\pi : C^* \to X$ denotes the projection map, we also have the identification

$$
T_{\pi(v)}(X) \simeq \mathrm{Hom}\,(\mathbb{R}v, v^\perp/\mathbb{R}v).
$$

The condition of vanishing scalar curvature is of course a local condition. Assume that X^* is an open set of X and let D^* be its inverse image in C^*. If $v \in D^*$, $\alpha, \beta \in T_{\pi(v)}(X^*)$ and $f \in \Gamma^{-2}(X^*)$ is a positive function, then for the corresponding metric g on X^*, at $\pi(v)$, one has

$$
g(\alpha, \beta) = f(v)(\alpha(v), \beta(v)).
$$

It follows immediately then that if $c : X^* \to D^*$ is a cross-section of $\pi|D^*$ such that $f \circ c = 1$ and V^* is the image of c then the restriction $\nu : V^* \to X^*$

of π to V^* is an isometry with respect to the metric g on X^* and the metric on V^* which is induced on it by the Euclidean pseudo-Riemannian metric B in $\mathbb{R}^8$. In fact if $x \in V^*$ then one has the identification $T_x(V^*) = x^\perp/\mathbb{R}x$ and

$$g(\overline{v}_1, \overline{v}_2) = f(x)\, B(v_1, v_2), \quad v_1, v_2 \in x^\perp$$

where $\overline{v}_1, \overline{v}_2$ are the images in $T_{\nu(x)}(X^*)$ of v_1, v_2 defined by the differential of ν.

Now given $u, w \in C^*$ such that $B(u, w) = 1/2$, let V be the orthogonal complement in $\mathbb{R}^8$ of $\mathbb{R}u + \mathbb{R}w$. Let $c : V \to C^*$ be defined by

$$c(v) = u + v - B(v, v)w\,.$$

Let V^* be the image of c and let $X^* = \pi(V^*)$. Then X^* is open in X and clearly X can be covered by open sets of this form. Furthermore the notation above concerning V^* applies. The differential of c at $v \in V$ is given by

$$dc(x) = -2B(v, x)w + x$$

for any $x \in V$. From this it follows that $\pi \circ C : V \to X^*$ is a diffeomorphism. Let $f \in \Gamma^{-2}(X^*)$ be the function which is identically 1 on V^*. Then the corresponding metric g on X^* at the point $\pi \circ c(v)$ is

$$\begin{aligned} g(d(\pi \circ c)(x),\, d(\pi \circ C)(y)) &= B(-2B(v, x)w + x,\, -2B(v, y)w + y) \\ &= B(x, y) \end{aligned}$$

for all $x, y \in V$.

Since the pseudo-Riemannian structure defined by B on V is clearly flat (zero curvature tensor) we see therefore that the metric g is flat.

Let g' be the metric which corresponds to any positive function $f' \in \Gamma^{-2}(X)$. Then, on X^* $g' = hg$ where $h \in C^\infty(X^*) = \Gamma^0(X^*)$ and $f' = hf$ on D^*. Now, on D^* it is not difficult to prove that

$$\Delta f' = \gamma \Delta_g h$$

where $\gamma \in \Gamma^{-4}(X^*)$ is nonzero everywhere. Thus by the previous proposition $g' \in \mathcal{M}_0 = \mathcal{M} \cap H^\infty$ if and only if scalar curvature of g' equals zero.

4. General Relativity

The mathematical formulation of the general theory of relativity begins with a pseudo-Riemannian manifold (M, g) where M is a 4-dimensional manifold and g is a Lorentzian metric tensor. In such a case one refers to (M, g) as a model of space-time. A major physical significance of the geometry on M induced by g is that the geodesics - both time-like and light-like - describe the movement of particles and light under the influence of gravity. The key question then is - what is the metric tensor g? Einstein has set up an equation for g in terms of an energy-momentum tensor T. To understand what T is, we first note that if $\mathcal{G}$ denotes the infinite dimensional "manifold" of all Lorentzian metric tensors on M and if $g \in \mathcal{G}$, then the tangent space $T_g(\mathcal{G})$ to $\mathcal{G}$ at g may be identified with the linear space of all symmetric tensor fields S on M of type $(2,0)$ (that is $S_{ij} = S_{ji}$). If $T_g(\mathcal{G})_o = \{T \in T_g(\mathcal{G}) | \operatorname{supp} T \text{ is compact}\}$ then $T_g(\mathcal{G})$ and $T_g(\mathcal{G})_o$ are in duality using the volume element on M associated to g. The energy-momentum tensor is the element $T \in T_g(\mathcal{G})$ corresponding to a particular linear functional l on $T_g(\mathcal{G})_o$ arising from a Lagrangian function $\mathcal{L}$. We will define $\mathcal{L}$ and l for a case which includes a number of physically interesting examples. Let $\mathcal{Z}$ be the direct product of the space of all tensor fields on M with itself some finite number of times. Consider the direct product $\mathcal{Z} \times \mathcal{G}$ and let $\gamma \colon \mathcal{Z} \times \mathcal{G} \to \mathcal{G}$ be the projection on the second factor. Assume that $\mathcal{X} \subset \mathcal{Z} \times \mathcal{G}$ is a submanifold such that if $\rho = \gamma | \mathcal{X}$ then

$$\rho \colon \mathcal{X} \to \mathcal{G}$$

is a fibration. A Lagrangian (chosen on the basis of some physical theory), $\mathcal{L}$, is a $C^\infty(M)$ valued function on $\mathcal{X}$ such that if $w \in \mathcal{X}$ and $p \in M$ then $(\mathcal{L}(w))(p)$ depends only on the 1-jet of w at p where the derivatives involved are covariant with repect to $\rho(w)$. If now $D \subset M$ is a domain with compact closure then one obtains a scalar valued function, ϕ, on $\mathcal{X}$ where, for $w \in \mathcal{X}$, $\phi(w)$ is the integral of $\mathcal{L}(w)$ over D with respect to the volume element on D defined by $\rho(w)$. We then look for a section — perhaps only locally — s of ρ such that, for $g \in \mathcal{G}$, the vertical differential $(d(\phi|\mathcal{X}_g))_{s(g)} = 0$ where $\mathcal{X}_g$ is the ρ-fiber over g. But since the vertical derivative vanishes differentiating ϕ horizontally at $s(g)$ well-defines the linear functional l on $T_g(\mathcal{G})_o$ - and consequently the energy momentum tensor T. In favorable circumstances this is independent of D since the condition for $(d(\phi|\mathcal{X}_g))_{s(g)} = 0$ is expressible point-wise on M by Euler-Lagrange equations for the tensor fields defining $s(g)$.

Now according to Einstein the metric $g \in \mathcal{G}$ whose geodesics define gravitional behavior for the physical situation associated to $\mathcal{L}$ is one such that (Einstein's equation)

$$R_{ij} - (\operatorname{Scal} g)/2\, g_{ij} = T_{ij} \tag{7}$$

where R_{ij} is the Ricci tensor associated to g and $\operatorname{Scal} g$ is the scalar curvature associated to g. The tensor $\operatorname{Ein} g \in T_g(\mathcal{G})$ where $(\operatorname{Ein} g)_{ij} = R_{ij} - (\operatorname{Scal} g)/2\, g_{ij}$ is called the Einstein tensor associated with g.

Now returning to our previous considerations we recall that X is a 6-dimensional projective variety, doubly covered by $S^3 \times S^3$, with a conformal structure of signature $(3,3)$, on which $SO(4,4)_e$ acts as a group of conformal transformations. We recall that $\mathcal{M} \subset \Gamma^{-2}(X)$ is the cone of conformal metrics in the this conformal class and that the subcone $\mathcal{M}_o = \{h \in \mathcal{M} | Scal\, h = 0\}$ is $SO(4,4)_e$ invariant and that the linear span of $\mathcal{M}_o$ is dense in a Hilbert space $\mathcal{H}$ on which one has an irreducible unitary representation π of minimal Gelfand-Kirillov dimension. Now let $h \in \mathcal{M}$ and assume that M is a 4-dimensional submanifold of X such that (M, g) is a model of space-time where $g = h|M$. According to the principles of general relativity one ought to have a solution of Einsteins equation (7) for an energy momemtum tensor T which, arise as described above, from a Lagrangian. In a number of important examples the Lagrangian is a natural function of the tensor (matter) fields involved. For example, the case where the matter field is an electro-magnetic field or the matter fields are the scalar functions pressure and density associated to a perfect fluid. We now observe that the embedding of (M, g) in (X, h) provide us with tensor fields of this kind, and in fact additional ones, on M. For example let N be the 2-plane on M defined as the normal bundle of M in X. Then h defines not only a Riemannian structure on N - so that N has the structure of a complex line bundle on M but also a connection in this complex line bundle. The curvature of this line bundle is a closed 2-form ω on M which can then be regarded as an electro-magnetic field on M. In addition one has 2 natural scalar functions on M. If $p \in M$ let

$$(8)$$

$$K(p) = \text{sectional curvature of } N_p$$

To define the second one, for any $x \in N_p$, let C_x be the symmetric operator on $T_p(M)$ defined so that if $u, v \in T_p(M)$ and η is a vector field on M whose value at p is v then - where ∇^h denotes covariant differentiation with respect to h -

$$h(C_x u, v) = h(x, \nabla_u^h \eta)$$

One then has a natural section H of N which at p is given by the equation

$$h(H(p), x) = tr\, C_x$$

The second scalar function L on M is defined at p by putting

$$(9)$$

$$L(p) = h(H(p), H(p))$$

If R denotes the curvature tensor on X with respect to h then R defines at p (or at any other point in X) an operator on the space S_p of symmetric operators on $T_p(X)$. Indeed in tensor notation one defines for $P \in S_p$ an element $V = R(P) \in S_p$ by the relation

$$V^i{}_j = R^i{}_k{}^l{}_j P^k{}_l$$

For the reason which will be made clear below Witold Biedrzycki has introduced the following element $T \in T_g(\mathcal{G})$ as an "energy-momentum" tensor — based on the embedding of M in X. In particular as arising from the normal bundle N. The value of T at p, given as a bilinear form on $T_p(M)$, is defined as

$$(10)$$

$$(Tu, v) = h(R(P)u, v) + (K(p) - 3/16\, L(p))h(u, v)$$

where P is the orthogonal projection of $T_p(X)$ onto N_p.

Now one has a natural embedding of the "conformal group" $SO(4,2)_e$ in $SO(4,4)_e$. If one considers the corresponding action of $SO(4,2)_e$ on X then it is easily seen that there exists a unique closed orbit M. In fact M is 4-dimensional and it admits compactified Minkowski space $S^3 \times S^1$ as a 2-fold covering. In addition for any $h \in \mathcal{M}$ one has that (M,g) is a model for space-time where $g = h|M$. Finally one has now the following result of Biedrzycki connecting Einstein's equation (7) with our representation π of $SO(4,4)_e$ on $\mathcal{H}$.

Theorem x. (Biedrzycki). *Let $h \in \mathcal{M}$ and let $g = h|M$ where M is the unique closed orbit of $SO(4,2)_e$ on X. Let T be defined by (10). Then for (M,g) Einstein's equation*

$$R_{ij} - Scal\, g/2\, g_{ij} = T_{ij} \tag{11}$$

is satisfied for h and all its $SO(4,4)_e$ transforms if and only if $h \in \mathcal{M}_o \subset \mathcal{H}$. That is if and only if $Scal\, h = 0$.

For a proof of Theorem x see [Biedrzycki, Proc. Nat. Acad. Sci 1991].

REFERENCES

[K]   Kostant, B., "The Principle of Triality and A Distinguished Unitary Representation of $SO(4,4)$," *Differential Geometrical Methods in Theoretical Physics*, edited by K. Bleuler and M. Werner, Kluwer Academic Publishers, 1988, 65–108

[K1]   Kostant, B., "The Vanishing of Scalar Curvature and the Minimal Representation of $SO(4,4)$," *Operator Algebras, Unitary Representations, Enveloping Algebras, and Invariant Theory*, Birkhäuser Boston: Prog. Math. **92** 1990, 85–124

[B]   Borho, W., *A Survey on Eveloping Algebras of Semisimple Lie Algebras, I*, CMS Conference Proceedings, **5** 1986, 19–50

[C]   Chevalley, C., *The Algebraic Theory of Spinors*, Columbia University Press, 1954

[HE]   Hawking, S. and Ellis, G., *The large scale structure of space–time*, Cambridge University Press, 1973

[V]   Vogan, D., *Gelfand-Kirillov dimension for Harish-Chandra modules*, Inventions Math, **48** (1978), 75–98

Department of Mathematics, MIT, Cambridge, MA 02139, USA

Unitary Representations
of Reductive Lie Groups and the Orbit Method

David A. Vogan, Jr.*

Based on notes prepared by Jorge Vargas

Let G be a Lie group acting on a manifold X and preserving a G-invariant measure. Abstract harmonic analysis seeks to understand the action of G on X by understanding the unitary representation of G on $L^2(X)$. This may be done by "decomposing" $L^2(X)$ (in an appropriate sense) into irreducible unitary representations. In order to do this, it is useful to have in hand a family of irreducible unitary representations; not necessarily exhaustive, but large enough to solve a range of interesting harmonic analysis problems. The Kirillov-Kostant philosophy of coadjoint orbits seeks to provide such a family. The purpose of these notes is to describe what is known about implementing that philosophy, particularly for reductive groups.

We begin with an abelian Lie group A having finitely many connected components. Every irreducible unitary representation of A is one-dimensional, and so may be regarded as a continuous group homomorphism from A to $\mathbb{T}$, the group of complex numbers of absolute value one. If we write $\hat{A}$ for the unitary dual of A, then

$$\hat{A} = \mathrm{Hom}(A, \mathbb{T}).$$

Any continuous homomorphism of Lie groups can be differentiated to get a Lie algebra homomorphism. This gives a map

$$\mathrm{Hom}(A, \mathbb{T}) \to \mathrm{Hom}_{\mathrm{Lie}}(\mathrm{Lie}(A), \mathrm{Lie}(\mathbb{T})).$$

This map is surjective if A is simply connected, and injective if A is connected. Write $\mathfrak{a}_0$ for the Lie algebra of A. The Lie algebra of $\mathbb{T}$ may be identified with $i\mathbb{R}$. Then

$$\mathrm{Hom}_{\mathrm{Lie}}(\mathrm{Lie}(A), \mathrm{Lie}(\mathbb{T})) = \mathrm{Hom}_{\mathbb{R}}(\mathfrak{a}_0, i\mathbb{R}) = i\mathfrak{a}_0^*.$$

Assembling these maps, we find a map

$$\hat{A} \to i\mathfrak{a}_0^*$$

---
* Supported in part by NSF grant DMS-8805665.

with finite fibers. The philosophy of coadjoint orbits seeks to establish analogous relationships for non-abelian groups.

Suppose now that G is a Lie group. Write $\mathfrak{g}_0$ for the Lie algebra of G, and $\mathfrak{g}$ for its complexification. Write Ad for the adjoint representation of G on $\mathfrak{g}_0$, and Ad^* for its contragredient, the *coadjoint representation* on the dual vector space $\mathfrak{g}_0^*$. The differentials of these representations are written ad and ad^* respectively. For each $\lambda \in \mathfrak{g}_0^*$, define

$$G(\lambda) = \{g \in G \mid \mathrm{Ad}^*(g)\lambda = \lambda\}$$
$$\mathfrak{g}_0(\lambda) = \{X \in \mathfrak{g}_0 \mid \mathrm{ad}^*(X)\lambda = 0\}$$
$$= \{X \in \mathfrak{g}_0 \mid \lambda([X, \mathfrak{g}_0]) = 0\}.$$

Notice in particular that the restriction of λ to $\mathfrak{g}_0(\lambda)$ is a Lie algebra homomorphism. The *coadjoint orbit* $\mathrm{Ad}^*(G) \cdot \lambda \subset \mathfrak{g}_0^*$ may be identified with $G/G(\lambda)$.

Definition 1. An element $\lambda \in \mathfrak{g}_0^*$ is called *integral* if there is a representation $\pi : G(\lambda) \to GL(V)$ with the property that

$$d\pi(X) = i\lambda(X) \cdot \mathrm{Id}_V \qquad (X \in \mathfrak{g}_0(\lambda).$$

Here Id_V is the identity operator on the vector space V. It is equivalent to require that $i\lambda$ should be the differential of a (unitary) character of the identity component $G_0(\lambda)$. Notice that the element $0 \in \mathfrak{g}_0^*$ is always integral.

An *integral orbit datum* is a pair (λ, π) with $\lambda \in \mathfrak{g}_0^*$ and π an irreducible unitary representation of $G(\lambda)$ on a Hilbert space V, such that $d\pi = i\lambda \cdot \mathrm{Id}_V$.

The group G acts on the set of integral orbit data in an obvious way. To a first approximation, the philosophy of coadjoint orbits seeks to attach to each integral orbit datum (λ, π) an irreducible unitary representation $\Pi_{int}(\lambda, \pi)$ of G, depending only on the G-conjugacy class of the datum. Such a map Π_{int} is called an *orbit correspondence*. Whenever the adjoint action of G is trivial (for example, if G is finite or abelian) Π_{int} exists for trivial reasons: the isotropy group $G(\lambda)$ is all of G, so we can define $\Pi_{int}(\lambda, \pi) = \pi$. Here are two more interesting examples.

Example 1. Let $G = O(2)$, the group of isometries of $\mathbb{R}^2$ fixing the origin. The Lie algebra $\mathfrak{g}_0$ consists of skew-symmetric two-by-two matrices:

$$\mathfrak{g}_0 = \left\{ \begin{pmatrix} 0 & \theta \\ -\theta & 0 \end{pmatrix} \mid \theta \in \mathbb{R} \right\}.$$

We identify $\mathfrak{g}_0^*$ with $\mathbb{R}$ by evaluation at $\begin{pmatrix} 0 & 1 \\ -1 & 0 \end{pmatrix}$. The adjoint and coadjoint actions of $SO(2)$ are trivial, and $\mathrm{Ad}^* \begin{pmatrix} 0 & 1 \\ 1 & 0 \end{pmatrix} = -\mathrm{Id}$. There are two kinds of coadjoint orbit.

1) $\lambda \neq 0$. In this case $G(\lambda) = SO(2)$. The element λ is integral if and only if $i\lambda$ is the differential of a character of $SO(2)$; that is, if and only if the formula

$$\chi_\lambda \left(\exp \begin{pmatrix} 0 & \theta \\ -\theta & 0 \end{pmatrix} \right) = e^{i\lambda\theta}$$

defines a character of $SO(2)$. So λ is integral if and only if $\lambda \in \mathbb{Z}$.

2) $\lambda = 0$. In this case $G(\lambda) = O(2)$, and λ is integral. There are two irreducible unitary representations of $O(2)$ having differential zero: the trivial representation π_+, and the determinant character π_-.

We therefore find the following integral orbit data: pairs (n, χ_n), with n a non-zero integer, and $(0, \pi_\pm)$. The group G acts on these by sending (n, χ_n) to $(-n, \chi_{-n})$. We define an orbit correspondence by

$$\Pi_{int}(n, \chi_n) = \text{Ind}_{SO(2)}^{O(2)}(\chi_n), \qquad \Pi_{int}(0, \pi_\pm) = \pi_\pm.$$

It is easy to check that this correspondence is actually a bijection from the set of G-conjugacy classes of orbit data onto the unitary dual of $O(2)$.

Example 2. Let G be any compact Lie group, and fix $\lambda \in \mathfrak{g}_0^*$. The isotropy group $G(\lambda)$ is compact, so we can find a maximal torus $T \subset G_0(\lambda)$. It turns out that T is also a maximal torus in G_0. Now a finite-dimensional representation of the Lie algebra of a compact connected group can be exponentiated if and only if its restriction to a maximal torus can be exponentiated. It follows that λ is integral for G if and only if $i\lambda|_{t_0}$ is the differential of a character of T. In this case we write χ_λ for the corresponding character of T.

Fix now an integral orbit datum (λ, π); the restriction of π to T is therefore a multiple of χ_λ. Define $G^\sharp$ to be the subgroup of G generated by $G(\lambda)$ and G_0; that is, the union of all the connected components meeting $G(\lambda)$. It turns out that there is a unique irreducible representation $\pi^\sharp$ of $G^\sharp$ containing χ_λ as an extremal weight, with the additional property that $G(\lambda)$ acts on the χ_λ weight space by the representation π. (If G is connected, this is just the assertion that there is a unique irreducible representation of G of extremal weight λ, and that this weight has multiplicity one.) Define

$$\Pi_{int}(\lambda, \pi) = \text{Ind}_{G^\sharp}^G(\pi^\sharp).$$

Theorem 1 (Cartan-Weyl-Kostant). *Suppose (λ, π) is an integral orbit datum for the compact Lie group G. Then the representation $\Pi_{int}(\lambda, \pi)$ defined above is irreducible. This orbit correspondence establishes a bijection between G-conjugacy classes of integral orbit data and irreducible unitary representations of G.*

More details, and an explicit construction of $\Pi_{int}(\lambda, \pi)$ (using the Borel-Weil theorem) may be found in the first chapter of [V2].

We turn now to an example in which a nice orbit correspondence cannot be defined on integral orbit data.

Example 3. In $\mathbb{R}^{2n}$, let ω be the symplectic form defined by

$$\omega((x, y), (z, w)) = x \cdot w - y \cdot z \qquad (x, y, z, w \in \mathbb{R}^n),$$

and let $G = Sp(2n, \mathbb{R})$ be the corresponding symplectic group. Its Lie algebra is

$$\mathfrak{g}_0 = \left\{ \begin{pmatrix} A & B \\ C & -A^t \end{pmatrix} \mid A, B, C \in M_n(\mathbb{R}), B^t = B, C^t = C \right\}.$$

For each $X \in \mathfrak{g}_0$ we define a linear functional $\lambda_X \in \mathfrak{g}_0^*$ by $\lambda_X(Y) = \operatorname{tr}(XY)$. The map $X \mapsto \lambda_X$ is an isomorphism from $\mathfrak{g}_0$ onto $\mathfrak{g}_0^*$, intertwining Ad and Ad*. (The adjoint action of G is given by conjugation of matrices: $\mathrm{Ad}(g)X = gXg^{-1}$.

Consider now the Lie algebra element $X_1 = \begin{pmatrix} 0 & E_{11} \\ 0 & 0 \end{pmatrix}$. Here E_{11} is the n by n matrix with a one in the upper left corner, and all other entries zero. Write $\lambda_1 = \lambda_{X_1}$. Then $G(\lambda_1) = (\mathbb{Z}/2\mathbb{Z} \times Sp(2n - 2, \mathbb{R})) \ltimes N$, with N a unipotent group.

We claim that the restriction of λ_1 to $\mathfrak{g}_0(\lambda_1)$ is zero. We will check this only for $n = 1$; the general case is similar. We have

$$X_1 = \begin{pmatrix} 0 & 1 \\ 0 & 0 \end{pmatrix}, \quad G(\lambda_1) = \left\{ \begin{pmatrix} \pm 1 & t \\ 0 & \pm 1 \end{pmatrix} \mid t \in \mathbb{R} \right\}, \quad \mathfrak{g}_0(\lambda_1) = \left\{ \begin{pmatrix} 0 & t \\ 0 & 0 \end{pmatrix} \right\}.$$

Applying λ_1 to a typical element of $\mathfrak{g}_0(\lambda_1)$, we get

$$\lambda_1 \begin{pmatrix} 0 & t \\ 0 & 0 \end{pmatrix} = \operatorname{tr}\left(\begin{pmatrix} 0 & 1 \\ 0 & 0 \end{pmatrix} \begin{pmatrix} 0 & t \\ 0 & 0 \end{pmatrix} \right) = 0.$$

It follows that λ_1 is integral. We can take π to be the trivial representation of $G(\lambda_1)$; then (λ_1, π) is an integral orbit datum. Nevertheless, it can be shown that for $n \geq 2$, there is no irreducible unitary representation of $Sp(2n, \mathbb{R})$ attached to (λ_1, π) in any reasonable sense. (For an explanation of "reasonable" and a proof, see [V0].)

We therefore need something slightly different from integral orbit data to get a nice orbit correspondence. It should not be *too* different, however, since integral orbit data work so well for compact and abelian groups. To get a hint about how to proceed, let us look again at compact groups.

So suppose G is a compact connected Lie group, $T \subset G$ is a maximal torus, and $W = N(T)/T$ is the Weyl group. Write

$$\Delta = \Delta(\mathfrak{g}, \mathfrak{t}) \subset i\mathfrak{t}_0^*$$

for the set of roots, and fix a system of positive roots $\Delta^+ \subset \Delta$. The description of $\hat{G}$ in terms of integral orbit data in Theorem 1 relies on the "theorem of the highest weight." Here is a weak formulation of it.

Proposition 1. *In the setting above, suppose e^μ is any character of T. Then there is a unique irreducible representation $\Pi_{ext}(\mu)$ of G having e^μ as an extremal weight. This correspondence defines a bijection from the set of W-orbits on $\hat{T}$ onto $\hat{G}$.*

Suppose (λ, π) is an integral orbit datum for G, and that $G(\lambda)$ is a torus. Then the orbit correspondence of Theorem 1 is given by

$$\Pi_{int}(\lambda, \pi) = \Pi_{ext}(\lambda).$$

The Weyl character formula for an irreducible representation π is of the form

$$\Theta_\pi(\exp X) = \sum_{w \in W} \epsilon(w) e^{w\lambda(X)}/(\text{Weyl denominator})(X).$$

Here Θ_π is the character of π (the function on G defined by $\Theta_\pi(g) = \operatorname{tr}\pi(g)$); X is an element of t_0; $\epsilon(w) = \pm 1$ is the sign of the Weyl group element w; $\lambda \in it_0^*$ depends on π. (A little more precisely, write μ for the highest weight of π (one of the extremal weights), and $\rho \in it_0^*$ for half the sum of the positive roots of T in G. Then $\lambda = \mu + \rho$.) This character formula suggests that it may also be reasonable to define an orbit correspondence by relating π to something like the orbit of λ. This will be accomplished by the theory of "admissible orbit data."

The first complication is that for many compact groups, the numerator and denominator of the Weyl character formula are multi-valued functions on T; only their quotient is well-defined. For example, suppose $G = SO(3)$, with $T = SO(2)$ embedded in the upper left corner. Define

$$k(\theta) = \begin{pmatrix} \cos\theta & \sin\theta & 0 \\ -\sin\theta & \cos\theta & 0 \\ 0 & 0 & 1 \end{pmatrix} \in T.$$

If π is the irreducible representation of $SO(3)$ of dimension $2j + 1$, then the Weyl character formula has the form

$$\Theta_\pi(k(\theta)) = (e^{i(2j+1)\theta/2} - e^{-i(2j+1)\theta/2})/(e^{i\theta/2} - e^{-i\theta/2})$$
$$= \sin((j + 1/2)\theta)/\sin(\theta/2).$$

The numerator and denominator are periodic of period 4π, but the quotient is periodic of period 2π (as it must be to define a function on T).

Because of this phenomenon, admissible orbit data involve characters not of subgroups of G, but rather of certain coverings of subgroups. We begin by introducing such a covering for our maximal torus T in the compact connected group G. Define

$$e^{2\rho}(t) = \prod_{\alpha \in \Delta^+} e^\alpha(t).$$

The differential of $e^{2\rho} \in \hat{T}$ is 2ρ. There may not exist a character of T with differential ρ, so we construct a covering of T on which such a character does exist. Define

$$\widetilde{T} = \{(t, z) \in T \times \mathbb{C}^\times \mid e^{2\rho}(t) = z^2\}.$$

This is a Lie group, a closed subgroup of $T \times \mathbb{C}^\times$. Projection on the first factor provides a short exact sequence

$$1 \to \{\pm 1\} \to \widetilde{T} \to T \to 1,$$

which we use to identify the Lie algebras of T and $\widetilde{T}$. (It is only in the sense of the existence of this exact sequence that we call $\widetilde{T}$ a covering of T; it may not be a connected covering group.) Projection on the second factor defines a character of $\widetilde{T}$, which we call e^ρ:

$$e^\rho(t, z) = z.$$

The differential of e^ρ is $\rho \in i\mathfrak{t}_0^*$, and e^ρ is non-trivial on the kernel of the covering map: $e^\rho(-1) = -1$. It turns out that the action of the Weyl group on T lifts to $\widetilde{T}$ ([V2], page 43).

More generally, by a *genuine character* of $\widetilde{T}$ we will mean any character e^λ satisfying $e^\lambda(-1) = -1$. We say that e^λ is *regular* if it is not fixed by any non-trivial element of W. In particular, e^ρ is a genuine regular character. The *Weyl numerator* attached to a genuine character e^λ is the function on $\widetilde{T}$

$$N_\lambda(\tilde{t}) = \sum_{w \in W} \epsilon(w) e^\lambda(w\tilde{t});$$

this is non-zero if and only if e^λ is regular, and it satisfies $N_\lambda(-\tilde{t}) = -N_\lambda(\tilde{t})$. The *Weyl denominator* is by definition N_ρ.

Proposition 2. *Suppose T is a maximal torus in a compact connected Lie group G, and Δ^+ is a set of positive roots. If e^λ is any genuine regular character of $\widetilde{T}$, then there is a unique irreducible representation $\Pi_{Weyl}(\lambda)$, characterized by*

$$\Theta_{\Pi_{Weyl}(\lambda)}(t) = \pm N_\lambda(\tilde{t})/N_\rho(\tilde{t}).$$

Here $t \in T$, and $\tilde{t}$ is any preimage of t in $\tilde{T}$. This defines a bijection from the W-orbits of genuine regular characters of T onto the irreducible representations of G.

To get a result along the lines of Theorem 1, we need to generalize the construction of $\tilde{T}$ slightly. So suppose H is any Lie group, and $\gamma : H \to \mathbb{C}^\times$ is a character of H. We want to construct a covering of H on which γ admits a square root. So define

$$\tilde{H} = \{(h, z) \in H \times \mathbb{C}^\times \mid \gamma(h) = z^2\}.$$

(Of course this group depends on the choice of γ, but we will not need to make this explicit in the notation.) Projection on the first factor provides a short exact sequence

$$1 \to \{\pm 1\} \to \tilde{H} \to H \to 1.$$

An irreducible representation (τ, V) of $\tilde{H}$ must satisfy $\tau(-1) = -\mathrm{Id}_V$ or $\tau(-1) = \mathrm{Id}_V$; τ is called *genuine* in the first case and *non-genuine* in the second. The non-genuine irreducible representations of $\tilde{H}$ may be identified naturally with the irreducible representations of H. Projection on the second factor defines a genuine character of $\tilde{H}$ that we call $\gamma^{1/2}$:

$$\gamma^{1/2} : \tilde{H} \to \mathbb{C}^\times, \qquad \gamma^{1/2}(h, z) = z.$$

The square of $\gamma^{1/2}$ is the non-genuine character of $\tilde{H}$ corresponding to our original character γ of H.

In addition to the identification between non-genuine representations of $\tilde{H}$ and representations of H, there is a bijection between irreducible representations of H and genuine irreducible representations of $\tilde{H}$, given by tensoring with the character $\gamma^{1/2}$. (A little more precisely, we first identify the representation of H with a non-genuine representation of $\tilde{H}$, and then tensor with $\gamma^{1/2}$.)

Definition 2. Suppose G is a compact Lie group, and $\lambda \in \mathfrak{g}_0^*$. Choose a $G(\lambda)$-invariant parabolic subalgebra $\mathfrak{q}$ of $\mathfrak{g}$ having $\mathfrak{g}(\lambda)$ as a Levi factor; this is always possible. Write $\mathfrak{u}$ for the nil radical of $\mathfrak{q}$. The adjoint action of the group $G(\lambda)$ must preserve $\mathfrak{u}$, so we can define $e^{2\rho(\mathfrak{u})}$ to be the character of $G(\lambda)$ by which it acts on the top exterior power of $\mathfrak{u}$:

$$e^{2\rho(\mathfrak{u})}(g) = \det(\mathrm{Ad}(g)|_{\mathfrak{u}}) \qquad (g \in G(\lambda)).$$

As above we can define a covering $\tilde{G}(\lambda)$ on which the character $e^{2\rho(\mathfrak{u})}$ has a square root. A representation (τ, V) of $\tilde{G}(\lambda)$ is called *admissible* if it is irreducible, genuine, and satisfies

$$d\tau(X) = i\lambda(X) \cdot \mathrm{Id}_V \qquad (X \in \mathfrak{g}_0(\lambda)).$$

An *admissible orbit datum* is a pair (λ, τ) with $\lambda \in \mathfrak{g}_0^*$ and τ an admissible representation of $\widetilde{G}(\lambda)$.

In order for this definition to make sense, one needs to know that if $\mathfrak{q}'$ is another $G(\lambda)$-invariant parabolic subalgebra of $\mathfrak{g}$ with Levi factor $\mathfrak{g}(\lambda)$, then the covering $(\widetilde{G}(\lambda))'$ constructed using $e^{2\rho(\mathfrak{u}')}$ is naturally isomorphic to $\widetilde{G}(\lambda)$. This is true and not difficult to verify (using Lemma 1.32 in [V2]); we omit the argument.

The only difference between Definitions 1 and 2 (other than the narrower hypotheses in Definition 2) is that in the latter we ask for genuine representations of $\widetilde{G}(\lambda)$ rather than representations of $G(\lambda)$. Now we have observed already that the genuine representations of $\widetilde{G}(\lambda)$ are in one-to-one correspondence with the representations of $G(\lambda)$; so it may appear that the two definitions are essentially identical. The reason they are not is that the one-to-one correspondence is implemented by tensoring with a character having non-zero differential.

We turn now to the construction of an admissible orbit orbit correspondence for a compact group G. Suppose (λ, τ) is an admissible orbit datum. Choose a maximal torus $T \subset G(\lambda)$, and write λ_t for the restriction of λ to the Lie algebra of T. Let S be the set of roots of T taking positive values on $i\lambda_t$:

$$S = \{\alpha \in \Delta(\mathfrak{g}, \mathfrak{t}) \mid i\lambda_t(H_\alpha) > 0\}.$$

Then the span $\mathfrak{u}$ of the corresponding root vectors is the nil radical of a parabolic subalgebra $\mathfrak{q} = \mathfrak{g}(\lambda) + \mathfrak{u}$. We can use this parabolic to define the covering $\widetilde{G}(\lambda)$. As in the remarks before Definition 2, we may regard $\tau \otimes e^{-\rho(\mathfrak{u})}$ as an irreducible representation of $G(\lambda)$. There is at most one irreducible representation (π, V) of G with the property that the representation of $G(\lambda)$ on the subspace $V^\mathfrak{u}$ of $\mathfrak{u}$-invariant vectors contains $\tau \otimes e^{-\rho(\mathfrak{u})}$. If such a representation π exists, we define

$$\Pi_{adm}(\lambda, \tau) = \pi;$$

otherwise we set $\Pi_{adm}(\lambda, \tau) = 0$. (The condition for π to exist can be made quite explicit: it is

$$(i\lambda_t - \rho(\mathfrak{u}))(H_\alpha) \geq 0, \quad \text{all } \alpha \in S.$$

This is automatic (because of the choice of S made above) if $G(\lambda)$ is a torus, but not in general.)

Theorem 2 (Cartan-Weyl-Kostant). *Suppose (λ, τ) is an admissible orbit datum for the compact Lie group G. Then the representation $\Pi_{adm}(\lambda, \tau)$ just defined is irreducible or zero, and depends only on the G-conjugacy class of (λ, τ). Every irreducible representation of G occurs in this way. The representations attached to more than one conjugacy class* *of admissible orbit data are precisely those with singular extremal weights. If G is connected and $G(\lambda)$ is a torus, then*

$$\Pi_{adm}(\lambda, \tau) = \Pi_{Weyl}(\lambda)$$

(cf. Proposition 2).

Example 4. Suppose $G = SO(3)$. The Lie algebra $\mathfrak{g}_0$ may be identified with the space of three by three real skew-symmetric matrices, and $\mathfrak{g}_0^*$ is isomorphic to $\mathfrak{g}_0$ by the map sending a Lie algebra element X to the linear functional $\lambda_X(Y) = \operatorname{tr} XY$. With this identification, the coadjoint action is conjugation of matrices. We want to determine the set of admissible orbit data. Write λ_t for the linear functional corresponding to the matrix $\begin{pmatrix} 0 & t & 0 \\ -t & 0 & 0 \\ 0 & 0 & 0 \end{pmatrix}$. Then every coadjoint orbit contains an element λ_t, for a unique $t \geq 0$.

If $t = 0$, then $G(\lambda_0) = G$, $\mathfrak{q} = \mathfrak{g}$, $\mathfrak{u} = 0$, $e^{2\rho(\mathfrak{u})}$ is the trivial character, and

$$\widetilde{G}(\lambda) = \{(g, z) \in SO(3) \times \mathbb{C}^\times \mid 1 = z^2\} = SO(3) \times \{\pm 1\}.$$

The character $e^{\rho(\mathfrak{u})}$ is trivial on $SO(3)$ and sends -1 to -1. Since $SO(3)$ is connected, there is a unique admissible orbit datum (λ_0, τ_0) attached to λ_0; the character τ_0 is $e^{\rho(\mathfrak{u})}$. According to the construction before Theorem 2,

$$\Pi_{adm}(\lambda_0, \tau_0) = \text{trivial representation of } SO(3).$$

Suppose $t > 0$. Then $G(\lambda_t)$ is the standard maximal torus of $SO(3)$, discussed after Proposition 1. There are two roots $\pm\alpha$ of T in G; in the coordinates used earlier, they are

$$\pm\alpha(k(\theta)) = e^{\pm i\theta}.$$

The coroots are

$$H_{\pm\alpha} = \begin{pmatrix} 0 & \pm i & 0 \\ \mp i & 0 & 0 \\ 0 & 0 & 0 \end{pmatrix}.$$

Consequently $i\lambda_t(H_{\pm\alpha}) = \pm 2t$, so the set S of roots defined by λ_t consists of α alone. Therefore $e^{2\rho(\mathfrak{u})} = \alpha$, and

$$\widetilde{G}(\lambda_t) = \{(k(\theta), z) \in SO(2) \times \mathbb{C}^\times \mid z^2 = e^{i\theta}\}.$$

This is a two-fold cover of $SO(2)$. We identify $\widetilde{G}(\lambda_t)$ with the unit circle $\mathbb{T}$ by sending $e^{i\phi} \in \mathbb{T}$ to $(k(2\phi), e^{i\phi})$. With this identification, the character $e^{\rho(\mathfrak{u})}$ becomes the tautological character of $\mathbb{T}$ (sending $e^{i\phi}$ to $e^{i\phi}$). The characters of $\mathbb{T}$ are parametrized by the integers (by sending m to the

mth power of the tautological character); the genuine characters of $\widetilde{G}(\lambda_t)$ are those corresponding to odd integers. The differential of the character corresponding to m is easily computed to be $i\lambda_{m/2}$. Consequently λ_t is admissible (still for $t > 0$) if and only if $2t = 2j + 1$ is an odd integer. In that case there is a unique admissible orbit datum $(\lambda_{j+1/2}, \tau_{j+1/2})$. The character $\tau_{j+1/2} \otimes e^{-\rho(u)}$ of $\widetilde{G}(\lambda_{j+1/2})$ corresponds to the integer $2j$, and therefore to the character of T sending $k(\theta)$ to $e^{ij\theta}$. According to the construction before Theorem 2,

$$\Pi_{adm}(\lambda_{j+1/2}, \tau_{j+1/2})$$
$$= \text{irreducible representation of } SO(3) \text{ of dimension } 2j + 1.$$

In particular, the orbits through λ_0 and $\lambda_{1/2}$ both correspond to the trivial representation of $SO(3)$.

We want to extend Theorem 2 to noncompact groups. Fix therefore a reductive Lie group G, a maximal compact subgroup K of G, and a Cartan decomposition $\mathfrak{g}_0 = \mathfrak{k}_0 + \mathfrak{p}_0$. Fix a non-degenerate symmetric G-invariant bilinear form $\langle , \rangle$ on $\mathfrak{g}_0$, negative definite on $\mathfrak{k}_0$, positive definite on $\mathfrak{p}_0$, and making the Cartan decomposition orthogonal. Using this form, we get an isomorphism of $\mathfrak{g}_0$ with $\mathfrak{g}_0^*$, sending X to the linear functional λ_X defined by $\lambda_X(Y) = \langle X, Y \rangle$. Write X_λ for the element of $\mathfrak{g}_0$ corresponding to the linear functional λ. Using this identification, we can transfer many standard definitions and results from $\mathfrak{g}_0$ to $\mathfrak{g}_0^*$. It is often useful to give direct definitions or proofs in $\mathfrak{g}_0^*$, however.

We begin our analysis of coadjoint orbits by recalling the Jordan decomposition.

Definition 3. Suppose G is reductive, and $\lambda \in \mathfrak{g}_0^*$. We say that λ is *nilpotent* if $\lambda|_{\mathfrak{g}_0(\lambda)} = 0$. (This is equivalent to the requirement that X_λ be an ad-nilpotent element of $[\mathfrak{g}_0, \mathfrak{g}_0]$.) We say that λ is *semisimple* if $\mathrm{Ad}^*(G) \cdot \lambda$ is closed in $\mathfrak{g}_0^*$; or, equivalently, if $G(\lambda)$ is reductive in G. (This is equivalent to the requirement that $\mathrm{ad}(X_\lambda)$ be semisimple.) We say that λ is *elliptic* if X_λ is G-conjugate to an element of $\mathfrak{k}_0$, and *hyperbolic* if X_λ is G-conjugate to an element of $\mathfrak{p}_0$. Elliptic and hyperbolic elements are automatically semisimple.

It would be interesting to give direct characterizations of elliptic and hyperbolic elements (not involving X_λ).

Proposition 3 (Jordan decomposition). *Suppose G is a reductive Lie group, and $\lambda \in \mathfrak{g}_0^*$. Then there is a unique decomposition $\lambda = \lambda_s + \lambda_n$, such that*

i) λ_s is semisimple;
ii) λ_n is nilpotent; and
iii) $[X_{\lambda_s}, X_{\lambda_n}] = 0$.

This decomposition has the following additional property. Write $\lambda_n^0 \in \mathfrak{g}_0(\lambda_s)^*$ *for the restriction of* λ_n *to* $\mathfrak{g}_0(\lambda_s)$. *Then*

iv) $G(\lambda) = G(\lambda_s) \cap G(\lambda_n) = G(\lambda_n^0)$.

Part (iv) says that the isotropy subgroup for λ in G may be computed as the isotropy subgroup for the nilpotent element λ_n^0 in the smaller reductive group $G(\lambda_s)$. Attempts to generalize Theorem 2 to reductive groups usually take advantage of this decomposition. One tries first to construct a representation τ_n of $G(\lambda_s)$ using λ_n^0. (This step is not understood in general.) Next, one uses τ_n on $G(\lambda_s)$ and some kind of parabolic induction to construct a representation of G; λ_s enters as a "central character" on $G(\lambda_s)$. This step *is* fairly well understood, and we will explain it in some detail. It can in turn be broken into two simpler constructions, using the following refinement of the Jordan decomposition (due perhaps to Kostant).

Proposition 4 (real Jordan decomposition). *Suppose G is a reductive Lie group, and $\lambda \in \mathfrak{g}_0^*$. Then there is a unique decomposition* $\lambda = \lambda_h + \lambda_e + \lambda_n$, *such that*

i) λ_h *is hyperbolic;*
ii) λ_e *is elliptic;*
iii) λ_n *is nilpotent; and*
iv) X_{λ_h}, X_{λ_e}, *and* X_{λ_n} *commute with each other.*

This decomposition has the following additional properties. Write $\lambda_{en}^0 \in \mathfrak{g}_0(\lambda_h)^*$ *for the restriction of* $\lambda_e + \lambda_n$ *to* $\mathfrak{g}_0(\lambda_h)$. *Then*

v) $\lambda_s = \lambda_h + \lambda_e$; *and*
vi) $G(\lambda) = G(\lambda_h) \cap G(\lambda_e) \cap G(\lambda_n) = G(\lambda_{en}^0)$.

Now the strategy is to use τ_n (a representation of $G(\lambda_s) = G(\lambda_h) \cap G(\lambda_e) \subset G(\lambda_h)$), and λ_e as a central character, to construct a representation τ_{en} of $G(\lambda_h)$. This step uses the "cohomological parabolic induction" functors of Zuckerman; $G(\lambda_s)$ will be a Levi factor for a "θ-stable parabolic subalgebra" of $G(\lambda_h)$. Finally, one uses τ_{en}, and λ_h as a central character, to construct a representation of G. This last step uses the familiar parabolic induction construction introduced by Gelfand and Harish-Chandra; $G(\lambda_h)$ is a Levi factor of a real parabolic subgroup of G.

In order to explain these last two steps more completely, we assume first that λ is semisimple. It follows that $\mathfrak{g}_0(\lambda)$ contains a Cartan subalgebra $\mathfrak{h}_0$ of $\mathfrak{g}_0$. Define a set of roots S by

$$S = \{\alpha \in \Delta(\mathfrak{g}, \mathfrak{h}) \mid \operatorname{Re}\lambda(H_\alpha) > 0, \text{ or } \operatorname{Re}\lambda(H_\alpha) = 0 \text{ and } \operatorname{Im}\lambda(H_\alpha) > 0\}.$$

The span $\mathfrak{u}$ of the corresponding root vectors is the nil radical of a $G(\lambda)$-invariant parabolic subalgebra $\mathfrak{q} = \mathfrak{g}(\lambda) + \mathfrak{u}$. Define $e^{2\rho(\mathfrak{u})}$ to be the character of $G(\lambda)$ on the top exterior power of $\mathfrak{u}$. Just as in the compact case, this leads to a double cover

$$1 \to \{\pm 1\} \to \widetilde{G}(\lambda) \to G(\lambda) \to 1.$$

Just as in the compact case, any other choice of $G(\lambda)$-invariant parabolic subalgebra with Levi subalgebra $\mathfrak{g}(\lambda)$ leads to a canonically isomorphic covering.

Definition 4. Suppose G is a reductive Lie group, and $\lambda \in \mathfrak{g}_0^*$ is semisimple. Define $\widetilde{G}(\lambda)$ as above. A representation (τ, V) of $\widetilde{G}(\lambda)$ is called *admissible* if it is irreducible, genuine, and satisfies

$$d\tau(X) = i\lambda(X) \cdot \mathrm{Id}_V \qquad (X \in \mathfrak{g}_0(\lambda)).$$

A *semisimple admissible orbit datum* is a pair (λ, τ) with λ semisimple and τ an admissible representation of $\widetilde{G}(\lambda)$.

In accordance with the remarks before Definition 2 above, one checks easily that the admissible representations of $\widetilde{G}(\lambda)$ are in one-to-one correspondence with the irreducible representations of $G(\lambda)$ having differential $i\lambda + \rho(\mathfrak{u})$.

Example 5. Suppose $G = GL(2m, \mathbb{R})$. We identify $\mathfrak{g}_0^*$ with $2m$ by $2m$ matrices using the trace form: if μ is a matrix in $\mathfrak{g}_0^*$ and X is a matrix in $\mathfrak{g}_0$, then $\mu(X) = \mathrm{tr}(\mu X)$. The coadjoint action is then by conjugation of matrices, so $G(\mu)$ consists of all invertible matrices commuting with μ. Define

$$J = \begin{pmatrix} O & I_m \\ -I_m & 0 \end{pmatrix} \in \mathfrak{g}_0^*, \qquad \lambda = tJ,$$

with t a positive real number. Because J is the matrix of multiplication by i (in an appropriate identification of $\mathbb{R}^{2m}$ with $\mathbb{C}^m$), the stabilizer of λ is $GL(m, \mathbb{C})$. Because λ is skew-symmetric, it is an elliptic element. The orbit $G \cdot \lambda \simeq G/G(\lambda)$ may be identified with the space of complex structures on $\mathbb{R}^{2m}$. These complex structures may in turn be identified with the m-dimensional planes $H \subset \mathbb{C}^{2m}$ such that $H \cap \overline{H} = \{0\}$. This exhibits $G/G(\lambda)$ as an open subset of the Grassmannian of m-dimensional planes in $\mathbb{C}^{2m}$. Consequently $G/G(\lambda)$ carries an invariant complex structure. (This is a general property of elliptic coadjoint orbits.) Define $\mathfrak{q}$ to be the inverse image in $\mathfrak{g}$ of the holomorphic tangent space at the identity coset to $G/G(\lambda)$. (Here we identify the full complex tangent space with $\mathfrak{g}/\mathfrak{g}(\lambda)$.) It turns out that $\mathfrak{q} = \mathfrak{g}(\lambda) + \mathfrak{u}$ is the parabolic subalgebra constructed before Definition 4 for λ, so $e^{2\rho(\mathfrak{u})}$ is the character by which $G(\lambda)$ acts on the top exterior power of the holomorphic tangent space. A straightforward computation gives

$$e^{2\rho(\mathfrak{u})}(g) = (\det g/|\det g|)^{2m} \qquad (g \in GL(m, \mathbb{C})),$$

so that

$$\widetilde{G}(\lambda) = \{(g, z) \in GL(m, \mathbb{C}) \times \mathbb{C}^\times \mid (\det g/|\det g|)^{2m}$$
$$= z^2\} \simeq GL(m, \mathbb{C}) \times \{\pm 1\}.$$

Finally, one can compute that the character $g \mapsto \det g / |\det g|$ of $G(\lambda$ has differential J. It follows easily that $\lambda = tJ$ is admissible if and only if t is an integer. The corresponding admissible representation of $\widetilde{G}(\lambda)$ (which is unique) is the character

$$\tau(g, \epsilon) = \epsilon(\det g / |\det g|)^t \qquad ((g, \epsilon) \in GL(m, \mathbb{C}) \times \{\pm 1\}).$$

Example 6. Suppose $G = SL(2, \mathbb{R})$. As for $GL(2m)$, we identify $\mathfrak{g}_0$ and $\mathfrak{g}_0^*$ with two by two matrices of trace 0. Set $\lambda = \begin{pmatrix} t & 0 \\ 0 & -t \end{pmatrix}$, with t positive; this is a hyperbolic element. We have

$$G(\lambda) = \left\{ \begin{pmatrix} a & 0 \\ 0 & a^{-1} \end{pmatrix} \mid a \in \mathbb{R}^\times \right\}, \qquad \mathfrak{u} = \left\{ \begin{pmatrix} 0 & x \\ 0 & 0 \end{pmatrix} \mid x \in \mathbb{C} \right\}.$$

We therefore identify $G(\lambda)$ with $\mathbb{R}^\times$. The character of $G(\lambda)$ on the top exterior power of $\mathfrak{u}$ is $e^{2\rho(\mathfrak{u})}(a) = a^2$. Consequently

$$\widetilde{G}(\lambda) = \{(a, z) \in \mathbb{R}^\times \times \mathbb{C}^\times \mid a^2 = z^2\} \simeq \mathbb{R}^\times \times \{\pm 1\};$$

the last isomorphism sends (a, z) to $(a, z/a)$. There are two admissible characters $\tau_\pm$ of $\widetilde{G}(\lambda)$, namely

$$\tau_+(a, \epsilon) = \epsilon |a|^{it}, \qquad \tau_-(a, \epsilon) = \epsilon |a|^{it} \mathrm{sgn}(a).$$

In particular, λ is admissible for all $t > 0$.

Example 7. Suppose $G = GL(2, \mathbb{R})$. As for $GL(2m)$, we identify $\mathfrak{g}_0$ and $\mathfrak{g}_0^*$ with two by two matrices. Set $\lambda = \begin{pmatrix} t_1 & 0 \\ 0 & t_2 \end{pmatrix}$, with $t_1 > t_2$; this is a hyperbolic element. We have

$$G(\lambda) = \left\{ \begin{pmatrix} a_1 & 0 \\ 0 & a_2 \end{pmatrix} \mid a_i \in \mathbb{R}^\times \right\}, \qquad \mathfrak{u} = \left\{ \begin{pmatrix} 0 & x \\ 0 & 0 \end{pmatrix} \mid x \in \mathbb{C} \right\}.$$

We therefore identify $G(\lambda)$ with $\mathbb{R}^\times \times \mathbb{R}^\times$. The character of $G(\lambda)$ on the top exterior power of $\mathfrak{u}$ is $e^{2\rho(\mathfrak{u})}(a_1, a_2) = a_1 a_2^{-1}$. Consequently

$$\widetilde{G}(\lambda) = \{(a_1, a_2, z) \in \mathbb{R}^\times \times \mathbb{R}^\times \times \mathbb{C}^\times \mid a_1 a_2^{-1} = z^2\}.$$

In this case $\widetilde{G}(\lambda)$ is *not* isomorphic to the direct product $G(\lambda) \times \{\pm 1\}$, because $\widetilde{G}(\lambda)$ contains the element $(1, -1, i)$, which has order 4. There are four admissible characters of $\widetilde{G}(\lambda)$, namely

$$\tau_{(u_1, u_2)}(a_1, a_2, z) = (\mathrm{sgn}\, a_1)^{u_1} (\mathrm{sgn}\, a_2)^{u_2} |a_1|^{it_1} |a_2|^{it_2} (z/|z|) \qquad (u_i \in \mathbb{Z}/2\mathbb{Z}).$$

In particular, λ is admissible for all $t_1 > t_2$. (In fact it is admissible for all t_i: the elements with $t_1 < t_2$ define exactly the coadjoint orbits we have just considered, and those with $t_1 = t_2$ can be dealt with by a simpler argument.)

Theorem 3. *Suppose G is a real reductive Lie group, and $\lambda \in \mathfrak{g}_0^*$ is a semisimple element. Assume that (λ, τ) is an admissible orbit datum (Definition 4). Then there is attached to this datum a unitary representation $\Pi_{adm}(\lambda, \tau)$, which may in general be reducible or zero. If λ is a regular element (that is, if $\mathfrak{g}_0(\lambda)$ is a Cartan subalgebra), then $\Pi_{adm}(\lambda, \tau)$ is an irreducible tempered representation having regular infinitesimal character. This establishes a bijection between the irreducible tempered representations of regular infinitesimal character, and the G-conjugacy classes of regular semisimple admissible orbit data.*

The last half of this theorem (concerning tempered representations) is due to Harish-Chandra. In this case the problem of constructing $\Pi_{adm}(\lambda, \tau)$ is essentially the problem (solved by Harish-Chandra) of constructing discrete series representations. The remaining assertions (still in the tempered case) are by now fairly well-known (see [KZ]), but are still very deep. What remains is therefore the construction of a unitary representation $\Pi_{adm}(\lambda, \tau)$ in general. This we now sketch.

Write $\lambda = \lambda_h + \lambda_e$ for the Jordan decomposition of Proposition 4. Write θ for the Cartan involution of G with fixed point set K. After replacing λ by a conjugate under G, we may assume that $\theta\lambda_h = -\lambda_h$ and $\theta\lambda_e = \lambda_e$. Then θ preserves $G(\lambda_e)$ and $G(\lambda_h)$, and therefore also their intersection $G(\lambda)$ (Proposition 4). It follows that the restriction of θ is a Cartan involution for the reductive group $G(\lambda)$. In particular, $G(\lambda) \cap K$ is a maximal compact subgroup $G(\lambda)$.

Now τ is an irreducible genuine representation of $\widetilde{G}(\lambda)$, and $e^{\rho(\mathfrak{u})}$ is a genuine character of the same group. Consequently $\tau' = \tau \otimes e^{\rho(\mathfrak{u})}$ is an irreducible representation of $\widetilde{G}(\lambda)$ trivial on $\{\pm 1\}$. We therefore regard it as an irreducible representation of $G(\lambda)$. We abuse notation by identifying τ' with the corresponding irreducible $(\mathfrak{g}(\lambda), G(\lambda) \cap K)$-module. (This is particularly harmless because the assumption on the differential of τ forces τ and τ' to be finite-dimensional.) Recalling the parabolic subalgebra $\mathfrak{q}$ constructed before Definition 4, we regard τ' as an irreducible $(\mathfrak{q}, G(\lambda) \cap K)$-module by making $\mathfrak{u}$ act trivially. Recall from [Green], Chapter 6 the construction of produced modules:

$$\mathrm{pro}_{\mathfrak{q}, G(\lambda) \cap K}^{\mathfrak{g}, G(\lambda) \cap K}(\tau') = \mathrm{Hom}_{\mathfrak{q}}(U(\mathfrak{g}), V_{\tau'})_{(G(\lambda) \cap K) - \mathrm{finite}}.$$

The Hom is defined using the left action of $\mathfrak{q}$ on $U(\mathfrak{g})$, and $U(\mathfrak{g})$ acts on the produced module through its right action on $U(\mathfrak{g})$. The produced module is therefore a $(\mathfrak{g}, G(\lambda) \cap K)$-module. (For the reader who is trying to keep track of normalizations, we record here the infinitesimal character of the produced module. Fix a Cartan subalgebra $\mathfrak{h}$ as in the construction of $\mathfrak{q}$, choose a system of positive roots for $\mathfrak{h}$ in $\mathfrak{l}$, and let $\rho(\mathfrak{l})$ be half the sum of these positive roots. Then the infinitesimal character corresponds under the Harish-Chandra map to $i\lambda|_{\mathfrak{h}} + \rho(\mathfrak{l})$.) Finally, we apply to the produced module the sth derived functor of the Zuckerman functor $\Gamma_{\mathfrak{g},G(\lambda)\cap K}^{\mathfrak{g},K}$ ([Green], Chapter 6), obtaining a $(\mathfrak{g}, K)$-module. (Here s is the dimension of $\mathfrak{u}\cap\mathfrak{k}$.) This module is the Harish-Chandra module of a unitary representation (more about this in a moment), and it is this unitary representation that we call $\Pi_{adm}(\lambda, \tau)$. To summarize,

$$\Pi_{adm}(\lambda, \tau) = (\Gamma_{\mathfrak{g},G(\lambda)\cap K}^{\mathfrak{g},K})^s(\mathrm{pro}_{\mathfrak{q},G(\lambda\cap K}^{\mathfrak{g},G(\lambda)\cap K}(\tau \otimes e^{\rho(\mathfrak{u})})).$$

The infinitesimal character is the same as the one for the produced module.

The most difficult step in this outline is perhaps the proof that the final $(\mathfrak{g}, K)$-module is unitary. This is best understood by breaking the construction into two steps, as outlined after Proposition 4. The first step (induction from $G(\lambda)$ to $G(\lambda_h)$ using the parabolic subalgebra $\mathfrak{q}\cap\mathfrak{g}(\lambda_h)$) is what is usually known as Zuckerman's cohomological induction construction, involving a parabolic subalgebra preserved by the Cartan involution. That this produces a unitary representation is proved in [V1] (see also [W]). The second step (induction from $G(\lambda_h)$ to G using the parabolic subalgebra $\mathfrak{q} + \bar{\mathfrak{q}} = \mathfrak{q} + \mathfrak{g}(\lambda_h)$) is parabolic induction from a real parabolic subgroup. That this step preserves unitarity is classical and easy. (Because of the form of the infinitesimal character, one can also check that this step preserves irreducibility; so the question of reducibility or vanishing of $\Pi_{adm}(\lambda, \tau)$ may be studied entirely in the setting of Zuckerman's cohomological induction.)) Finally, one must check that this two-step construction gives the same result as the original one-step definition. This is a consequence of the induction by stages formula in [Green], Proposition 6.3.6.

Example 8. We describe here the representations attached to the various admissible orbit data in Examples 5–7. In Example 5, the maximal compact subgroup K is $O(2m)$, which has a maximal torus $T = SO(2)^m$. The characters of T are therefore naturally parametrized by $\mathbb{Z}^m$. The representation π_k of $GL(2m, \mathbb{R})$ attached to λ in the case $t = k$ (a positive integer) was first constructed by Speh (see [S]). The lowest K-type of π_k is the representation μ_{k+1} of $O(2m)$ having extremal weight $(k+1, \ldots, k+1)$. All of the K-types have multiplicity one, and have extremal weights of the form $(k + 1 + 2a_1, \ldots, k + 1 + 2a_m)$, with (a_i) a decreasing sequence of non-negative integers. For $m = 1$, π_k is a (relative) discrete series representation; this corresponds to the fact that λ is regular in that case. For $m > 1$, π_k is non-tempered.

In Example 6, $\Pi_{adm}(\lambda, \tau_+)$ is a unitary spherical principal series representation of $SL(2, \mathbb{R})$; $\Pi_{adm}(\lambda, \tau_-)$ is a unitary non-spherical principal series representation. In both cases the principal series representation with parameter 0 is omitted. (This is the reducible one in the non-spherical case.) This omission is not rectified by including the case $\lambda = 0$: that orbit corresponds by the construction of Theorem 3 to the trivial representation of G. Instead, the singular principal series representations should be regarded as attached to nilpotent coadjoint orbits.

In Example 7, the representations $\Pi_{adm}(\lambda, \tau_{(u_1,u_2)})$ are unitarily induced principal series representations of $GL(2, \mathbb{R})$. The coordinates t_1 and t_2 of λ give the continuous parameter, and the $u_i \in \mathbb{Z}/2\mathbb{Z}$ give the four possible discrete parameters in an obvious way. As the parameters vary, we get all the unitary principal series with regular infinitesimal character.

Here is an example in which the representation associated to an admissible orbit datum is reducible.

Example 9. Suppose G is a complex simple group of type G_2. Let λ be a non-zero singular elliptic element chosen so that $G(\lambda)$ contains a long root $SL(2)$, λ is admissible, and λ is as small as possible subject to the first two conditions. For complex reductive groups admissibility is equivalent to integrality, so $i\lambda$ is the differential of a unitary character τ of $G(\lambda)$. There is a real parabolic subgroup $P = G(\lambda)N$ with Levi factor $G(\lambda)$. (If J is the complex structure on $\mathfrak{g}_0$, then $J\lambda$ is hyperbolic, and $G(J\lambda) = G(\lambda)$. One can take for $\mathfrak{p}$ the parabolic subalgebra constructed for $J\lambda$ after Proposition 4.) It turns out that

$$\Pi_{adm}(\lambda, \tau) \simeq \operatorname{Ind}_P^G(\tau),$$

a degenerate principal series representation. (One can show that for G complex reductive and λ semisimple, the representations $\Pi_{adm}(\lambda, \tau)$ are always degenerate principal series representations.) This particular degenerate principal series representation is a direct sum of two irreducible representations. (This can happen in the classical groups as well, but all the known examples involve singular parameters; they do not give rise to reducible $\Pi_{adm}(\lambda, \tau)$.)

The representations in question are related to those discussed in Kostant's lectures. Recall that $G \subset SO(8, \mathbb{C})$, and that Kostant described a certain very singular unitary representation π of $SO(8, \mathbb{C})$. The restriction of π to G is a direct sum of four irreducible representations, two of which are isomorphic. The constituents of $\Pi_{adm}(\lambda, \tau)$ are two of these four representations.

The construction of $\Pi_{adm}(\lambda, \tau)$, although it is rather long and difficult, is at least philosophically quite natural and geometric; it amounts to looking at sections of a vector bundle on a flag manifold. From this point of view the enveloping algebra of G acts by differential operators on sections. For most λ, one gets in some sense "all" differential operators in this way, and the representation $\Pi_{adm}(\lambda, \tau)$ is irreducible. It turns out that the failure of irreducibility in general can be attributed to the failure of the enveloping algebra of G to provide enough differential operators. The next theorem makes this statement precise.

Theorem 4. *Suppose G is a real reductive Lie group, $\lambda \in \mathfrak{g}_0^*$ is a semisimple element, and (λ, τ) is an admissible orbit datum. Let $\mathfrak{q}$ be the parabolic subalgebra of $\mathfrak{g}$ constructed from λ after Proposition 4, and let X be the (projective algebraic) variety of parabolic subalgebras of $\mathfrak{g}$ conjugate to $\mathfrak{q}$. Let $\mathcal{D}_\lambda$ be the sheaf of twisted differential operators on X associated to λ, and D_λ its algebra of global sections.*

a) *There is a natural algebra homomorphism $U(\mathfrak{g}) \to D_\lambda$. It is surjective if $(i\lambda + \rho(\mathfrak{l})) (H_\alpha)$ is not a negative integer for any root α in $\mathfrak{u}$.*

b) *The algebra D_λ is finitely generated as $U(\mathfrak{g})$-bimodule for all λ.*

c) *There is a natural action of D_λ on (the Harish-Chandra module) $\Pi_{adm}(\lambda, \tau)$ extending the action of $U(\mathfrak{g})$ (by means of the homomorphism in (a)).*

d) *The action in (c) makes $\Pi_{adm}(\lambda, \tau)$ an irreducible (D_λ, K)-module (or zero).*

Parts (a) and (b) may essentially be found in [BB] (see for example Theorem 3.8). The produced module appearing in the construction of $\Pi_{adm}(\lambda, \tau)$ may be thought of as a space of formal power series sections of a bundle on X ([V2], Proposition 6.13), so the differential operator algebra D_λ acts on it. Applying the Zuckerman functor Γ to this action is not difficult, and we get (c). For (d), one can first prove that $\Pi_{adm}(\lambda, \tau)$ is irreducible as a $(\mathfrak{g}, K)$-module for λ sufficiently regular, and then apply a translation functor as in [VJ]. (The point is that the algebras D_λ behave much better under translation functors than $U(\mathfrak{g})$.)

How large a family of representations does Theorem 3 provide? Certainly one gets almost all (with respect to the Plancherel measure) tempered representations; so one has enough representations to decompose $L^2(G)$. If H is a symmetric subgroup of G, Oshima's Plancherel formula involves non-tempered representations in general. Almost all of them will be of the form $\Pi_{adm}(\lambda, \tau)$, however; so we have enough representations to decompose $L^2(G/H)$. If $G = SL(2, \mathbb{R})$, the representations in Theorem 3 are the tempered representations with non-singular infinitesimal character, and the trivial representation. Missing, therefore, are the three tempered representations of infinitesimal character 0, and the complementary series (parametrized by the open interval $(0, 1)$). The absence of the complementary series is not too distressing, since they seem to arise less often in interesting harmonic analysis problems. (The *fact* that they do not arise is often extremely interesting and difficult, but that is another story.) The absence of the three tempered representations is more serious, as we will soon see.

The first class of harmonic analysis problems for which we encounter serious difficulties is $L^2(G/\Gamma)$, with Γ an arithmetic subgroup of G. In this case the decomposition of L^2 is very far from understood. Nevertheless it is clear that the representations in Theorem 3 will not suffice; more precisely, that there will be irreducible subrepresentations of $L^2(G/\Gamma)$ (for appropriate G and congruence subgroups Γ) not of the form $\Pi_{adm}(\lambda, \tau)$ for any semisimple λ (see [A]). The three tempered representations of $SL(2, \mathbb{R})$ of infinitesimal character 0 are the first examples. For groups not of type A_n, one expects to find such representations that are entirely unlike any of the ones constructed in Theorem 3. This suggests that our assumption in Theorem 3 that λ is semisimple is too restrictive. For the balance of this paper, we consider the (largely untouched) problem of weakening that assumption.

The first problem is the definition of "admissible orbit datum" in general; recall that Definition 4 made use of a covering constructed using the hypothesis of semisimplicity. Fortunately Duflo has shown how to make such a definition in great generality. Suppose for a moment that G is any Lie group, and $\lambda \in \mathfrak{g}_0^*$, and $\Lambda = G \cdot \lambda$ is the corresponding coadjoint orbit. The tangent space to Λ at λ may be identified as

$$T_\lambda(\Lambda) \simeq \mathfrak{g}_0/\mathfrak{g}_0(\lambda).$$

The element λ defines a skew-symmetric bilinear form on $\mathfrak{g}_0$ by

$$\omega_\lambda(X, Y) = \lambda([X, Y]).$$

One checks easily that the radical of this form is exactly $\mathfrak{g}_0(\lambda)$, so ω_λ may be regarded as a non-degenerate symplectic form on $\mathfrak{g}_0/\mathfrak{g}_0(\lambda)$; or, equivalently, on $T_\lambda(\Lambda)$. These forms depend smoothly on λ (within a fixed orbit λ), and so define a two-form ω_Λ on Λ. A short calculation shows that ω_Λ is closed, so it defines a symplectic structure on Λ. By its definition ω_Λ is obviously G-invariant; in particular, ω_λ is $G(\lambda)$-invariant.

Definition 5 (Duflo - see [D] or [V2], Definition 10.16). Suppose G is a Lie group, and $\lambda \in \mathfrak{g}_0^*$. Define $Sp(\omega_\lambda)$ to be the group of linear transformations of $\mathfrak{g}_0/\mathfrak{g}_0(\lambda)$ preserving the form ω_λ just defined. According to the last remark, the isotropy action defines a homomorphism

$$i_\lambda : G(\lambda) \to Sp(\omega_\lambda).$$

The symplectic group has a distinguished two-fold cover $Mp(\omega_\lambda)$:

$$1 \to \{1, \zeta\} \to Mp(\omega_\lambda) \xrightarrow{\pi} Sp(\omega_\lambda) \to 1;$$

if $\mathfrak{g}_0/\mathfrak{g}_0(\lambda) \neq 0$, then $Mp(\omega_\lambda)$ is the unique connected two-fold cover. We can pull this covering back to $G(\lambda)$ using i_λ, setting

$$\widetilde{G}(\lambda) = \{(g, m) \in G(\lambda) \times Mp(\omega_\lambda) \mid i_\lambda(g) = \pi(m)\},$$

the *metaplectic covering of* $G(\lambda)$. Then

$$1 \to \{1, \zeta\} \to \widetilde{G}(\lambda) \to G(\lambda) \to 1.$$

A representation (τ, V) of $\widetilde{G}(\lambda)$ is called *genuine* if $\tau(\zeta) = -\mathrm{Id}_V$. It is called *admissible* if it is irreducible, unitary, genuine, and satisfies

$$d\tau(X) = i\lambda(X) \cdot \mathrm{Id}_V \qquad (X \in \mathfrak{g}_0(\lambda)).$$

An *admissible orbit datum* is a pair (λ, τ) with $\lambda \in \mathfrak{g}_0^*$ and τ an admissible representation of $\widetilde{G}(\lambda)$.

Formally this is rather similar to Definitions 2 and 4. One important technical difference is that $\widetilde{G}(\lambda)$ is not equipped in general with a one-dimensional genuine representation. Consequently the admissible representations are not so simply related to representations of $G(\lambda)$ (see the discussion before Definition 2).

There is a consistency question to be addressed at this point: one must check that if G is reductive and λ is semisimple, then the covering $\widetilde{G}(\lambda)$ constructed here is isomorphic to the one constructed before Definition 4 (so that the two definitions of admissible agree). This is true and not too difficult to prove, once one knows something about the metaplectic representation; we will not discuss the argument.

An *admissible orbit correspondence* would attach to each admissible orbit datum a unitary representation (possibly reducible or zero). It turns out that this is not the best thing to ask for in general; we will introduce refinements in Definition 6 below.

Example 10. We use the notation of Example 3, with $G = Sp(2n, \mathbb{R})$. Write λ for the element λ_1 considered there. The orbit $\Lambda = G \cdot \lambda$ consists of all the non-zero linear transformations in $\mathfrak{g}_0^*$ of the form $w \mapsto \omega(v, w) \cdot v$. The linear transformation determines v up to sign; in this way Λ may be identified with $(\mathbb{R}^{2n} - \{0\})/\{\pm 1\}$. The tangent space $\mathfrak{g}_0/\mathfrak{g}_0(\lambda)$ may therefore be naturally identified with $\mathbb{R}^{2n}$, and the group $Sp(\omega_\lambda)$ with $Sp(2n, \mathbb{R})$. There is a subtle point about the isotropy map i_λ. If $H \subset GL(V)$ fixes a point $v \in V$, then the tangent space at v may be identified with V, and the isotropy action of H on the tangent space is just given by the inclusion $H \hookrightarrow GL(V)$. This suggests that in our case the isotropy action i_λ should be given by the inclusion $G(\lambda) \subset G$. This is almost but not quite correct. What happens is that since Λ sits in $\mathbb{R}^{2n}/\{\pm 1\}$, not all elements of $G(\lambda)$ actually fix the corresponding vector v: those in the the identity component do, but the non-identity component sends v to $-v$. We can write

$$G(\lambda) \simeq G(\lambda)_0 \times \{\pm \mathrm{Id}\},$$

the latter factor being the center of G. The isotropy map i_λ is then the identity on $G(\lambda)_0$, and trivial on $\{\pm \mathrm{Id}\}$. The metaplectic covering of $G(\lambda)$

is therefore just the preimage of $G(\lambda)_0$ in $Mp(2n, \mathbb{R})$, times $\{\pm \mathrm{Id}\}$. (This is *not* true for other λ in $Sp(2n, \mathbb{R})$.) We saw in Example 3 that the restriction of λ to $\mathfrak{g}_0(\lambda)$ is zero. It follows that an irreducible representation τ of $\widetilde{G}(\lambda)$ is admissible if and only if $\tau(\zeta) = -1$, and τ is trivial on the identity component of $\widetilde{G}(\lambda)$.

If $n = 1$ (so that $G(\lambda)_0 \simeq \mathbb{R}$) then $\widetilde{G}(\lambda) \simeq \mathbb{R} \times \{1, \zeta\} \times \{\pm \mathrm{Id}\}$. Consequently λ is admissible, and there are exactly two admissible orbit data $(\lambda, \tau_\pm)$ (with $\tau_\pm$ sending $-\mathrm{Id}$ to ± 1). These orbit data are related (although not in a completely straightforward way — see Examples 13 and 14 below) to two of the three tempered representations of $SL(2, \mathbb{R}) \simeq Sp(2, \mathbb{R})$ of infinitesimal character 0.

If $n > 1$, then $G(\lambda)_0$ contains $Sp(2n - 2, \mathbb{R})$. The preimage of this subgroup in $Mp(2n, \mathbb{R})$ is naturally isomorphic to the connected group $Mp(2n-2, \mathbb{R})$. It follows that ζ belongs to the identity component of $\widetilde{G}(\lambda)$, and therefore that λ is not admissible. This is consistent with the claim in Example 3 that there is no irreducible unitary representation attached to λ.

Finally, this example can be repeated with G replaced by its double cover $Mp(2n, \mathbb{R})$. In that case one finds (after some difficulties sorting out the various coverings) that $\widetilde{G}(\lambda)/\widetilde{G}(\lambda)_0 \simeq (G(\lambda)/G(\lambda)_0) \times \{1, \zeta\}$, so that the admissible representations of $\widetilde{G}(\lambda)$ correspond naturally to the representations of $G(\lambda)/G(\lambda)_0$. This component group is $\mathbb{Z}/4\mathbb{Z}$ if $n = 1$ and $\mathbb{Z}/2\mathbb{Z}$ if $n > 1$; so in any case one finds two new admissible orbit data. If $n > 1$, these data should be attached to the two irreducible components of the metaplectic representation of G. For $n = 1$ the situation is less straightforward; probably the components of the metaplectic representation and certain tempered representations should all be attached to the new admissible orbit data.

We would like an orbit correspondence similar to Theorem 3 for general admissible orbit data for reductive groups. Using the ideas in the proof of Theorem 3, the construction of such a correspondence can be reduced to the case of nilpotent orbit data. Since we have no good general theorem in any case, we will confine our attention to nilpotent λ.

One useful observation about the semisimple case is that Theorem 4 is in some respects easier than Theorem 3: that is, that the description of the differential operator algebra D_λ is easier than the description of the modules over it. One might even claim that Theorem 4 helps to suggest how the representations in Theorem 3 should be constructed. (This is historically inaccurate, but logically reasonable.) We will therefore try to begin not by constructing representations attached to real coadjoint orbits, but by attaching algebras to complex coadjoint orbits.

Recall that we have fixed before Definition 3 a real reductive group G and maximal compact subgroup K. It is convenient to fix also a complex reductive group $G_{\mathbb{C}}$ having Lie algebra $\mathfrak{g}$ (the complexification of the Lie algebra of G). We assume that the adjoint action of G on its Lie algebra factors through a homomorphism

$$i : G \rightarrow G_{\mathbb{C}};$$

we assume that the differential of i is the inclusion $\mathfrak{g}_0 \hookrightarrow \mathfrak{g}$. (Because we want to allow G to be non-linear, we do not assume that i is injective.)

Fix a complex nilpotent coadjoint orbit

$$\mathcal{O}_{\mathbb{C}} = G_{\mathbb{C}} \cdot \lambda \subset \mathfrak{g}^*.$$

An *equivariant covering for* $\mathcal{O}_{\mathbb{C}}$ is a $G_{\mathbb{C}}$-homogeneous space $\tilde{\mathcal{O}}_{\mathbb{C}}$ that is a finite cover of $\mathcal{O}_{\mathbb{C}}$. Explicitly,

$$\tilde{\mathcal{O}}_{\mathbb{C}} \simeq G_{\mathbb{C}}/G_{\mathbb{C}}(\lambda)_1, \qquad G_{\mathbb{C}}(\lambda)_0 \subset G_{\mathbb{C}}(\lambda)_1 \subset G_{\mathbb{C}}(\lambda).$$

Let $R_{\tilde{\mathcal{O}}_{\mathbb{C}}}$ be the algebra of regular functions on the variety $\tilde{\mathcal{O}}_{\mathbb{C}}$. (This is the same as the algebra of regular functions on $G_{\mathbb{C}}$ invariant on the right by $G_{\mathbb{C}}(\lambda)_1$.) The finitely generated algebra $R_{\tilde{\mathcal{O}}_{\mathbb{C}}}$ carries a natural action of $G_{\mathbb{C}}$ by automorphisms. The morphism of algebraic varieties

$$\tilde{\mathcal{O}}_{\mathbb{C}} \rightarrow \mathcal{O}_{\mathbb{C}} \hookrightarrow \mathfrak{g}^*$$

gives rise to a homomorphism of algebras

$$\xi : S(\mathfrak{g}) \rightarrow R_{\tilde{\mathcal{O}}_{\mathbb{C}}}.$$

The algebra $R_{\tilde{\mathcal{O}}_{\mathbb{C}}}$ has a natural grading by $1/2\mathbb{N}$ that makes ξ a graded homomorphism. It is finite (in the sense that it makes $R_{\tilde{\mathcal{O}}_{\mathbb{C}}}$ a finitely generated $S(\mathfrak{g})$-module). Set $X = \operatorname{Spec} R_{\tilde{\mathcal{O}}_{\mathbb{C}}}$, an affine algebraic variety. Then $G_{\mathbb{C}}$ acts on X with finitely many orbits, all of which have even dimension; the unique open orbit is $\tilde{\mathcal{O}}_{\mathbb{C}}$. The homomorphism ξ corresponds to a finite morphism of varieties $X \rightarrow \mathfrak{g}^*$; its image is the closure of the orbit $\mathcal{O}_{\mathbb{C}}$. (If $\tilde{\mathcal{O}}_{\mathbb{C}} = \mathcal{O}_{\mathbb{C}}$, then X is just the normalization of the orbit closure.)

Theorem 5 (W. Graham). *Fix an equivariant covering $\tilde{\mathcal{O}}_{\mathbb{C}}$ for the complex nilpotent coadjoint orbit $\mathcal{O}_{\mathbb{C}}$. Assume in addition that $X - \tilde{\mathcal{O}}_{\mathbb{C}}$ has codimension strictly greater than 2 in X. (It is equivalent to assume that no orbit in the closure of $\mathcal{O}_{\mathbb{C}}$ has codimension 2.) Then there is an algebra $A_{\tilde{\mathcal{O}}_{\mathbb{C}}}$ having the following properties.*

a) $A_{\tilde{\mathcal{O}}_{\mathbb{C}}}$ is filtered by $1/2\mathbb{N}$, and the associated graded algebra is naturally isomorphic to $R_{\tilde{\mathcal{O}}_{\mathbb{C}}}$.

b) There is an action of $G_{\mathbb{C}}$ on $A_{\tilde{\mathcal{O}}_{\mathbb{C}}}$ by algebra automorphisms, and a filtered $G_{\mathbb{C}}$-equivariant homomorphism $U(\mathfrak{g}) \rightarrow A_{\tilde{\mathcal{O}}_{\mathbb{C}}}$. The associated graded homomorphism is the map $\xi : S(\mathfrak{g}) \rightarrow R_{\tilde{\mathcal{O}}_{\mathbb{C}}}$ defined above.

Graham constructs the algebra $A_{\widetilde{\mathcal{O}}_{\mathbb{C}}}$ as a deformation of $R_{\widetilde{\mathcal{O}}_{\mathbb{C}}}$ (see [V2], Chapter 9, and [V3]). It is characterized uniquely by some natural hypotheses on the deformation. The properties stated in Theorem 5 already suffice to show that $A_{\widetilde{\mathcal{O}}_{\mathbb{C}}}$ must be finitely generated as a bimodule for $U(\mathfrak{g})$, as the analogy with Theorem 4 suggests it ought to be.

There is an enormous amount of evidence to suggest that the analogue of Theorem 5 holds without the codimension hypothesis on $\mathcal{O}_{\mathbb{C}}$. It is not yet clear what kind of uniqueness statement to expect in general, however; we return to this point after the next example.

Example 11. We use again the notation of Example 3, with $G = Sp(2n, \mathbb{R})$ and $G_{\mathbb{C}} = Sp(2n, \mathbb{C})$. Write λ for the element λ_1 considered there. Every non-zero element v of $\mathbb{C}^{2n}$ defines a linear transformation T_v in $\mathfrak{g}^*$ by $T_v(w) = \omega(v, w)v$. As in Example 10, we see that the map $v \mapsto T_v$ is a two-to-one morphism from $\mathbb{C}^{2n} - \{0\}$ onto $\mathcal{O}_{\mathbb{C}}$. We may therefore take $\widetilde{\mathcal{O}}_{\mathbb{C}} = \mathbb{C}^{2n} - \{0\}$. Since the origin has codimension at least 2, the algebra $R_{\widetilde{\mathcal{O}}_{\mathbb{C}}}$ of regular functions is just the polynomial algebra in $2n$ variables. It turns out that the natural grading of $R_{\widetilde{\mathcal{O}}_{\mathbb{C}}}$ (which we have not explained) is half the usual one on polynomials. The elements of "degree" one in $R_{\widetilde{\mathcal{O}}_{\mathbb{C}}}$ are therefore the quadratic polynomials. These may be identified with $\mathfrak{g}$; this gives the homomorphism $\xi : S(\mathfrak{g}) \to R_{\widetilde{\mathcal{O}}_{\mathbb{C}}}$, with image the even polynomials.

For the algebra $A_{\widetilde{\mathcal{O}}_{\mathbb{C}}}$ of Theorem 5, we may take the Weyl algebra A_n, with generators $p_1, \ldots, p_n$ and $q_1, \ldots, q_n$. (One can think of A_n as polynomial coefficient differential operators on $\mathbb{C}^n$. Then p_i is the operator of multiplication by x_i, and q_j is $\partial/\partial x_j$.) We filter A_n by declaring the generators to have order $1/2$. The symmetric quadratic polynomials in the generators (spanned by $p_i p_j + p_j p_i$, $p_i q_j + q_j p_i$, etc.) form a Lie algebra (under commutator of the associative algebra structure) isomorphic to $\mathfrak{g}$. By the universal property of $U(\mathfrak{g})$, we get a filtered algebra homomorphism $U(\mathfrak{g}) \to A_{\widetilde{\mathcal{O}}_{\mathbb{C}}}$; it is easy to check that the associated graded homomorphism is ξ. (The image of $U(\mathfrak{g})$ consists of all differential operators preserving the even polynomial functions.)

Notice that the construction in Example 11 works perfectly well even if $n = 1$, although the hypotheses of Theorem 5 fail in that case. What is (perhaps) surprising and (certainly) inconvenient is that the three tempered representations of infinitesimal character 0 (which we would like to relate to this coadjoint orbit) cannot be made into modules for the Weyl algebra: the infinitesimal characters do not agree. It seems that we must allow for several different algebras $A_{\widetilde{\mathcal{O}}_{\mathbb{C}}}$ to be attached to a single coadjoint orbit. This difficulty appears to arise only when the hypotheses of Theorem 5 fail; that is, in the presence of orbits of codimension 2 in the closure of $\mathcal{O}_{\mathbb{C}}$.

Definition 6. Suppose G is a reductive group as above, $\lambda \in \mathfrak{g}_0^*$

is a nilpotent element, and $\widetilde{\mathcal{O}}_{\mathbb{C}}$ is an equivariant covering for the orbit $\mathcal{O}_{\mathbb{C}} = G_{\mathbb{C}} \cdot \lambda$. We have natural homomorphisms

$$\widetilde{G}(\lambda) \to G(\lambda) \xrightarrow{i} G_{\mathbb{C}}(\lambda).$$

Define $\widetilde{G}(\lambda)_1$ to be the preimage of $G_{\mathbb{C}}(\lambda)_1$ (notation as before Theorem 5). This is a subgroup of $\widetilde{G}(\lambda)$ of finite index, containing the distinguished central element ζ. It is therefore natural to speak of *genuine* and *admissible* representations of $\widetilde{G}(\lambda)_1$, exactly as in Definition 5. A *nilpotent admissible covering datum* is a triple $(\lambda, \widetilde{\mathcal{O}}_{\mathbb{C}}, \tau_1)$, with λ and $\widetilde{\mathcal{O}}_{\mathbb{C}}$ as above, and τ_1 an admissible representation of $\widetilde{G}(\lambda)_1$. (We may sometimes omit the covering $\widetilde{\mathcal{O}}_{\mathbb{C}}$ from the notation if no confusion can result.)

A *nilpotent admissible orbit correspondence* would attach to each covering datum $(\lambda, \widetilde{\mathcal{O}}_{\mathbb{C}}, \tau_1)$ first an algebra $A_{\widetilde{\mathcal{O}}_{\mathbb{C}}}$ satisfying the conclusions of Theorem 5, and then an $(A_{\widetilde{\mathcal{O}}_{\mathbb{C}}}, K)$-module $\Pi_{adm}(\lambda, \tau_1)$. This module should be irreducible (as suggested by Theorem 4), and the underlying $(\mathfrak{g}, K)$-module should be the Harish-Chandra module of a unitary representation (as suggested by Theorem 3). (For technical reasons connected with the theory of primitive ideals, it seems unreasonable to allow $\Pi_{adm}(\lambda, \tau_1) = 0$, as we did in the case of semisimple orbits.) It will turn out that no nice correspondence exists with these properties, even under the hypotheses of Theorem 5. Before turning to counterexamples, we consider a trivial case to illustrate some of the formal aspects of such a correspondence.

Example 12. Suppose G is a finite group. We can take $K = G_{\mathbb{C}} = G$. The only complex coadjoint orbit is $0 \simeq G/G$. An equivariant covering of this orbit is therefore an arbitrary homogeneous space G/G_1; here G_1 is our group $G_{\mathbb{C}}(\lambda)_1$. The metaplectic cover $\widetilde{G}(\lambda)$ is $G \times \mathbb{Z}/2\mathbb{Z}$, and its subgroup $\widetilde{G}_1$ is $G_1 \times \mathbb{Z}/2\mathbb{Z}$. Once the equivariant covering is fixed, an admissible covering datum therefore amounts to an irreducible representation τ_1 of G_1. We define

$$\Pi_{adm}(\tau_1) = \mathrm{Ind}_{G_1}^G (\tau_1).$$

The most immediate objection is that this representation is usually not irreducible. But we asked not that it should be irreducible under G, but that it should be an irreducible (A, G)-module, for an appropriate algebra A. So we need to understand A. The variety X constructed before Theorem 5 is just G/G_1, and the algebra R is the algebra of functions on X with the trivial grading. If the requirements in the conclusion of Theorem 5 are to be satisfied, we must define A also to be the algebra of functions on G/G_1, with the trivial filtration and the natural action of G (by translation of functions). The induced representation defined above may be regarded as the space of sections of a vector bundle on G/G_1, so A certainly acts on it. It is an elementary exercise to verify that the induced representation is indeed irreducible as an (A, G)-module.

Example 13. We work in the setting of Example 11, except that we take $G = Mp(2n, \mathbb{R})$ so that λ will be admissible. The group $\widetilde{G}(\lambda)_1$ of Definition 6 is of the form

$$(Mp(2n - 2, \mathbb{R}) \ltimes N) \times \{1, \zeta\},$$

with N a connected unipotent group. If $n > 1$ there is exactly one admissible representation τ_1 of $\widetilde{G}(\lambda)_1$ (trivial on the identity component, and sending ζ to -1). The corresponding representation $\Pi_{adm}(\lambda, \tau_1)$ is the full (reducible) metaplectic representation, on which the Weyl algebra does indeed act irreducibly.

If $n = 1$, τ_1 still makes sense, and we should still define $\Pi_{adm}(\lambda, \tau_1)$ to be the metaplectic representation. However, the metaplectic group $Mp(2n - 2, \mathbb{R})$ is just $\mathbb{Z}/2\mathbb{Z}$, so there is a second admissible representation $\tau_{1,-}$ (which acts by the non-trivial character of the $Mp(0, \mathbb{R})$ factor). Somewhat paradoxically, it is the character $\tau_{1,-}$ that comes from an admissible covering datum for $Sp(2, \mathbb{R})$. (Recall from Example 10 that such admissible data exist for this orbit only for $n = 1$. Of course the condition for descending to $Sp(2, \mathbb{R})$ is being trivial on an appropriate central element, and one would think that τ_1 rather than $\tau_{1,-}$ satisfies this condition. The subtle point is the construction of the isomorphism of $\widetilde{G}(\lambda)_1$ with $G(\lambda)_1 \times \{1, \zeta\}$. What emerges is that the condition to descend is triviality on (ϵ, ζ), with ϵ the central element of $Mp(2n - 2, \mathbb{R})$.)

At any rate, we conclude that $\Pi_{adm}(\lambda, \tau_{1,-})$ should be a representation of $Sp(2, \mathbb{R}) \simeq SL(2, \mathbb{R})$ on which the Weyl algebra acts irreducibly. There is only one such representation: the direct sum of the two non-unitary principal series representations (spherical and non-spherical) having infinitesimal character $\frac{1}{2}\rho$. The sphericial component is at the midpoint of the complementary series, and so corresponds to a unitary representation; but the non-spherical component does not. We conclude that there is no representation $\Pi_{adm}(\lambda, \tau_{1,-})$ in this case.

There is another kind of difficulty in constructing an orbit correspondence. In the setting of Definition 6, the intersection of $\mathcal{O}_{\mathbb{C}}$ with $\mathfrak{g}_0^*$ will be the union of a finite number of coadjoint orbits $\mathcal{O}^1, \ldots, \mathcal{O}^r$. Choose representatives $\lambda^1, \ldots, \lambda^r$ for these orbits. Fix an algebra $A_{\widetilde{\mathcal{O}}_{\mathbb{C}}}$ having the properties in the conclusion of Theorem 5, and an $(A_{\widetilde{\mathcal{O}}_{\mathbb{C}}}, K)$-module X. In [V4] one can find a procedure (under a mild "self-adjointness" hypothesis on $A_{\widetilde{\mathcal{O}}_{\mathbb{C}}}$, satisfied for the algebras constructed in Theorem 5) for constructing from X a sum $T(X, \lambda^i)$ of admissible representations of $\widetilde{G}(\lambda^i)_1$ (for each i). (The main result is Theorem 8.7, but one needs also Theorems 2.13 and 7.14. The theorems are stated only for $(\mathfrak{g}, K)$-modules, but the proofs apply to $(A_{\widetilde{\mathcal{O}}_{\mathbb{C}}}, K)$-modules.) This procedure ought to be an inverse of the orbit correspondence we want. Our example in which the orbit correspondence does not exist showed that a certain admissible representation did not appear in any $T(X, \lambda^i)$ with X unitary. The second kind of diffi-culty is that even if X is irreducible, $T(X, \lambda^i)$ may be non-zero for several different i. This suggests that such a representation X is attached not to a single coadjoint orbit, but rather to some combination of several orbits.

Example 14. Suppose $G = SL(2, \mathbb{R})$, $G_\mathbb{C} = SL(2, \mathbb{C})$, and $\mathcal{O}_\mathbb{C}$ is the principal nilpotent orbit. We can take for $A_{\widetilde{\mathcal{O}}_\mathbb{C}}$ the quotient of $U(\mathfrak{g})$ by the ideal generated by the kernel of the infinitesimal character 0. This is the algebra that acts on the two principal series with parameter 0. There are exactly three irreducible $(A_{\widetilde{\mathcal{O}}_\mathbb{C}}, K)$-modules: a holomorphic limit of discrete series representation π^+; an anti-holomorphic limit of discrete series rep-resentation π^-; and a spherical principal series representation π^0. These are the three tempered representations with singular infinitesimal character mentioned after Theorem 4.

The intersection $\mathcal{O}_\mathbb{C} \cap \mathfrak{g}_0^*$ is a union of exactly two orbits, which may be represented by the elements $\lambda = \begin{pmatrix} 0 & 1 \\ 0 & 0 \end{pmatrix}$ (cf. Examples 3 and 10) and $-\lambda$. Attached to each are two admissible representations, which we now write as $\tau_\pm(\lambda)$ and $\tau_\pm(-\lambda)$ respectively (Example 10). Here is how the map T acts.

$$\begin{aligned} T(\pi^+, \lambda) &= \tau_-(\lambda), & T(\pi^+, -\lambda) &= 0 \\ T(\pi^-, \lambda) &= 0, & T(\pi^-, -\lambda) &= \tau_-(-\lambda) \\ T(\pi^0, \lambda) &= \tau_+(\lambda), & T(\pi^0, -\lambda) &= \tau_+(-\lambda). \end{aligned}$$

From this we conclude that the representations $\pi^\pm$ may be thought of as attached to the admissible orbit data $(\pm\lambda, \tau_-(\pm\lambda))$; but that π^0 ought to be attached to a formal "sum" of the two orbit data $(\lambda, \tau_+(\lambda))$ and $(-\lambda, \tau_+(-\lambda))$.

The correspondence T evidently encodes precisely the possibilities for a nilpotent admissible orbit correspondence; so we can phrase questions about such a correspondence in terms of T. In [V4] (Theorem 4.6) there is one positive result.

Theorem 6. *Suppose we are in the setting of Theorem 5 (so that the closure of $\mathcal{O}_\mathbb{C}$ contains no orbits of codimension 2). Let $A_{\widetilde{\mathcal{O}}_\mathbb{C}}$ be as in that theorem, and suppose X is an irreducible $(A_{\widetilde{\mathcal{O}}_\mathbb{C}}, K)$-module. Then there is at most one G-orbit $\mathcal{O}^i \subset \mathcal{O}_\mathbb{C} \cap \mathfrak{g}_0^*$ with the property that $T(X, \lambda) \neq 0$ for $\lambda \in \mathcal{O}^i$. If $A_{\widetilde{\mathcal{O}}_\mathbb{C}}$ has no quotient of smaller Gelfand-Kirillov dimension, then there is exactly one such orbit.*

The last hypothesis on $A_{\widetilde{\mathcal{O}}_\mathbb{C}}$ is probably always satisfied by the algebra Graham constructs.

Theorem 6 says that under the codimension hypothesis, we can expect to attach representations to single coadjoint orbits. Here are some natural further questions.

Open Problems. Suppose we are in the setting of Definition 6, and that $A_{\widetilde{\mathcal{O}}_{\mathbb{C}}}$ is an algebra satisfying the conclusions of Theorem 5. We assume also the technical conditions needed to define the representations $T(\lambda^i, X)$ discussed before Example 14.

1) Suppose we are in the setting of Theorem 6. Show that the representation $T(\lambda^i, X)$ is necessarily irreducible or zero.

In the presence of codimension 2 orbits, there are counterexamples to the analogous assertion. The simplest has G the double cover of $SL(3, \mathbb{R})$; $\mathcal{O}_{\mathbb{C}}$ the principal nilpotent orbit; $A_{\widetilde{\mathcal{O}}_{\mathbb{C}}}$ the quotient of $U(\mathfrak{g})$ by the ideal generated by the infinitesimal character 0; and X the irreducible principal series representation of infinitesimal character 0 for the non-linear group. Then X is induced from a two-dimensional representation of the Borel subgroup, and it follows (not trivially) that $T(\lambda, X)$ also has dimension 2. But one can check that the admissible representations have dimension 1; $T(\lambda, X)$ consists of two copies of one of them.

2) Suppose we are in the setting of Theorem 6. Show that X is uniquely determined by $T(\lambda^i, X)$.

Again there are counterexamples in the presence of codimension 2 orbits. If $\mathcal{O}$ is principal, then $T(\lambda^i, X)$ sees essentially only the action of $Z(G)$ on X. Any two irreducible principal series X and X' for a split linear group having the same central and infinitesimal character will have $T(\lambda^i, X) \simeq T(\lambda^i, X')$. Non-isomorphic pairs (X, X') with these properties exist for $SL(3, \mathbb{R})$ (and indeed for almost all split groups except $SL(2, \mathbb{R})$).

3) Describe the range of T. More precisely, if we are given for each i a sum ρ^i of admissible representations, give criteria for the existence of an $(A_{\widetilde{\mathcal{O}}_{\mathbb{C}}}, K)$-module X such that $T(\lambda^i, X) = \rho^i$. When can X be chosen unitarizable?

Here the main difficulty is probably in the presence of codimension 2 orbits, where the ρ^i are not nearly enough information to specify an X. But [V4] also provides an example (Example 12.4) in the setting of Theorem 5, in which T is not surjective (so that an orbit correspondence cannot be defined on all admissible covering data). On the other hand, Theorem 4.11 of [V4] does provide some (very weak) information about how the ρ^i must be related (say for an irreducible X to exist).

In addition to [V4], the reader may consult [V5] for some related material, particularly on the passage from nilpotent to general coadjoint orbits. Of course we have said essentially nothing about the most interesting and fundamental question, which is the (preferably direct and geometric) construction of unitary $(A_{\widetilde{\mathcal{O}}_{\mathbb{C}}}, K)$-modules. Many of these modules should be found among the "singular unitary representations" of classical groups studied by Li in [L]. All of the "special unipotent representations" defined in [ABV] should also appear in the image of appropriate nilpotent orbit correspondences. No satisfactory general relationship is known among the constructions of [L] and [ABV] and the coadjoint orbit philosophy outlined here. Connecting any two of these (even by a detailed conjecture) would be very useful.

References

[ABV]   J. Adams, D. Barbasch, and D. Vogan, "The Langlands classification and irreducible characters for real reductive groups," to appear.

  [A]   J. Arthur, "Unipotent automorphic representations: conjectures," 13–71 in *Orbites Unipotentes et Représentations II. Groupes p-adiques et Réels*, Astérisque **171–172** (1989).

 [BB]   W. Borho and J.-L. Brylinski, "Differential operators on homogeneous spaces I," Invent. Math. **69** (1982), 437–476.

  [D]   M. Duflo, "Théorie de Mackey pour les groupes de Lie algébriques," Acta Math. **149** (1982), 153–213.

 [KZ]   A. Knapp and G. Zuckerman, "Classification of irreducible tempered representations of semisimple Lie groups," Ann. of Math. **116** (1982), 389–501. (See also Ann. of Math. **119** (1984), 639.)

  [L]   J.-S. Li, "Singular unitary representations of classical groups," Invent. Math. **97** (1989), 237–255.

  [S]   B. Speh, "Unitary representations of SL(n,$\mathbb{R}$) and the cohomology of congruence subgroups," 483–505 in *Non-commutative Harmonic Analysis and Lie groups*, J. Carmona and M. Vergne, eds., Lecture Notes in Mathematics **880**. Springer-Verlag, Berlin-Heidelberg-New York, 1981.

 [V0]   D. Vogan, "Singular unitary representations," 506–535, in *Non-commutative Harmonic Analysis and Lie groups*, J. Carmona and M. Vergne, eds., Lecture Notes in Mathematics **880**. Springer-Verlag, Berlin-Heidelberg-New York, 1981.

[Green] D. Vogan, *Representations of Real Reductive Lie Groups*. Birkhäuser, Boston-Basel-Stuttgart, 1981.

 [V1]   D. Vogan, "Unitarizability of certain series of representations," Ann. of Math. **120** (1984), 141–187.

 [V2]   D. Vogan, *Unitary Representations of Reductive Lie Groups*. Annals of Mathematics Studies, Princeton University Press, Princeton, New Jersey, 1987.

 [VJ]   D. Vogan, "Irreducibility of discrete series representations for semisimple symmetric spaces," 191–221 in *Representations of Lie groups, Kyoto, Hiroshima, 1986*, K. Okamoto and T. Oshima, editors. Advanced Studies in Pure Mathematics, volume 14. Kinokuniya Company, Ltd., Tokyo, 1988.

[V3] D. Vogan, "Noncommutative algebras and unitary representations," in *The Mathematical Heritage of Hermann Weyl*, R. O. Wells, Jr., ed., Proceedings of Symposia in Pure Mathematics, **48**. American Mathematical Society, Providence, Rhode Island, 1988.

[V4] D. Vogan, "Associated varieties and unipotent representations," preprint.

[V5] D. Vogan, "Dixmier algebras, sheets, and representation theory," in *Operator Algebras, Unitary Representations, Enveloping Algebras, and Invariant Theory*, A. Connes, M. Duflo, A. Joseph, and R. Rentschler, eds. Birkhäuser Boston, 1990.

[W] N. Wallach, "On the unitarizability of derived functor modules," Invent. Math. **78** (1984), 131–141.

Department of Mathematics
MIT
Cambridge, MA 02139

Twistor Theory for Riemannian Manifolds

John Rawnsley

Introduction

I should begin by saying a little about the origin of the word "twistor". Robinson [27] parameterised some null solutions of Maxwell's equations by means of shear-free twisting null congruences of geodesics in Minkowski space. Penrose [24] gave an interpretation in terms of incidence geometry of space-time and an associated complex manifold. Robinson's solutions can then be generalised [25] by means of contour integrals on this complex manifold to yield all real analytic solutions. The fact that Robinson's solutions were in terms of twisting congruences led Penrose to coin the name "twistors" for the points of the space and "twistor space" for the three-dimensional complex manifold. Later Penrose extended the method to all the zero mass linear field equations and to various non-linear situations.

In the context of Riemannian geometry where the metric is positive definite, a fibration $\pi\colon \mathbb{C}P^3 \to S^4$ was used in [2] to translate the problem of finding self-dual connections on bundles over S^4 into one of finding certain algebraic bundles on $\mathbb{C}P^3$. Moreover they show how to obtain similar fibrations for any four-manifold whose Weyl curvature is (anti-) self-dual.

All the above situations take a problem in real differential geometry and translate it into a problem in complex analysis. We shall use the term *twistor method* for such a translation and *twistor space* for the complex manifold involved.

There are many examples predating Penrose's work, which can now also be seen as twistor methods, such as the Weierstrass correspondence between minimal surfaces in $\mathbb{R}^3$ and null curves in $\mathbb{C}^3$. See the lectures on twistor theory in [17] for a survey. In these lectures we shall see that generalized flag manifolds can be interpreted as twistor spaces for a large class of symmetric spaces. For another approach to flag manifolds as twistor spaces and the relationship with representation theory, see the book by Baston and Eastwood [4].

The bundle of complex structures as a twistor space

We begin by considering the question of the integrability of natural almost complex structures in certain fibre bundles over even-dimensional manifolds.

Given a differentiable manifold M, a *complex twistor space* over M consists of a complex manifold Z and a C^∞ submersion $\pi : Z \to M$ whose

---

Lecture notes by Isabel Dotti-Miatello

fibres are complex submanifolds. For each z in the fibre $\pi^{-1}(x)$, $x \in M$ the quotient $T_z M / \mathcal{V}_z$ is isomorphic to $T_x M$ (here $\mathcal{V}_z$ denotes the vertical tangent space $\ker d\pi_z$). If J is the endomorphism of the tangent bundle TZ which satsifies $J^2 = -I$ then the fibre $\pi^{-1}(x)$ will be a complex submanifold provided that $\mathcal{V}_z$ is J-stable for each z in the fibre. Hence J gives rise to a family of endomorphisms $j(z)$ of $T_x M$, $(j(z))^2 = -I$, $z \in \pi^{-1}(x)$.

We denote by $J(M)$ the bundle of all complex structures on the tangent spaces of M. In other words, the fibre of $J(M)$ at $x \in M$ consists of all endomorphisms of $T_x M$ with square $-I$. A complex twistor space Z then gives rise to a diagram

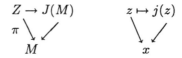

The fibres $J(M)_x$ of the bundle $J(M) \to M$ are complex manifolds. In fact, if $\dim M = 2n$, and on $\mathbb{R}^{2n}$ we pick an endomorphism J_o satisfying $J_o^2 = -I$ then every other endomorphism with $J^2 = -I$ is conjugate to J_o by an element $g \in GL(2n, \mathbb{R})$. The stabilizer of $GL(2n, \mathbb{R})$ at J_o is given by

$$\{g \in GL(2n, \mathbb{R}) | g J_o g^{-1} = J_o\} \cong GL(n, \mathbb{C})$$

so we can identify $J(M)_x$ with the complex manifold $GL(2n, \mathbb{R})/GL(n, \mathbb{C})$.

The following natural question now arises: Is $J(M)$ itself a candidate for a twistor space? That is, does $J(M)$ have a natural complex structure with respect to which the fibres are complex submanifolds? To answer this question, we note that if $J(M)$ has such a complex structure, then in the fibre directions it is already known since it must coincide with the complex manifold structure we just described. If we denote the bundle of tangents to the fibres by $\mathcal{V}$, then in order to give $J(M)$ a complex structure we need to extend the complex structure we have on $\mathcal{V}$ in complementary directions. Thus we have to choose a horizontal bundle $\mathcal{H}$ with $TJ(M) = \mathcal{V} \oplus \mathcal{H}$. Then the differential of the bundle projection π maps $\mathcal{H}_z$ isomorphically onto $T_x M$ where $x = \pi(z)$. But the point z in $J(M)$ is a complex structure $j(z)$ on $T_x M$, so we can use the isomorphism to transfer it to $\mathcal{H}_z$ and thus make $\mathcal{H}_z$ into a complex vector space too. Thus, in fact, any horizontal bundle automatically has a complex structure. We can then combine the complex vector bundle structures on $\mathcal{H}$ and $\mathcal{V}$ to give an almost complex structure on $J(M)$. We call this almost complex structure J_1. We get a second almost complex structure J_2 by reversing the sign of J_1 on the fibres.

Although they look very similar, these two almost complex structures behave quite differently. For instance, we shall see that, under restrictions on the curvature of M, J_1 can be integrable whilst J_2 is never integrable (except in the case of zero-dimensional fibre in which case M has dimension two).

The construction of the almost complex structures J_1, J_2 depends on the choice of a horizontal distribution $\mathcal{H}$. One way to get such a distribution is by means of a connection in the tangent bundle of M. If M is a Riemannian manifold it would be natural to use the Levi-Civita connection determined by the Riemannian metric. In this setting it is also natural to look at just those endomorphisms j which are compatible with the metric.

Thus suppose we have a Riemannian manifold (M, g) then we define a bundle $J(M, g)$ over M with fibres

$$J(M, g)_x = \{j \in J(M)_x | g_x(jX, jY) = g_x(X, Y), \forall X, Y \in TM_x\}.$$

Here the fibre is obviously isomorphic to $O(2n)/U(n)$ if M has dimension $2n$. If $O(M, g)$ denotes the principal $O(2n)$-bundle of orthonormal frames of TM then $J(M, g)$ is the associated bundle

$$O(M, g) \times_{O(2n)} O(2n)/U(n) \cong O(M, g)/U(n)$$

and the horizontal distribution on $O(M, g)$ determined by the Levi-Civita connection obviously projects to $J(M, g)$. In what follows we shall restrict our attention to this case. The case of $J(M)$ is described in [22]. By making this restriction we can expect that the structure of J_1 and J_2 will be intimately connected with the Riemannian geometry of (M, g).

To study the integrability question for J_1 on $J(M, g)$ one can use the Nirenberg-Newlander Theorem which tells us that J_1 is integrable if and only if the $-i$-eigenbundle of J_1 on $TJ(M, g)^{\mathbb{C}}$ is closed under Lie brackets. Dually it is enough that the ideal of differential forms generated by the $(1, 0)$-forms is closed under exterior differentiation. In this second version we can use the submersion $O(M, g) \to J(M, g)$ to lift the problem to the frame bundle. It is easy to see that an equivalent condition is that the ideal of forms generated by pull-backs to $O(M, g)$ of $(1, 0)$-forms on $J(M, g)$ should also be d-closed.

Working on $O(M, g)$ has the advantage that the tangent bundle is trivial, and in fact the geometry gives us natural globally defined forms which form a basis at every point, so we are able to check the condition by explicit computation. To see this we first choose a base-point J_0 for the orthogonal almost-complex structures on $\mathbb{R}^{2n}$. Then we identify $U(n)$ with the centralizer of J_0 in $O(2n)$ and its Lie algebra $\mathfrak{u}(n)$ with the centralizer of J_0 in $\mathfrak{o}(2n)$.

Any $A \in \mathfrak{o}(2n)$ can be decomposed into a piece which commutes with J_0 and a piece which anti-commutes with J_0:

$$A = \tfrac{1}{2}(A - J_0 A J_0) + \tfrac{1}{2}(A + J_0 A J_0).$$

If we denote by

$$\mathfrak{m} = \{A \in \mathfrak{o}(2n) | A J_0 = -J_0 A\}$$

then we see we have a direct sum decomposition

$$\mathfrak{o}(2n) = \mathfrak{u}(n) \oplus \mathfrak{m}$$

which is $U(n)$-invariant. Furthermore $A \mapsto J_0 A$, $A \in \mathfrak{m}$, gives $\mathfrak{m}$ a $U(n)$-invariant complex structure and, by translation by $O(2n)$, an invariant complex structure on $O(2n)/U(n)$. This will be the complex structure on the fibres of $J(M, g)$.

As usual, left multiplication by J_0 on $\mathfrak{m}$ is extended to the complexification $\mathfrak{m}^{\mathbb{C}} = \mathfrak{m} \otimes_{\mathbb{R}} \mathbb{C}$ to be $\mathbb{C}$-linear. Consequently the eigenvalues are $\pm i$ and the corresponding eigenspaces are denoted by $\mathfrak{m}^+$, $\mathfrak{m}^-$. One verifies easily that $\mathfrak{m}^+$ and $\mathfrak{m}^-$ are abelian subalgebras. Let $J^+ = \frac{1}{2}(I - iJ_0)$, $J^- = \frac{1}{2}(I + iJ_0)$ be the projections onto the $\pm i$-eigenspaces of J_0 on $\mathbb{C}^{2n}$. Then

$$\mathfrak{m}^+ = J^+ \mathfrak{m}, \qquad \mathfrak{m}^- = J^- \mathfrak{m}.$$

The trivialization mentioned above can be given by a pair of natural 1-forms θ and α. θ is the $\mathbb{R}^{2n}$-valued form (the soldering form) which realizes the isomorphism $TM \cong O(M, g) \times_{O(2n)} \mathbb{R}^{2n}$ and α the connection form of the Levi-Civita connection which is $\mathfrak{o}(2n)$-valued. There are $2n + n(2n - 1)$ independent components of these forms which is exactly the dimension of $O(M, g)$ as a manifold.

Set $\theta^+ = J^+ \theta$ and $\alpha^+ = J^+ \alpha J^-$. Then it is easy to see that pointwise θ^+ and α^+ span the values of the pull-back to $O(M, g)$ of the $(1,0)$-forms of J_1 on $J(M, g)$. To check integrability for J_1 it is then enough to verify that θ^+ and α^+ generate a d-closed ideal.

First observe that the Levi-Civita connection having no torsion implies that

$$0 = d\theta + \alpha \wedge \theta.$$

Similarly, the curvature R of the Levi-Civita connection is represented by an $\mathfrak{o}(2n)$-valued 2-form ρ on $O(M, g)$ given by

$$\rho = d\alpha + \tfrac{1}{2}\alpha \wedge \alpha.$$

Thus

$$
\begin{aligned}
d\theta^+ &= d(J^+ \theta) \\
&= J^+ d\theta \\
&= -J^+ (\alpha \wedge \theta) \\
&= -J^+ \alpha \wedge J^+ \theta - J^+ \alpha J^- \wedge \theta \\
&= -J^+ \alpha \theta^+ - \alpha^+ \wedge \theta.
\end{aligned}
$$

Likewise

$$
\begin{aligned}
d\alpha^+ &= d(J^+ \alpha J^-) \\
&= J^+ d\alpha J^- \\
&= J^+ \rho J^- - J^+ \alpha \wedge \alpha J^- \\
&= J^+ \rho J^- - J^+ \alpha \wedge J^+ \alpha J^- - J^+ \alpha J^- \wedge \alpha J^-.
\end{aligned}
$$

This shows that J_1 will be integrable if and only if $J^+ \rho J^-$ belongs to the ideal generated by θ^+ and α^+. This translates in terms of the Riemann curvature tensor R on M to the condition that

$$j^+ R(j^- X, j^- Y) j^- = 0$$

for every j in $J(M, g)$ (here $j^\pm$ denote the projections onto the $\pm i$-eigenspaces of j).

To interpret this condition on the curvature tensor R, suppose that it holds for a particular $j_0 \in J(M, g)$ which we can suppose corresponds with J_0 relative to a frame b_0 under the isomorphism $J(M, g) \cong O(M, g) \times_{O(2n)} O(2n)/U(n)$. Then we want it also to hold for every $j = b J_0 b^{-1}$ with b any orthonormal frame of TM_x. As a tensor on M, the curvature R corresponds with a function $\widehat{R} : O(M, g) \to \mathfrak{o}(2n) \otimes (\wedge^2 \mathbb{R}^{2n})^*$ on the frame bundle given by

$$\widehat{R}(b)(u, v)w = b^{-1} R(bu, bv) bw$$

Hence $\widehat{R}$ satisfies the condition

$$J_0^+ \widehat{R}(b)(J_0^- u, J_0^- v) j_0^+ = 0 \qquad (*)$$

not only for $b = b_0$ but for every orthonormal frame b. Since we can get from one frame to another by the action of the orthogonal group, this says that $\widehat{R}$ must take its values in the largest $O(2n)$-invariant subspace of $\mathfrak{o}(2n) \otimes (\wedge^2 \mathbb{R}^{2n})^*$ satisfying $(*)$.

If we view J_0 as an element of the Lie algebra $\mathfrak{o}(2n)$, then differentiating the natural action of $O(2n)$, J_0 acts on $\mathfrak{o}(2n) \otimes (\wedge^2 \mathbb{R}^{2n})^*$. The tensor $J_0^+ A(J_0^- \cdot, J_0^- \cdot) J_0^-$ is then the projection of a tensor $A(\cdot, \cdot)$ into the $4i$-eigenspace of J_0. Thus integrability of J_1 can be restated as saying that the Riemann curvature tensor can have no non-zero component in any $4i$-eigenspace of any $j \in J(M, g)$. Note that on tensors of rank four the possible eigenvalues are $\{0, \pm 2i, \pm 4i\}$, whilst on tensors of rank two they are $\{0, \pm 2i\}$. We know the $4i$ eigenvalue does occur if the dimension of M is at least four, and under the action of the orthogonal group the curvature tensors split into three irreducible pieces, the scalar curvature, the traceless Ricci curvature (a rank two tensor) and the Weyl curvature. It follows that the $4i$-eigenvalue can only occur in the Weyl curvature and hence that integrability of J_1 is the vanishing of this component. Thus we have the following theorem due independently to several people, see for example [22] or [12]:

Theorem. J_1 on $J(M, g)$ is integrable if and only if (M, g) is locally conformally flat.

Examples. The standard metric g_0 on S^{2n} is conformally flat. In this case

$$J(S^{2n}, g_0) = O(2n + 1)/U(n)$$

with its usual complex structure. A more interesting example is provided by the $2n$-torus with flat metric g_0. In this case $J(T^{2n}, g_0)$ is an example of a non-Kählerian complex manifold due to Kodaira.

Since there are not many conformally flat manifolds, our first attempt to associate a twistor space to a manifold does not yield many examples, so it needs some modification. To see how to do this consider the case where M is oriented. We can then restrict attention to those endomorphisms j which induce the given orientation. Call the bundle of these $J_+(M, g)$. As before we give $J_+(M, g)$ an almost complex structure J_1 using the Levi-Civita connection. If we repeat the previous argument to get the condition for integrability, then it is clear that everything goes as before except that we conclude that the Riemann curvature tensor cannot have any non-zero component in any $4i$-eigenspace of any j in $J_+(M, g)$. In dimension six or higher, the space of curvature tensors has the same decomposition into irreducible subspaces under $SO(2n)$ as under $O(2n)$, so the result is the same. However in dimension four the Weyl tensor splits into two pieces on only one of which (the self-dual part) the $4i$-eigenvalue can occur. Hence we recover the result of Atiyah-Hitchin-Singer which forms the basis for their study of Yang-Mills equations on four-dimensional manifolds in terms of holomorphic bundles over associated three-dimensional complex manifolds:

Theorem. [2] *If (M, g) is an oriented Riemannian four-manifold then the almost-complex structure J_1 on $J_+(M, g)$ is integrable if and only if the Weyl curvature tensor is anti-self-dual.*

It is not hard to see that the projective spin bundle they use is isomorphic to $J_+(M, g)$.

Before going on to see how we can modify $J(M, g)$ further to obtain more cases where J_1 is integrable, it is worth commenting on the other almost complex structure J_2 which we get by reversing the sign on the fibres. The question of integrability for J_2 would involve replacing α^+ by α^- in the above calculations and one quickly sees in dimension four and larger that J_2 can never be integrable. On the other hand it turns out to be of importance in the twistor theory of harmonic maps of Riemann surfaces as was discovered by Eells and Salamon [13].

If Σ is a Riemann surface and $\phi : \Sigma \to M$ a map into the Riemannian manifold (M, g), we denote by ∇^ϕ the pull-back of the Levi-Civita connection as a covariant derivative in $\phi^{-1}TM^{\mathbb{C}}$. Then we have the following special case of a theorem of Koszul and Malgrange:

Theorem. [19] *There is a unique structure of a holomorphic vector bundle in $\phi^{-1}TM^{\mathbb{C}}$ such that the $\bar{\partial}$-operator coincides with the $(0,1)$ part of ∇^ϕ. That is: A local section s of $\phi^{-1}TM^{\mathbb{C}}$ is holomorphic if and only if $\nabla^\phi_{\frac{\partial}{\partial \bar{z}}} s = 0$ for any holomorphic coordinate z on Σ.*

When the domain is two-dimensional the harmonic map equation depends only on the cconformal class of the domain metric and can be written in terms of a holomorphic corrdinate for the conformal structure as

$$\nabla^{\phi}_{\frac{\partial}{\partial \bar{z}}} d\phi(\frac{\partial}{\partial z}) = 0.$$

This can be interpreted in terms of the Koszul-Malgrange holomorphic structure as saying that the $(1,0)$-part of the differential $d\phi$ of ϕ is holomorphic when viewed as a 1-form with values in $\phi^{-1}TM^{\mathbb{C}}$. There is, however, a second much more twistor theoretic holomorphic characterization of a harmonic map of a surface:

Theorem. [8] *A map $\phi: \Sigma \rightarrow M$, Σ a Riemann surface, is conformal, harmonic with $\phi^* w_1(M) = 0$ if and only if there exists a J_2-holomorphic map $\psi: \Sigma \rightarrow J(M,g)$ with $\phi = \pi \circ \psi$ where J_2-holomorphic means $d\phi J_{\Sigma} = J_2 d\phi$.*

The neccessity of the conditions in the theorem is easy to understand. If $\phi = \pi \circ \psi$ and ψ is J_2-holomorphic then $d\phi(\frac{\partial}{\partial z})$ lies in the $(1,0)$ vectors for the complex structure $\psi(z)$ on $TM_{\phi(z)}$. The space of $(1,0)$-vectors is isotropic for the complex bilinear extension of the metric, and hence so is $d\phi(\frac{\partial}{\partial z})$. But this is just the condition that ϕ be conformal. A map ψ with $\phi = \pi \circ \psi$ can be viewed as a complex structure on the real vector bundle $\phi^{-1}TM$ and so if ψ exists $\phi^{-1}TM$ is orientable and hence has vanishing first Stiefel-Whitney class. Hence the topological condition. The hard part of the theorem lies in showing that these condtions are also sufficient.

In many cases such a ψ can be found so that $d\phi(T\Sigma)$ consists entirely of horizontal vectors on which J_1 and J_2 agree, so that ψ becomes holomorphic for an integrable complex structure (when J_1 is integrable!) and so is a complex curve in the usual sense. This explains our interest in the case where J_1 is integrable even though when studying harmonic maps it is J_2 which at first appears to play the important role. In general it is hard to find explicit examples of holomorphic curves for a non-integrable almost complex structure, so we try to express things in terms of the integrable structure J_1. In the next lecture we shall try to find more general situations where J_1 is integrable on suitably chosen subsets of $J(M,g)$.

FLAG MANIFOLDS AS TWISTOR SPACES

In the previous section we considered an almost complex structure J_1 on $J(M,g)$ and showed that only for locally conformally flat manifolds is it integrable. The problem is that integrability imposes a lot of conditions on the curvature coming from the fibres of $J(M,g)$, so to find more plentiful examples of twistor spaces we have to find fibrations with smaller fibres. We start by showing that flag manifolds provide a natural class of examples.

Let G be a compact semi-simple Lie group, $\mathfrak{g}_0$ its Lie algebra, $\mathfrak{g} = \mathfrak{g}_0^{\mathbb{C}}$ its complexification and $\mathfrak{q} \subset \mathfrak{g}$ a parabolic subalgebra. If $\mathfrak{h}_0 = \mathfrak{g}_0 \cap \mathfrak{q}$ and $\mathfrak{h} = \mathfrak{h}_0^{\mathbb{C}}$ then $\mathfrak{q} = \mathfrak{h} \oplus \mathfrak{n}$ where $\mathfrak{n}$ is the nilradical of $\mathfrak{q}$. The key observation is the following:

Proposition. *There exists a unique $\xi \in Z(\mathfrak{h}_0)$ such that*

i) ad_ξ has eigenvalues in $i\mathbb{Z}$,

ii) if $\mathfrak{g}_{(r)}$ is the ir-eigenspace of ad_ξ and $\mathfrak{n}^{(s)}$ is the s-th step in the descending central series of $\mathfrak{n}$ then

$$\mathfrak{q} = \sum_{r \geq 0} \mathfrak{g}_{(r)}, \qquad \mathfrak{n}^{(s)} = \sum_{r \geq s} \mathfrak{g}_{(r)}.$$

We call ξ the *canonical element* of the parabolic subalgebra $\mathfrak{q}$.

If we set $\tau = Ad \exp \pi \xi = e^{\pi ad_\xi}$ then $\tau^2 = e^{2\pi ad_\xi} = id$. Thus τ is an involution on $\mathfrak{g}_0$ with eigenspace decomposition $\mathfrak{g}_0 = \mathfrak{k}_0 \oplus \mathfrak{p}_0$ where $\mathfrak{k}_0 = \mathfrak{g}_0^\tau$, $\mathfrak{p}_0 = \mathfrak{g}_0^{-\tau}$. Furthermore

$$\mathfrak{k} = \mathfrak{k}_0^{\mathbb{C}} = \sum_{r \in 2\mathbb{Z}} \mathfrak{g}_{(r)}, \qquad \mathfrak{p} = \sum_{r \in 2\mathbb{Z}+1} \mathfrak{g}_{(r)}.$$

Since $\mathfrak{g}_{(0)} = \mathfrak{h}$ we have $\mathfrak{h}_0 \subset \mathfrak{k}_0$ and $H \subset K \subset G$ for the corresponding Lie subgroups of G. This gives us a fibration $G/H \to G/K$ with G/K an inner symmetric space (by construction) and G/H a generalized flag manifold. It can be shown that every inner symmetric space arises this way. We claim that each of these fibrations exhibits G/H as a twistor space over G/K.

Set $\mathfrak{p}^+ = \mathfrak{q} \cap \mathfrak{p} = \sum_{r \geq 0} \mathfrak{g}_{(2r+1)}$. Then $\mathfrak{p}^+ \subset \mathfrak{p}$ is maximally isotropic with respect to the Killing form B hence is the $+i$-eigenspace of some complex structure j on $\mathfrak{p}_0$. Since the latter is the tangent space to G/K at the identity coset, $j \in J(G/K, B)$ and so we have a G-invariant diagram:

$$G/H \hookrightarrow J(G/K, B)$$
$$\searrow \qquad \swarrow$$
$$G/K$$

which shows G/H is a twistor space for G/K as claimed. Moreover it is clear from the construction that the complex structure J_1 coincides with the standard invariant complex structure on G/H given by the parabolic subalgebra $\mathfrak{q}$. This gives an abundant supply of twistor spaces for inner symmetric spaces with integrable J_1. In fact the following theorem shows that these are essentially the only possibilities for inner symmetric spaces.

Theorem. *Let G/K be an inner compact semisimple symmetric space and N_1 the Nijenhuis tensor of J_1 on $J(G/K, g)$, g any invariant metric. The zero set of N_1 is a finite union of G-orbits which are generalized flag manifolds fibered canonically over G/K.*

Sketch of proof. If j is in the zero set of N_1 then $j^- R(j^+ X, j^+ Y) j^+) = 0$. Setting $T^+ = j^+ (T_x G/K)$ the previous condition says $R(T^+, T^+,) T^+ \subset T^+$. Thus if T^+ corresponds to $\mathfrak{p}^+$ in $\mathfrak{p}_0^{\mathbb{C}} \cong T_x(G/K)^{\mathbb{C}}$, we want solutions of the condition $[[\mathfrak{p}^+, \mathfrak{p}^+], \mathfrak{p}^+] \subset \mathfrak{p}^+$ with $\mathfrak{p}^+$ maximally isotropic. This makes $\mathfrak{p}^+ + [\mathfrak{p}^+, \mathfrak{p}^+]$ into a Lie algebra. If one sets $\mathfrak{h}_0 = \{\eta \in \mathfrak{k}_0 | [\eta, \mathfrak{p}^+] \subset \mathfrak{p}^+\}$ and

$\mathfrak{h} = \mathfrak{h}_0^{\mathbb{C}}$ then one easily sees that the K-orbit of $\mathfrak{p}^+$ in the set of maximally isotropic subspaces of $\mathfrak{p}_0^{\mathbb{C}}$ is complex, and thus $\mathfrak{h}$ has maximal rank in $\mathfrak{k}$ with $[\mathfrak{p}^+, \mathfrak{p}^+]$ is a nilpotent subalgebra of $\mathfrak{k}$. It is then easy to see that $\mathfrak{q} = \mathfrak{h} + \mathfrak{p}^+ + [\mathfrak{p}^+, \mathfrak{p}^+]$ is a parabolic subalgebra of $\mathfrak{g}$ whose nilradical is $\mathfrak{p}^+ + [\mathfrak{p}^+, \mathfrak{p}^+]$. See [8] for more details.

An important consequence of the previous theorem is that G acts transitively on connected components of the zero set of N_1. In particular each component is a submanifold of $J(G/K, g)$.

Bryant [5] also considered twistor spaces over inner symmetric spaces, but imposed an extra condition – that the horizontal $(1,0)$ bundle be holomorphic – before classifying them. Our methods can thus be used to give an alternative derivation of his classification. His condition amounts to the requirement that $[[\mathfrak{p}^+, \mathfrak{p}^+], \mathfrak{p}^+] = 0$ instead of the weaker condition $[[\mathfrak{p}^+, \mathfrak{p}^+], \mathfrak{p}^+] \subset \mathfrak{p}^+$, and says that the canonical element of the parabolic $\mathfrak{q}$ has eigenvalues $\{0, \pm i, \pm 2i\}$.

At the end of the first lecture we stated a result exhibiting conformal harmonic maps of a Riemann surface Σ as projections of J_2-holomorphic curves in $J(M, g)$. We have just seen that when $M = G/K$ we have certain flag manifolds sitting inside $J(M, g)$ on which J_1 is integrable. We can ask if it is possible to choose the lift so that it lands inside one of these submanifolds. Since we can regard the flag manifold as the conjugacy class of parabolic subalgebras, and since there are only a finite number of conjugacy classes of parabolic subalgebras any two nearby parabolic subalgebras are necessarily conjugate, we can equivalently view maps into a (generalized) flag manifold as a bundle of parabolic subalgebras which is a subbundle of the trivial bundle over Σ with fibre $\mathfrak{g}$. All we now need to do is translate the condition of being J_2-holomorphic into these Lie algebra terms.

If we have a map $\psi : \Sigma \to G/H$ we shall denote the corresponding bundle of parabolic subalgebras by $\underline{\psi}$. Obviously we have a similar bundles $\underline{\mathfrak{h}}, \underline{\mathfrak{k}}$ of isotropy subalgebras and and complementary bundle $\underline{\mathfrak{p}}$ which is the pullback to Σ of the tangent bundle of G/K by the composition of ψ with projection from G/H to G/K. We denote this composition by ϕ, and view $\phi_*(\frac{\partial}{\partial z})$ as a (local) section δ of $\underline{\mathfrak{p}}$. Then $\underline{\psi}$ is J_2-holomorphic provided that $\underline{\psi}$ is a holomorphic sub-bundle of the trivial bundle $\underline{\mathfrak{g}}$ for the Koszul-Malgrange holomorphic structure coming from the Levi-Civita connection of G/K and δ lies in the nilradical sub-bundle of $\underline{\psi}$.

Thus to build lifts into flag manifolds we need techniques for building bundles of parabolic subalgebras. A general technique is provided by the Harder-Narasimhan filtration of a holomorphic bundle over a Riemann surface. If E is any holomorphic vector bundle over Σ there is a unique holomorphic filtration by sub-bundles $\ldots \subset E_{-1} \subset E_0 \subset E_1 \subset \ldots \subset E$ such that each quotient E_i / E_{i-1} is either trivial or semi-stable of slope i. See [1] for an account in the context of Yang-Mills theory on Riemann surfaces. If E has a holomorphic non-degenerate bilinear form, then uniqueness of the filtration forces the orthogonal of E_i to be E_{1-i}. Similarly if E is a holomorphic bundle of Lie algebras then $[E_i, E_j] \subset E_{i+j}$ and so E_0 is a bundle of subalgebras with E_i a nilpotent ideal. It follows at once that if E is a bundle of semisimple Lie algebras then the fiberwise Killing form gives a holomorphic nondegenerate bilinear form to E. Thus E_0 is a bundle of subalgebras which is orthogonal to the bundle E_1 of nilpotent Lie algebras. It follows that each fiber of E_0 is a parabolic subalgebra.

In the case Σ is the two-sphere an even stronger structure theorem due to Birkhoff and Grothendieck is available which allows us to choose a holomorphically trivial sub-bundle of Cartan subalgebras and hence bundle of Borel subalgebras in such a way that we have control of the topological degrees of the resulting root line bundles. We apply the above arguments to the trivial bundle $\mathfrak{g}$ with the holomorphic structure coming from the Levi-Civita connection on an inner symmetric space. In order to force the differential of the map to Lie in E_1 we need some extra work which is described in detail in [8]. The result is given in the following theorem:

Theorem. Let $\phi\colon S^2 \to G/K$ be a harmonic map into an inner symmetric space. Then there exists a generalized flag manifold G/H canonically embedded in the $J(G/K,g)$ bundle and a J_2-holomorphic map $\psi\colon S^2 \to J(G/K,g)$ whose image is in G/H.

This result is not terribly useful as it stands, but forms a starting point for an analysis of harmonic maps of two-spheres by holomorphic methods. For example, if we measure the height of a twistor space by the length of the central series of the nilradical of the parabolic, then one can show that in the height two case there are certain holomorphic differentials which are obstructions to the lift being horizontal. When the domain is a two-sphere such differentials must vanish, and so the lift will be horizontal and hence also J_1-holomorphic. This means it is then a genuine holomorphic curve in a complex manifold. One can then ask: For which inner symmetric spaces are all the components of the zero set of the Nijenhuis tensor height two flag manifolds? The answer is: Just for the rank one case, S^{2n} and $\mathbb{C}P^n$. This explains why Calabi [10] and Eells and Wood [15] are able to obtain such a complete classification for minimal two-spheres in these two cases (see also partial results obtained by Barbosa [3], Burns [6], Catenacci et al. [11] and Glaser and Stora [16]). It is much harder to manipulate the twistor lifts into genuine holomorphic data when the flag manifolds occuring in $J(M,g)$ are of height bigger than two. But see Wolfson [30] and Wood [31] for the case of complex Grassmannian targets using more geometrical methods.

A more striking result is obtained when we look at a more restrictive class of maps, namely those which are stable harmonic. We recall that given $\phi\colon M \to N$ a smooth map, the energy $E(\phi)$ of ϕ is defined by

$$E(\phi) = \int_M \tfrac{1}{2}|d\phi|^2 vol_g.$$

The map ϕ is harmonic if it extremises the energy on all compact subdomains of M. Moreover, as in all variational problems, we can look at the second variation. If ϕ is harmonic it is said to be *stable* if $\frac{d^2}{dt^2}E(\phi_t)|_{t=0} \geq 0$ for all variations ϕ_t of ϕ.

If $\phi: S^2 \to G/K$ is a stable harmonic map into an inner symmetric space G/K of compact type, then a form of the second variation due to Moore and Micallef [21] shows that if $\psi: S^2 \to G/H$ is the twistor lift given by the previous theorem, then $\delta = \phi_*(\frac{\partial}{\partial z})$ is in the centre of the nilradical of the bundle of parabolics ψ. This gives an easy proof of the following theorem:

Theorem. *If $\phi: S^2 \to G/K$ is a stable harmonic map into an irreducible Hermitian symmetric space then ϕ is holomorphic or antiholomorphic ($\pm$holomorphic).*

The reason for such a theorem is that when G/K is irreducible, $\mathfrak{g}$ is simple and so the centre $\mathfrak{z}$ of $\mathfrak{n}$ is irreducible under $\mathfrak{q}$ for any parabolic subalgebra $\mathfrak{q}$. On the other hand, for the parabolic subalgebras we construct as twistor lifts we have an $\mathfrak{h}$-invariant splitting

$$\mathfrak{z} = \mathfrak{p}' \cap \mathfrak{z} + \mathfrak{p}'' \cap \mathfrak{z}$$

Irreducibility means that one of these two components must vanish at each point. But δ lies in $\mathfrak{z}$ and hence lies in either the $(1, 0)$ or $(0, 1)$ vectors at each point. Thus ϕ is either holomorphic or antiholomorphic at each point. A regularity theorem of Siu says that then only one of these possibility holds for all points.

The converse of the previous theorem holds in an even stronger form. Eells and Sampson (see also Lichnerowicz [20]) observed that holomorphic maps between Kähler manifolds are absolute minima of the energy in their homotopy class and hence are stable harmonic maps.

When $G/K = \mathbb{C}P^n$ the previous theorem was proved by Siu and Yau [29] in their work on the Frankel conjecture. For targets other than $\mathbb{C}P^n$ Siu [28] used a curvature condition to show that all stable harmonic maps into the classical irreducible Hermitian symmetric spaces are $\pm$holomorphic. The remaining excpetional spaces were dealt with by Zhang [33]. However our proof is uniform without any case-by-case analysis.

Burstall, Rawnsley and Salamon [9] have generalized this result to all Riemannian symmetric spaces as targets. We can restrict attention to the compact type since harmonic maps from S^2 into non-positively curved manifolds are necessarily constant by a theorem of Eells and Sampson [14].

Theorem. *Let $\phi: S^2 \to G/K$ be a stable harmonic map into a Riemannian symmetric space of compact type. Then there is a totally geodesically embedded submanifold $G_1/K_1 \subset G/K$ such that G_1/K_1 is Hermitian symmetric and ϕ is a holomorphic map into G_1/K_1.*

From the point of view of applying this theorem, there are three classes of irreducible symmetric spaces of compact type. Namely, $\pi_2(G/K)$ can be one of $\mathbb{Z}$, 0 or $\mathbb{Z}_2$. The first of these is the Hermitian symmetric case which we have already dealt with, and in that case the subspace G_1/K_1 is $\pm$holomorphic, agreeing with the previous theorem.

In the case $\pi_2(G/K) = 0$ the subspaces G_1/K_1 are zero-dimensional so we deduce:

Theorem. *Let $\phi: S^2 \to G/K$ be a stable harmonic map into a simply-connected Riemannian symmetric space with $\pi_2(G/K) = 0$ then ϕ is constant.*

Howard and Wei in a joint paper [18], and, independently, Ohnita [23] and Pluzhnikov [26] have shown that if the identity map of a manifold is unstable then any stable harmonic map is necessarily constant. An examination of their list of symmetric spaces with unstable identity map shows it to be the same as the list of spaces with vanishing π_2 and so the above theorem is a special case of theirs.

In the case where π_2 is $\mathbb{Z}_2$ we have many non-trivial examples of the subspaces G_1/K_1. Moreover in this case they have the property that any $\pm$holomporhic map of a Riemann surface into any G_1/K_1 is automatically stable as a harmonic map into G/K. It is interesting to observe in a case-by-case analysis that the maximal G_1/K_1 are all copies of $\mathbb{C}P^n$. The holomorphic sectional curvatures of these $\mathbb{C}P^n$ coincide with the maximum of the sectional curvatures of two-planes in the tangent spaces to G/K. In this sense they would appear to be analogous to Helgason's spheres of maximal sectional curvatures in symmetric spaces.

So far we have only been considering the twistor space of the target of a harmonic map. There is no time to go into any details, but it is possible to use twistor theory also for the domain, and in a joint work with Burns, Burstall and de Bartolomeis [7] we have obtained the following results (we shall say a symmetric space is *simple* if the Lie algebra of Killing fields is simple):

Theorem. *Let $\phi: \Sigma \to G/K$ be a stable harmonic map of a closed Riemann surface into a compact simple Hermitian symmetric space. Then ϕ is $\pm$holomorphic.*

Note that using the twistor space of the range we were only able to obtain this result for the two-sphere as domain.

Theorem. *Let $\phi: M \to G/K$ be a stable harmonic map, where M is a four-dimensional real analytic Riemannian manifold and G/K is a compact simple Hermitian symmetric space. If there exists a point where $d\phi$ has rank at least three then M has a unique Kähler structure with respect to which ϕ is holomorphic.*

Theorem. *Let $\phi: \mathbb{C}P^n \to G/K$ be a stable harmonic map into a compact simple Hermitian symmetric space. Then ϕ is $\pm$holomorphic.*

References

1. M. F. Atiyah and R. Bott, *Yang-Mills equations over Riemann surfaces*, Phil. Trans. R. Soc. Lond. A **308** (1982), 523–615.
2. M. F. Atiyah, N. J. Hitchin and I. M. Singer, *Self-duality in four dimensional Riemannian Geometry*, Proc. Roy. Soc. London A **362** (1978), 425–461.
3. J. Barbosa, *On minimal immersions of S^2 into S^{2m}*, Trans. Amer Math. Soc. **210** (1975), 75–106.

4. R. J. Baston and M. G. Eastwood, *The Penrose transform: its interaction with representation theory*, Oxford University Press, Oxford, 1989.

5. R. Bryant, *Lie groups and twistor spaces*, Duke math. J. **52** (1985), 223–261.

6. D. Burns, *Harmonic maps from* $\mathbb{C}P^1$ *to* $\mathbb{C}P^n$, Proc. Tulane Conf. (R. Knill, ed.), Lect. Notes in Math. vol 1263, Springer-Verlag, Berlin, Heidelberg, New York, 1982.

7. D. Burns, F. Burstall, P. de Bartolomeis and J. Rawnsley, *Stability of harmonic maps of Kähler manifolds*, J. Diff. Geometry **30** (1989), 579–594.

8. F. E. Burstall and J. H. Rawnsley, *Twistor theory for Riemannian symmetric spaces*, Lecture Notes in Mathematics, vol. 1424, Springer-Verlag, Berlin, Heidelberg, New York, 1990.

9. F. Burstall, J. Rawnsley and S. Salamon, *Stable harmonic 2-spheres in symmetric spaces*, Bull. Amer. Math. Soc. **16** (1987), 274–278.

10. E. Calabi, *Minimal immersions of surfaces in Euclidean spheres*, J. Diff. Geom. **1** (1967), 111–125.

11. R. Catenacci, M. Martenllini and C. Reina, *On the classical energy spectrum of* $\mathbb{C}P^2$ *models*, Phys. Lett. B **115** (1982), 461–462.

12. M. Dubois-Violette, *Structures complexes au-dessus des variétés*, Mathématiques et Physique. Progress in math., 37, Birkhäuser, Boston, 1983.

13. J. Eells and S. M. Salamon, *Twistorial construction of harmonic maps of surfaces into four-manifolds*, Ann. Scuola Norm. Sup. Pisa **12** (1985), 589–640.

14. J. Eells and J. Sampson, *Harmonic mappings of Riemannian manifolds*, Amer. J. Math. **86** (1964), 109–160.

15. J. Eells and J. C. Wood, *Harmonic maps from surfaces into projective spaces*, Adv. in Math. **49** (1983), 217–263.

16. V. Glaser and R. Stora, *Regular solutions of the* $\mathbb{C}P^n$ *model and further generalizations*, CERN preprint.

17. V. L. Hansen (ed.), *Differential Geometry*, Nordic Summerschool, Lyngby, Springer-Verlag, Berlin, Heidelberg, New York, 1987.

18. R. Howard and W. Wei, *Nonexistence of stable harmonic maps to and from certain homogeneous spaces and submanifolds of Euclidean spaces*, Trans. Amer. Math. Soc. **294** (1986), 319–331.

19. J. L. Koszul and B. Malgrange, *Sur certaines structures fibrées complexes*, Arch. Math. **9** (1958), 102–109.

20. A. Lichnerowicz, *Applications harmoniques et variétés kählériennes*, Rend. Sem. Mat. Fis. Milano **39** (1969), 186–195.

21. M. Micallef and J. D. Moore, *Minimal two-spheres and the topology of manifolds with positive curvature on totally isotropic two-planes*, Ann. of math. **127** (1988), 199–227.

22. N. R. O'Brian and J. H. Rawnsley, *Twistor spaces*, Ann. Glob. Anal. and Geom. **3** (1985), 29–58.

23. Y. Ohnita, *Stability of harmonic maps and standard minimal immersions*, Tôhoku Math. J. **38** (1986), 259–267.

24. R. Penrose, *Twistor algebra*, J. Math. Phys. **8** (1967), 345–366.

25. R. Penrose, *Twistor theory: an approach to the quantisation of fields and space-time*, Physics Reports **6** (1972), 241-316.

26. A. I. Pluzhnikov, *On the minimum of the Dirichlet functional*, Sov. math. Dokl. **34** (281-284), 1987.

27. I. Robinson, *Null electromagnetic fields*, J. Math. Phys. **2** (1961), 290–291.

28. Y. T. Siu, *A curvature characterization of the hyperquadrics*, Duke Math. J. **47** (1980), 641–654.

29. Y. T. Siu and S. T. Yau, *Compact Kähler manifolds of positive bisectional curvature*, Inv. Math. **59** (1980), 189–204.

30. J. Wolfson, *Harmonic sequences and harmonic maps of surfaces into complex Grassmanians*, J. Diff. Geom. **27** (1988), 161–178.

31. J. C. Wood, *Explicit construction and parametrisation of harmonic two-spheres in the unitary group*, Proc. Lond. Math. Soc. **58** (1989), 608–624.

32. W. J. Zakrzewski, *Classical solutions of two-dimensional Grassmanian models*, J. Geom. Phys. **1** (1984), 39–63.

33. J. Q. Zhong, *The degree of strong nondegeneracy of the bisectional curvature of bounded symmetric domains*, Proc. 1981 Hangzhore Conf., Birkhäuser, Boston, 1984, pp. 127–139.

Mathematics Institute
University of Warwick
Coventry CV4 7AL, UK

You Can't Hear the Shape
of a Manifold

Carolyn S. Gordon*

Two compact Riemannian manifolds are said to be isospectral if the associated Laplace-Beltrami operators have the same eigenvalue spectrum. The question, often posed as "Can you hear the shape of a drum," asks whether isospectral manifolds are necessarily isometric. The answer is now well-known to be negative; the first counterexample was a pair of isospectral tori given by Milnor in 1964. Until 1980, however, the only other examples discovered were a few additional pairs of tori or twisted products with tori.

In 1980, new examples began to appear, and the past decade has seen a flood of examples as well as fairly general techniques for constructing examples. Among these examples are pairs of manifolds with non-isomorphic fundamental groups, locally symmetric spaces both of rank one and higher rank, Riemann surfaces of every genus ≥ 4, continuous families of isospectral manifolds, lens spaces, and other examples. However, all the known examples have one property in common: the isospectral manifolds have a common Riemannian covering (M, g) which admits an isometric action by a (possibly finite) Lie group G. The isospectral manifolds are quotients of M by discrete subgroups Γ_i of G. Thus the manifolds in the examples are locally isometric.

The purpose of this article is to describe the various examples, primarily from a representation theoretic point of view. In almost every case, the isospectrality of each pair of manifolds arises from an eigenvalue between the actions of G on $L^2(\Gamma_1 \backslash G)$ and $L^2(\Gamma_2 \backslash G)$. In section 1, after first looking at the isospectral condition on tori, we look at manifolds which are compact quotients $\Gamma \backslash G$ of Lie groups (i.e., in the notation above, $M = G$ with a left-invariant metric). In section 2, we consider continuous families of isospectral manifolds of this type. In sections 3 and 4, we generalize to the case where G does not act transitively on M; here the main technique is a theorem of T. Sunada [S] for actions of finite groups G and a generalization [DG] for arbitrary G. Next in section 5 we give a modification of these techniques, developed by R. Brooks, to construct isospectral conformally equivalent Riemmannian metrics.

To the author's knowledge, all known examples of isospectral manifolds can be constructed by the techniques of sections 1–5 with the exception of

---

* partially supported by National Science Foundation grant DMS-8601966

certain isospectral Heisenberg manifolds, described in section 6, and lens spaces, described in section 7.

The fact that the known examples have so much in common may only reflect the difficulty in constructing examples without a good deal of symmetry. Major open questions remain. Must every pair of isospectral manifolds be locally isometric? In particular, do there exist isospectral, simply-connected manifolds?

The book [Be1], [BGM] and [C] contain good introductions to spectral geometry and references; in particular, [Be1] contains an extrememly extensive and beautifully organized bibliography. There have been several recent expository articles on isospectral manifolds ([Be2], [Br1], [D], [G1]). The current article overlaps with these other expositions but adopts a more representation theoretic point of view.

The author would like to express her gratitude to the organizing committe of the Workshop on Representation Theory of Lie Groups and Their Applications, and especially to Roberto Miatello and Juan Tirao, who prepared a first manuscript of these lecture notes.

1. Lie Group Quotients

The first example of isospectral, non-isometric manifolds – a pair of 16-dimensional flat tori – was given by Milnor [M] in 1964 (see also BGM]). A flat torus is a quotient $\Gamma\backslash\mathbb{R}^n$ of $\mathbb{R}^n$ by a lattice Γ of maximal rank. The Laplace Beltrami operator is just the Euclidean Laplacian $\Delta = -\sum \frac{\partial^2}{\partial^2 x_i}$. Letting Γ^* be the dual lattice, i.e.,

$$\Gamma^* = \{\tau \in \mathbb{R}^n : \tau \cdot \gamma \in \mathbb{Z} \quad \text{for all} \quad \gamma \in \Gamma\},$$

then an orthonormal basis of $L^2(\Gamma\backslash\mathbb{R}^n$ is given by $\{f_r : \tau \in \Gamma^*\}$, where $f_\tau(x) = \exp(2\pi i\tau \cdot x)$ is an eigenfunction of the Laplacian with eigenvalue $4\pi^2\|\tau\|^2$.

Thus two tori, $\Gamma_1\backslash\mathbb{R}^n$ and $\Gamma_2\backslash\mathbb{R}^n$, are isospectral if and only if there exists a bijective norm-preserving correspondence $\theta : \Gamma_1^* \to \Gamma_2^*$, or equivalently, there exists a bijective norm-preserving correspondence $\theta : \Gamma_1 \to \Gamma_2$ (i.e., Γ_1 and Γ_2 have the same "length spectrum"). The tori are isometric if and only if some such choice of θ extends to a linear map $\theta \in O(n, \mathbb{R})$, i.e., the lattices are congruent. Milnor's example of isospectral, non-isometric tori arose from the construction by Witt of a pair of non-congruent lattices with the same length spectra.

To what extent does the isospectral torus construction extend to compact quotients of arbitrary Lie groups? Let G be a connected Lie group, Γ a co-compact discrete subgroup (i.e., $\Gamma\backslash G$ is compact), and g a left-invariant Riemannian metric on G. Note that g induces a metric, which we again denote by g, on $\Gamma\backslash G$. We will denote by ρ_Γ the right action of G on $L^2(\Gamma\backslash G)$. Given two co-compact discrete subgroups Γ_1 and Γ_2, consider the conditions:

(1) $\rho_{\Gamma_1} \cong \rho_{\Gamma_2}$ (where ρ_Γ denotes the right action of G on $L^2(\Gamma\backslash G)$)

(2) $\rho_{\Gamma_1} \cong \rho_{\Gamma_2}$ "module orthogonal automorphisms" (i.e., there exists decompositions, $\rho_{\Gamma_1} \cong \oplus_{\alpha \in A}\sigma_\alpha$ and $\rho_{\Gamma_2} \cong \oplus_{\alpha \in A}\sigma_\alpha \circ \Phi_\alpha$ where, for each α, Φ_α is a g-orthogonal automorphism of G.

(3) $spec\ \Gamma_1\backslash G, g =spec\ \Gamma_2\backslash G, g$. (Here spec (M, g) denotes the Laplace spectrum of the manifold (M, g).) Then:

Theorem A. *Conditon (1) implies (3).*

Theorem A'. *Condition (2) implies (3).*

While Theorem A' obviously implies Theorem A, we are stating them separately here for later reference.

Proof. First observe that (1) implies (3). Indeed, the Laplacian of the left-invariant metric g is given by $\Delta = -\sum_{j=1}^n X_j^2$, where $\{X_j\}_1^n$ is any g-orthogonal basis of the Lie algebra. The operator extends to $L^2(\Gamma_i\backslash G)$ as $\Delta = -\sum_{j=1}^n (\rho_{\Gamma_{i*}}(X_j))^2$. Thus if $\rho_{\Gamma_1} \cong \rho_{\Gamma_2}$, then the Laplacians on $L^2(\Gamma_1\backslash G)$ and $L^2(\Gamma_2\backslash G)$ are unitarily equivalent.

Now assume (2). If Φ is any orthogonal automorphism of (G, g), then $\{\Phi_*X_1, \ldots, \Phi_*X_n\}$ is also an orthonormal basis. Thus for any α, the Laplacian can be expressed on $L^2(\Gamma_1\backslash G)$ as $\Delta = -\sum_{j=1}^n ((\rho_{\Gamma_1} \circ \Phi_\alpha)_*(X_j))^2$. It again follows that the Laplacians of $\Gamma_1\backslash G$ and $\Gamma_2\backslash G$ are unitarily equivalent.

Thus $(1) \Rightarrow (2) \Rightarrow (3)$. Example (1) below shows that (2) is strictly weaker than (1). An example in section 6 (Heisenberg example) will show that (3) is strictly weaker than (2).

Example 1. For flat tori $\Gamma_i\backslash\mathbb{R}^n$ $(i = 1, 2)$, conditions (2) and (3) are equivalent. Indeed, as discussed above, (3) implies the existence of a norm preserving bijection $\theta : \Gamma_1^* \to \Gamma_2^*$. We then have

$$L^2(\Gamma_1\backslash\mathbb{R}^n) = \oplus_{\tau \in \Gamma_1^*} <f_\tau>, \quad L^2(\Gamma_2\backslash\mathbb{R}^n) = \oplus_{\tau \in \Gamma_1^*} <f_{\theta(\tau)}> .$$

Since $\|\tau\| = \|\theta\tau\|$, there exist $A = A_\tau \in SO(n)$ such that $\theta(\tau) = A(\tau)$. Since the characters $f_{\theta(\tau)} \circ A$ and f_τ of $\mathbb{R}^n$ are equivalent, condition (2) holds.

Notice, however, that $\rho_{\Gamma_1} \cong \rho_{\Gamma_2}$ only if the tori are isometeric.

Example 2. For the Lie groups $SL(2, \mathbb{R})$ and $SL(2, \mathbb{C})$, Vignéras [V] constructed examples of pairs of non-conjugate co-compact discrete subgroups Γ_1 and Γ_2 for which the representation ρ_{Γ_1} and ρ_{Γ_2} are equivalent. In the case of $G = SL(2, \mathbb{C})$, the groups Γ_1 and Γ_2 are not even abstractly isomorphic, so the isospectral manifolds $\Gamma_1 \backslash G$ and $\Gamma_2 \backslash G$ have non-isomorphic fundamental groups.

Vignéras used her construction of equivalent representations to produce the first examples of isospectral Riemann surfaces and hyperbolic manifolds. Given a compact subgroup K of G, let g be a left-invariant metric on G which is also right K-invariant. Then g induces a left-invariant metric $\bar{g}$ on G/K so that the projection $\pi : G \to G/K$ is a Riemannian submersion. (In other words, the differential π_* maps the orthogonal complement of the fibre at any point x of G isometrically to the tangent space of G/K at $\pi(x)$.) The Laplacian Δ of (G, g) commutes with the right K-action, and the Laplacian of $(G/K, \bar{g})$ is just the restriction of Δ to the right K-invariant functions. Thus if ρ_{Γ_1} is equivalent to ρ_{Γ_2} so that $\Gamma_1 \backslash G, g)$ is isospectral to $(\Gamma_2 \backslash G, g)$, then $(\Gamma_1 \backslash G/K, \bar{g})$ and $(\Gamma_2 \backslash G/K, \bar{g})$ are also isospectral.

Taking $G = SL(2, \mathbb{R})$ and $K = SO(2, \mathbb{R})$, Vignéras obtained isospectral non-isometric Riemann surfaces $\Gamma_1 \backslash SL(2, \mathbb{R})/SO(2)$ and $\Gamma_2 \backslash SL(2, \mathbb{R})/SO(2)$. Taking $G = SL(2, \mathbb{C})$, she obtained isospectral three-dimensional hyperbolic manifolds with non-isomorphic fundamental groups.

2. Isospectral Deformations on Nilmanifolds

We now apply the methods of section 1 to construct continuous families of isospectral, non-isometric Riemannian metrics. Let G be a simply-connected nilpotent Lie group and $\mathfrak{g}$ its Lie algebra.

According to the Kirillov theory, equivalence classes of irreducible unitary representations of G are in one-to-one correspondence with co-adjoint orbits of $\mathfrak{g}^*$. In other words, each linear functional $\lambda \in \mathfrak{g}^*$ gives rise to an irreducible unitary representation π_λ of G, and two such representations π_λ and π_μ are equivalent if and only if $\mu \in Ad(G) \cdot \lambda$. Recall also that if $\Phi \in \operatorname{Aut}(G)$, then $\pi_\lambda \circ \Phi \cong \pi_{\Phi_*\lambda}$.

Definition. We say that $\phi \in \operatorname{Aut}(G)$ is almost inner if $\phi(x)$ is conugate to x for all $x \in G$.

Proposition. [GW1]. *Let G be a simply-connected nilpotent Lie group, Φ an almost inner automorphism of G and Γ a co-compact discrete subgroup of G. Then the representations ρ_Γ and $\rho_{\Phi(\Gamma)}$ are equivalent. In particular, if g is a left-invariant Riemannian metric on G, then spec $(\Gamma \backslash G, g) = spec(\Phi(\Gamma) \backslash G, g)$.*

Proof. The condition that Φ is almost inner is equivalent to the condition $\Phi^*(A) \in Ad(G) \cdot \lambda$ for all $\lambda \in \mathfrak{g}^*$. (See [G2]). Thus by the Kirillov theory, if $\rho_\gamma \cong \oplus \pi_\lambda$, then $\rho_{\Phi(\Gamma)} \cong \oplus \pi_\lambda \circ \Phi^{-1} \oplus \pi_\lambda \cong \rho_\Gamma$. The final statement follows from Theorem A.

Remark.

(1) The proposition remains true if we only assume that $\Phi(\gamma)$ is conjugate to γ in G for all $\Gamma \in \Gamma$. This condition gurantees that $\pi_\lambda \sim \pi_\lambda \circ \Phi^{-1}$ for all the π_λ which occur in ρ_Γ. We say that Φ is almost inner relative to Γ in this case.

(2) The Riemannian manifold $(\Phi(\Gamma)\backslash G, g)$ is isometric to $(\Gamma\backslash, \Phi^* g)$, so g and $\Phi^* g$ are isospectral metrics on the same manifold $\Gamma\backslash G$.

Proposition [GW1]. *The almost inner automorphisms of G form a simply-connected nilpotent subgroup $AIA(G)$ of $Aut(G)$, frequently of dimension greater than that of the inner automorphism group $Inn(G)$. If $\{\Phi_t\}$ is a one-parameter subgroup of $AIA(G)$ not contained in $Inn(G)$, then for any choice of Γ and g, resulting isospectral deformation $\Gamma\backslash G, \Phi_t^* g)$ is non-trivial.*

Example 3. Let

$$
G = \left\{ \begin{pmatrix} 1 & x_1 & x_2 & z_1 & & & \\ & 1 & 0 & y_1 & & & \\ & & 1 & y_2 & & 0 & \\ & & & 1 & & & \\ & & 0 & & 1 & x_1 & z_2 \\ & & & & & 1 & y_2 \\ & & & & & & 1 \end{pmatrix} \right\}
$$

Denote the elements of G by $h = (x_1, x_2, y_1, y_2, z_1, z_2)$. Then the multiplication is given by

$$
hh' = (x_1 + x_1', \ldots, y_2 + y_2', z_1 + z_1' + x_1 y_1' + x_2 y_2', z_2 + z_2' + x_1 y_2')
$$

G admits a 2-parameter family of almost inner, non-inner automorphisms given by

$$
\Phi_{s,t}(h) = (x_1, x_2, y_1, y_2, z_1, z_2 + sx_1 + ty_2).
$$

To see that these are almost inner, note that conjugation in G is given by

$$
h'h(h')^{-1} = (x_1, x_2, y_1, y_2, z_1 + x_1' y_1 + x_2' y_2 - x_1 y_1'
$$
$$
- x_2 y_2', z_2 + x_1' y_2 - x_1 y_2').
$$

Thus $\Phi_{s,t}(h) = h'h(h')^{-1}$ with $h' = (x_1', x_2', y_1', y_2', 0, 0)$ where

$$
(x_1', x_2', y_1', y_2') = \begin{cases} (0,0,0) & \text{if } x_1 = y_2 = 0 \\ (t, -ty_1/y_2, 0, 0) & \text{if } x_1 = 0, y_2 \neq 0 \\ (0, 0, sx_1/x_2, -s) & \text{if } x_1 \neq 0, y_2 = 0 \\ (t, -ty_1/y_2, sx_1/x_2, -s) & \text{if } x_1 \neq 0, y_2 \neq 0. \end{cases}
$$

Choosing any co-compact discrete subgroup Γ, e.g., the subgroup of matrices with integer entries, and any left-invariant Riemannian metric g on G, we obtain a non-trivial 2-parameter isospectral deformation $(\Gamma\backslash, (g_{s,t})$ where $g_{s,t} = \Phi_{s,t}^* g$.

In the paper [DGGW1], we examine the changing geometry of the metrics $g_{0,t}$ for a particular choice of initial metric g. In [DGGW2,3,4,5], we investigate how the geometry changes in general for deformations constructed by almost inner automorphisms.

Remark. He Ouyang and Hubert Pesce have very recently (and independently) proven a converse to the propositions above in case G is two-step nilpotent: If $\{g_t\}_t$ is a continuous family of left-invariant metrics on G such that spec $(\Gamma\backslash G, g_0)$ for all t, then there exists a continuous family $\{\Phi_t\}_t$ of almost inner automorphisms relative to Γ such that $g_t = \Phi_t^*(g_0)$. As a corollary, if $\{\Gamma_t\}_t$ is a continuous family of co-compact discrete subgroups of a two-step nilpotent Lie group G such that $\rho_{\Gamma_t} \sim \rho_{\Gamma_0}$ for all t, then there exists a continuous family $\{\Phi_t\}_t$ of almost inner automorphisms relative to Γ such that $\Gamma_t = \Phi_t(\Gamma_0)$. They expect to eliminate the "two-step" hypothesis. However, the continuity assumption cannot be dropped. Ruth Gornet has constructed a pair of non-isomorphic co-compact discrete subgroups Γ_1 and Γ_2 of a three-step nilpotent Lie group G with $\rho_{\Gamma_1} \sim \rho_{\Gamma_2}$.

3. Sunada's Technique

In the previous sections we have restricted our attention to locally homogenous spaces, i.e., Riemannian manifolds covered by homogeneous Riemannian manifolds. We now look at more general manifolds although we will still require that the covering manifold exhibit a certain amount of symmetry.

T. Sunada gave an elegant technique for constructing isospectral manifolds with a common finite Riemannian covering.

Theorem B. [S]. *Let G be a finite group and Γ_1 and Γ_2 subgroups. Suppose each conjugacy class of G intersects Γ_1 and Γ_2 in the same number of elements. (Note that this hypothesis is equivalent to $L^2(\Gamma_1\backslash G) \cong L^2(\Gamma_2\backslash G)$ as G-modules; i.e., $\rho_{\Gamma_1} \sim \rho_{\Gamma_2}$). Let (M, g) be a compact Riemannian manifold on which G acts freely on the left by isometries. Then*

$$\text{spec } (\Gamma_1\backslash M, g) = \text{spec } (\Gamma_2\backslash M, g).$$

Sunada proved this theorem by showing that the heat kernels of $(\Gamma_1\backslash M, g)$ and $(\Gamma_2\backslash M, g)$ have the same trace. Since the trace of the heat kernels of a compact Riemannian manifold is given by $\zeta(t) = \sum_{j=0}^{\infty} e^{-\lambda_j t}$ where $\lambda_0 \leq \lambda_1 \leq \lambda_2 \leq \dots$ is the Laplace spectrum, the heat trace determines the spectrum.

P. Berard [Be3] very recently gave a new proof of this theorem using representation theoretic techniques. His proof is actually more general; he can allow M to be non-compact and establish isospectrality of the discrete part of the spectrum.

The beauty of Sunada's technique is its simplicity. To apply the technique, begin with a finite group G and non-conjugate subgroups Γ_1 and Γ_2 satisfying Sunada's conjugacy condition. (There is an abundance of such triples (G, Γ_1, Γ_2). One example will be given below. See [S] for additional examples.) Let $\overline{M}$ be a manifold whose fundamental group surjects onto G, and let M be a normal covering group G. We can then give M a G-invariant Riemannian metric g and construct a tower of Riemannian coverings

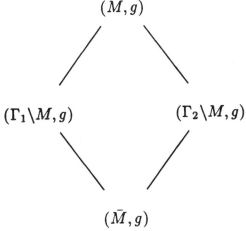

Applying Sunada's theorem, we see that

$$\text{spec}\,(\Gamma_1 \backslash M, g) = \text{spec}\,\Gamma_2 \backslash M, g).$$

Since Γ_1 is not conjugate to Γ_2, any isometry between the two manifolds will be accidental. By choosing g carefully, one can obtain non-isometric examples.

Examples. The many examples of isospectral, non-isometric manifolds constructed via Sunada's method include: (this is not a complete list)

(a) Riemann surfaces in every genus ≥ 4 and genus 3 surfaces with non-standard metrics (Buser [Bu], Brooks - Tse[BrT]);

(b) higher rank locally symmetric spaces (Spatzier [Sp]);

(c) flat 2-dimensional manifolds with boundary embedded in $\mathbb{R}^3$ (Buser [Bu]);

(d) isospectral manifolds with simple spectra (i.e., no multiple eigenvalues). (Zelditch [Z]).

We describe two of these examples in detail. Let $G = SL(3, Z_2)$ and

$$\Gamma_1 = \left\{ \begin{pmatrix} 1 & * & * \\ 0 & * & * \\ 0 & * & * \end{pmatrix} \right\} \gamma_2 = \left\{ \begin{pmatrix} 1 & 0 & 0 \\ * & * & * \\ * & * & * \end{pmatrix} \right\}.$$

Then (G, Γ_1, Γ_2) satisfies Sunada's hypothesis.

Each of $\Gamma_i \backslash G$ has order 7. The group G is generated by two elements A, B of order 7. The commutator C of A and B also has order 7.

Brooks-Tse (BrT) (see also Brooks [Br1]) used this triple to obtain isospectral genus 4 surfaces as follows. Let $\overline{M}$ be the Riemann surface with fundamental polygon,

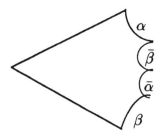

The fundamental group of $\overline{M}$ surjects onto G, so as described above, we can construct a pair of isospectral 7-fold coverings $\Gamma_1 \backslash M$ and $\Gamma_2 \backslash M$ of $\overline{M}$. (it is not necessary to construct the common covering M explicitly.) To obtain $M_1 = \Gamma_1 \backslash M$, we take 7 copies of the fundamental polygon for $\overline{M}$, label the copies by the elements of $\Gamma_1 \backslash G$ and glue them together according to the action of G. (For example, if the generator A of G maps the coset $\Gamma_1 x$ to the coset $\Gamma_1 y$, then glue side $\bar{\alpha}$ of the polygon labeled $\Gamma_1 x$ to side α of the polygon labeled $\Gamma_1 y$. Similarly, glue the β and γ edges according to the action of B and C respectively.) The gluing patterns for M_1 and M_2 are illustrated in figure 1.

To obtain Buser's example of isospectral flat manifolds with boundary, apply the same procedure to the title below.

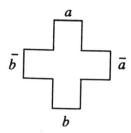

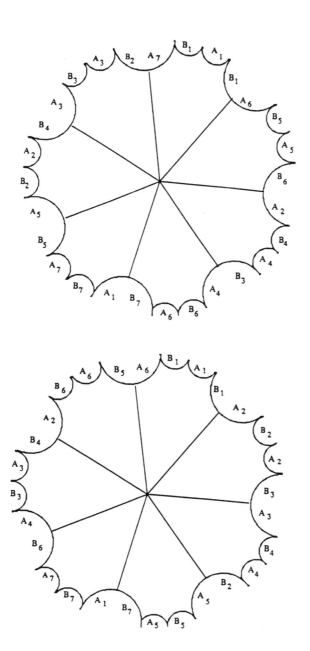

4. Generalized Technique

As discussed in section 2, Sunada's theorem (Theorem B) says that if G is a finite group with subgroups Γ_1 and Γ_2 such that $\rho_{\Gamma_1} \sim \rho_{\Gamma_2}$ (in the notation of section 1), then given any compact Riemannian manifold (M, g) on which G acts in a nice way by isometries, the quotient manifolds $(\Gamma_1 \backslash M, g)$ and $(\Gamma_2 \backslash M, g)$ are isospectral. In Theorem A, on the other hand, we showed that if G is a Lie group and Γ_1 and Γ_2 co-compact discrete subgroups with $\rho_{\Gamma_1} \sim \rho_{\Gamma_2}$, then spec $(\Gamma_1 \backslash G, g) = $ spec $\Gamma_2 \backslash, g)$ for any left-invariant metric.

Theorem B is less restrictive in the sense that M need not be homogeneous; however, it requires G to be finite.

The following theorem simultaneously generalizes Theorems A and B.

Theorem C. [DG]. *Let G be a Lie group and Γ_1 and Γ_2 co-compact discrete subgroups with $\rho_{\Gamma_1} \sim \rho_{\Gamma_2}$. Let (M, g) be a Riemannian manifold on which G acts on the left by isometries, and suppose that the subgroups Γ_1 and Γ_2 act freely and properly discontinuously with compact quotients $\Gamma_i \backslash M$. Then*

$$\mathrm{Spec}\,(\Gamma_1 \backslash M, g) = \mathrm{spec}\,(\Gamma_2 \backslash M, g)\,.$$

The condition $\rho_{\Gamma_1} \sim \rho_{\Gamma_2}$ is equivalent to a certain conjugacy condition. The proof in [DG] is generalization of Sunada's proof using the heat trace. Again Berard has recently given a new representation theoretic proof [Be4].

Example. Let G be the six-dimensional Lie group in Example 3, and let Γ be the integer lattice. Note that the center $Z(G)$ is isomorphic to $\mathbb{R}^2$ and $\Gamma \cap (G) = \mathbb{Z}^2$. Let $\overline{G} = G/(\Gamma \cap Z(G))$ be a nilpotent Lie group covered by G, and $\overline{\Gamma} = \Gamma/(\Gamma \cap Z(G))$. Let $\{\Phi_{s,t}\}_{s,t}$ be the family of almost inner automorphisms constructedas in Example 3. Note that $\Phi_{s,t}$ fixes $Z(G)$ pointwise and thus induces an almost inner automorphism $\overline{\Phi}_{s,t}$ of $\overline{G}$. Set $\overline{\Gamma}_{s,t} = \overline{\Phi}_{s,t}(\overline{\Gamma})$. Then for each (s,t) the triple $(\overline{G}, \overline{\Gamma}, \overline{\gamma}_{s,t})$ satisfies the hypothesis of Theorem C.

Let B be a principal $T^2 = R^2/Z^2$ bundle, e.g.,

$$S^3 \times S^3 \rightarrow S^2 \times S^2$$

(take two copies of the Hopf fibration). Note that $T^2 \cong Z(\overline{G})$ acts by translation on $\overline{G}$. Form the associated $\overline{G}$-bundle

$$M = B \times_{T^2} \overline{G}\,.$$

Note that $\overline{G}$ acts on M; in fact M is a principal $\overline{G}$-bundle.

In case $B = S^3 \times S^3$, as above M is diffeomorphic to $S^3 \times S^3 \times R^4$ (Since $\overline{G}$ contracts to T^2).

Given any $\overline{G}$-invariant Riemannian metric on M, Theorem C implies that

$$spec\,(\Gamma\backslash M, g) = spec\,(\Gamma_{s,t}\backslash M, g)\,.$$

If there exists a diffeomorphism $\Psi_{s,t}$ of M such that

$$\Psi_{s,t}(\gamma \cdot x) = \overline{\Phi}_{s,t}(\gamma)\Psi_{s,t}(x)$$

for all $\gamma \in \Gamma$, $x \in M$, then $(\overline{\Gamma}_{s,t}\backslash M, g) \cong (\overline{\Gamma}\backslash M, g_{s,t})$ where $g_{s,t} =)\Psi_{s,t}^{-1})^* g$. We then obtain an isospectral family of metrics $\{g_{s,t}\}_{s,t}$ on $\overline{\Gamma}\backslash M$. This is the case for example when $B = S^3 \times S^3$ as above. To correct an error in [DG], we remark that $\overline{\Gamma}\backslash M$ is *not* diffeomorphic to $S^3 \times S^3 \times T^4$. The author is grateful to Dorothee Schüth for pointing out this error.

5. Isospectral Conformally Equivalent Metrics

R. Brooks applied Theorems B and C to given a method for constructing isospectral, conformally equivalent metrics.

Theorem ([BPY, BG]). *Let (M, g) be a Riemannian manifold, G a Lie group acting on M on the left by isometries, and Γ_1 and Γ_2 discrete subgroups of G. Assume,*
(1) $(M, g, G, \Gamma_1, \Gamma_2)$ satisfies the hypotheses of Theorem B or C.
(2) There exists an isometry τ of M, not in G, such that $\Gamma_2 = \tau\Gamma_1\tau^{-1}$.
Let $u \in C^\infty(M)$ be G-invariant but not τ invariant. Then

$$spec\,(\Gamma_1\backslash M, e^u g) = spec\,(\Gamma_1\backslash(M, e^{u\circ\tau} g)\,.$$

Proof. Observe that G acts by isometries on $(M, e^u g)$, so Theorem B or C implies spec $\Gamma_1\backslash M, e^u g) =$ spec $(\Gamma_2\backslash M, e^u g)$. But τ induces a diffeomorphism from $\Gamma_1\backslash M$ to $\Gamma_2\backslash M$ and, since $\tau^* g = g$, it gives an isometry from $(\Gamma_1\backslash M, e^{u\circ\tau} g)$ to $(\Gamma_2, \backslash M, e^u g)$.

Example. [BG] Let G be the six-dimensional nilpotent Lie group of Example 3 and Γ the integer lattice. Choose a one-parameter family Φ_t of almost inner automorphisms Φ_t, say $\Phi_t = \Phi_{0,t}$ where $\Phi_{s,t}$ is given in Example 3. Let $F = G \rtimes \mathbb{R}$, where $t \in \mathbb{R}$ acts on G as Φ_t; i.e., in F,

$$(h, t) \cdot (h', t') = (h\Phi_t(h'), t + t')\,.$$

Note that Φ_t extends to an inner automorphism of F, namely conjugation by $(0, t)$.

Let $M = F$ with a left-invariant metric G. Set $\widetilde{G} = G \rtimes \mathbb{Z} \subset F, \widetilde{\Gamma} = \Gamma \rtimes \mathbb{Z}$ and $\widetilde{\Gamma}_t = \Phi_t(\Gamma) \rtimes \mathbb{Z}$. Then for each t, $(M, g, \widetilde{G}, \widetilde{\Gamma}, \widetilde{\Gamma}_t)$ satisfies the hypotheses of Theorem C. Moreover, $\widetilde{\Gamma}_t = \tau_t\widetilde{\Gamma}\tau_t^{-1}$ where τ_t is left translation by $(0, t)$. (Note that $\tau_t \notin \widetilde{G}$ when $t \notin \mathbb{Z}$). Thus hypotheses (i) and (ii) are satisfied.

Letting v be any non-constant periodic function on $\mathbb{R}$ of period one, then the function $u \in C^\infty(M)$ given by $u(h, s) = v(s)$ is $\widetilde{G}$-invariant but not τ_t-invariant. Note that

$$u \circ \tau_t(h, s) = u(h, t + s).$$

Set $u_t = u \circ \tau_t$. By the theorem, $\{e^{u_t}g\}_t$ is an isospectral family of metrics on $\widetilde{\Gamma}\backslash M$.

Since at first glance one might think that these metrics are isometric, we briefly indicate why they are not. For details, see [DGGW6]. Recall that $M = F$, an extension of G. Let $\bar{g}$ be the left-invariant metric on G induced by the metric g on F. As in section 2, we obtain a non-trivial isospectral family of metric $\bar{g}_t = \Phi_t^* \bar{g}$ on $\Gamma\backslash G$.

The manifold $\widetilde{\Gamma}\backslash M = \widetilde{\Gamma}\backslash F$ is a bundle over S^1 with fibre $S\backslash G$, and one can show that the metric g restricts on the fibre over $[s] \in S^1$ to $\bar{g}_s$. (We note that $\bar{g}_{s+1} \cong \bar{g}_s$ so this is well-defined.) In particular, all the fibres have the same volume, since isospectral metrics always have equal volumes.

The effect of the conformal factor e^u is to rescale the metric on the fibre over $[s] \in S^1$ by the factor $e^{v(s)}$. As we deform u (recall $u_t(h, s) = u(h, s+t)$), the scaling factors translate around the circle; i.e., $e^{u_t}g$ restricts on the sth fibre to $e^{v(s+t)}\bar{g}_s$. One can show that fibres all lie in the same homology class of M, and the smallest fibre is volume-minimizing in that homology class. If a is the minimal value of v, obtained say at $v(s_o)$, then at time t, the fibre over $[s_o - t]$ is volume minimizing and has metric $e^a \bar{g}_{s_o-t}$. Since the metrics $\{\bar{g}_{s_o-t}\}_t$ are isospectral but not isometric, the geometry of the volume-minimizing fibre changes, results in a change in the geometry of M.

6. Heisenberg Manifolds

All isospectral manifolds constructed via any of Theorems A, A$'$, B, or C are actually strongly isospectral (see [DG]); in particular, the Laplacians acting on p-forms are isospectral for all p. In this section we give examples of two types of isospectral left-invariant metrics on Heisenberg manifolds. One type does not satisfy any of the theorems above; for at least some (and probably all) of the examples of this type, the Laplacians acting on 1-forms are not isospectral. The second type satisfies Theorem A$'$ but not Theorem A.

By Heisenberg manifold, we mean a compact quotient $\Gamma\backslash H_n$ of the $(2n + 1)$-dimensional Heisenberg group H_n by a discrete subgroup Γ, together with a metric g whose lift to H_n is left-invariant. Recall that H_n is the set of all real $(n+2) \times (n+2)$ matrices of the form

$$\begin{pmatrix} 1 & x_1 & \cdots & x_n & z \\ & 1 & & 0 & y_1 \\ & & \ddots & & \vdots \\ & 0 & & \ddots & y_n \end{pmatrix}$$

We denote elements of H_n by $h = (x,t,y)$, with $x, y \in \mathbb{R}^n$ and $t \in \mathbb{R}$. Note that multiplication is given by

$$hh' = (x+x', y+y', t+t'+x \cdot y').$$

Up to conjugacy, all co-compact discrete subgroups of H_n are obtained as follows. Let $c \in \mathbb{R}^+$ and let $\mathcal{L}$ be a lattice of maximal rank in $\mathbb{R}^{2n}$ with the property that $x \cdot y' \in c\mathbb{Z}$ whenever (x,y) and $(x',y') \in \mathcal{L}$. Then $\Gamma(\mathcal{L}, c) = \{(x,y,t) : (x,y) \in \mathcal{L}, t \in c\mathbb{Z}\}$ is a co-compact discrete subgroup of H_n.

Let h be an inner product on $\mathbb{R}^{2n}$ and let I denote the standard inner product on $\mathbb{R}$. Then $h \times I$ defnies an inner product on the tangent space to H_n at the identity element, and hence extends to a left-invariant Riemannian metric on g. Up to automorphism, all left-invariant metrics are of this form.

Given a Heisenberg manifold $(\Gamma \backslash H_n, g)$ with $\Gamma = \Gamma(\mathcal{L}, c)$ and g as above, the inner product h defines a flat metric on the torus $\mathcal{L} \backslash \mathbb{R}^{2n}$; we refer to $(\mathcal{L} \backslash \mathbb{R}^{2n}, h)$ as the associated torus.

Theorem [GW2]. *Fix a metric g as above and fix $c \in \mathbb{R}^+$. Let $\Gamma_1 = \Gamma(\mathcal{L}_1, c)$ and $\Gamma_2 = \Gamma(\mathcal{L}_2, c)$ be a two co-compact discrete subgroups of H_n. Then $\mathrm{spec}(\Gamma_1 \backslash H_n, g) = \mathrm{spec}(\Gamma_2 \backslash H_n, g)$ if and only if the associated tori $(\mathcal{L}_1 \backslash \mathbb{R}^{2n}, h)$ and $(\mathcal{L}_2 \backslash \mathbb{R}^{2n}, h)$ are isospectral.*

We remark that in all the examples we've been able to construct, the tori are actually isometric.

In [GW2], we proved this theorem by explicitly calculating the eigenvalues of an arbitrary Heisenberg manifold. The paper [G3] gives another proof based on the Kirillov theory of representations of a nilpotent Lie group G (applied to $G = H_n$) and on Richardson's formula for the decomposition of $L^2(\Gamma \backslash H)$ into irreducible invariant subspaces. We sketch the latter proof below.

Proof. Write

$$L^2(\Gamma_i \backslash H_n) = \mathcal{H}_i^{(1)} \bigoplus \mathcal{H}_i^{(2)},$$

direct sum of H_n-modules where

$\mathcal{H}_i^{(2)}$ = direct sum of irreducible infinite dimensional H_n-invariant subspaces

$\mathcal{H}_i^{(1)}$ = direct sum of 1-dimensional invariant subspaces.

Using the Kirillov theory, one shows that in order to determine which irreducible representations of H_n occur in $\mathcal{H}_i^{(2)}$, it is enough to know c, and in order to determine their multiplicities, it is enough to know the volume of $\mathcal{L}_i\backslash\mathbb{R}^{2n}$. Thus if the associated tori are isospectral (and hence have the same volume), then $\mathcal{H}_1^{(2)} \cong \mathcal{H}_2^{(2)}$ as H_n-modules and, as in section 1, the Laplacians restricted to these subspaces are isospectral.

Next $\mathcal{H}_i^{(1)}$ consists of L^2-functions on $\Gamma_i\backslash H_n$ which do not depend on the final variable t. Thus $\mathcal{H}_i^{(1)}$ may be identified with $L^2\mathcal{L}_i\backslash\mathbb{R}^{2n})$. One checks that the Laplacian restricted to $\mathcal{H}_i^{(1)}$ agrees, under this identification, with the Laplacian of the torus metric h. Thus, by hypothesis, the Laplacians acting on $\mathcal{H}_1^{(1)}$ and $\mathcal{H}_2^{(1)}$ are isospectral, and the theorem is proved.

Note. The proof of the theorem shows that $\mathcal{H}_1^{(2)} \cong \mathcal{H}_2^{(2)}$ as H_n modules. Moreover, the proof together with Example (1) of section 1 shows that, $\mathbb{R}^{2n}$-modules, $\mathcal{H}_1^{(1)} \cong \mathcal{H}_2^{(1)}$ modulo h-orthogonal automorphisms. If these orthogonal automorphisms extend to g-orthogonal automorphisms of H_n, (this happens sometimes, but not always), then $\mathcal{H}_1^{(1)}$ will be isomorphic to $\mathcal{H}_2^{(2)}$ as H_n-modules, modulo orthogonal automorphisms. In that case, the isospectral pair of manifold satisfies Theorem A'.

Let us now see some concrete examples. Consider $\Gamma(\mathcal{L}, c)$ with $c = 1$ and $\mathcal{L} = r_1\mathbb{Z} \times \ldots \times r_n\mathbb{Z} \times s_1\mathbb{Z} \times \ldots \times s_n\mathbb{Z}$. Let h be of the form

$$h = \begin{pmatrix} a_1 & & & & & & & \\ & \ddots & & & & & 0 & \\ & & a_n & & & & & \\ & & & a_n & & & & \\ & & & & \ddots & & & \\ 0 & & & & & & a_n \end{pmatrix} \tag{1}$$

The torus $\mathcal{L}\backslash\mathbb{R}^{2n}, h)$ is isometric to $\mathbb{Z}^{2n}\backslash\mathbb{R}^{2n}$ with the flat metric

$$\begin{pmatrix} a_1 r_1^2 & & & & & \\ & \ddots & & & 0 & \\ & & a_n r_n^2 & & & \\ & & & a_1 s_1^2 & & \\ & 0 & & & \ddots & \\ & & & & & a_n s_n^2 \end{pmatrix}$$

For $\mathcal{L} = r_1\mathbb{Z} \times \ldots \times s_n\mathbb{Z}$ and $\mathcal{L}' = r_1'\mathbb{Z} \times \ldots \times s_n'\mathbb{Z}$, the tori $(\mathcal{L}\backslash\mathbb{R}^{2n}, h)$ and $(\mathcal{L}'\backslash\mathbb{R}^{2n}, h)$ have the same spectrum $\Leftrightarrow (a_1 r_1^2, \ldots, a_n r_n^2, a_1 s_1^2, \ldots, a_n s_n^2)$ is a permutation of $(a_1(r_1')^2, \ldots, a_n(r_n'), a_1(s_1')^2, \ldots, a_n(s_n')^2)$. For instance, if we specify h by $(a_1, \ldots, a_n)$ and $\mathcal{L}$ by $(r_1, \ldots, r_n; s_1, \ldots, s_n)$, then the chart below gives four isospectral pairs.

n	h	$\mathcal{L}$	$\mathcal{L}'$	Isomorphic fundamental groups?	isospectral on forms?
2	(1,4)	(4,2;1,2)	(4,1/2; 4,2)	no	no
2	(1,1)	(2,1;2,1)	(2,2;1,1)	no	yes
2	(1,4)	(4,2;1,1/2)	(1,2;1,2)	yes	no
3	(1,1,1)	(2,4,3;6,5,10)	(3,2,5;4,10,6)	yes	yes

One can show that $\Gamma(\mathcal{L}, 1) \cong \Gamma(\mathcal{L}', 1)$ if and only if $(r_1 s_1, \ldots, r_n s_n)$ is a permutation of $(r_1' s_1', \ldots, r_n' s_n')$; this gives the information in the fifth column.

How do we determine whether the Laplacians are isospectral on forms? As discussed in the note above, we first determine whether, as H_n-modules as opposed to just $\mathbb{R}^{2n}$-modules, $\mathcal{H}_1^{(1)} \cong \mathcal{H}_2^{(1)}$ modulo g-orthogonal automorphisms. For h as in (1), and $\mathcal{L}$ and $\mathcal{L}'$ rectangular lattices, a computation of the g-orthogonal automorphisms shows that this condition is equivalent to the following:

Let $V_1, \ldots, V_k$ be the eigenspaces of h in $\mathbb{R}^{2n}$. Thus, in the first example in the chart above, $V_1 = \{x_2 = y_2 = 0\}$ and $V_2 = \{x_1 = y_1 = 0\}$, while the second example h has only one eigenspace $\mathbb{R}^4$. Each of $\mathcal{L}$ and $\mathcal{L}'$ intersects each V_i in a lattice of maximal rank. The metric L restricts on V_i to a scalar multiple of the standard Euclidean metric. As shown in [G3], $\mathcal{H}_1^{(1)} \cong \mathcal{H}_2^{(2)}$ modulo orthogonal automorphisms if and only if the tori $V_j/(V_j \cap \mathcal{L})$ and $V_j/(V_j \cap \mathcal{L}')$ are isometric, $1 \leq j \leq k$. (Here $\mathcal{L}$ plays the role of $\mathcal{L}_1$ and $\mathcal{L}'$ the role of $\mathcal{L}_2$. Note that for pairs of rectangular tori, isospectral is equivalent to isometric.) Thus is each of these pairs of subtori are isometric, then the Laplacians acting on p-forms are isospectral for all p. Conversely, a direct computation shows that if any of these pairs are not isometric, then the Laplacians acting on 1-forms are not isospectral.

In the second and fourth examples in the chart, this condition is trivially satisfied, so that Laplacians are isospectral on forms. In the first example, $V_1 \cap \mathcal{L} = 4\mathbb{Z} \times \mathbb{Z}$ while $V_1 \cap \mathcal{L}' = 4\mathbb{Z} \times 4\mathbb{Z}$, thus the first pair of subtori (and also the second pair) are non-isometric. Hence the Laplacians are not isospectral on 1-forms. The third example is similar.

Remark. Very recently, Ruth Gornet generalized the Heisenberg construction to obtain isospectral higher step nilmanifolds for which the Laplacians on 1-forms are not isospectral.

7. Lens Spaces

In addition to the examples of section 6, there is one other known class of examples which do not satisfy any of Theorems A, A', B, or C. The manifolds involved are lens spaces. These examples, constructed by Ikeda, are especially exciting, as they establish the following theorem.

Theorem [I2]. *Given $p \in \mathbb{Z}^+$, there exists pairs of manifolds whose Laplacians are isospectral on functions and on k-forms for all $k < p$ but not on p-forms.*

The lens spaces are constructed as follows: Let

$$R(\theta) = \begin{pmatrix} \cos 2\pi\theta & \sin 2\pi\theta \\ -\sin 2\pi\theta & \cos 2\pi\theta \end{pmatrix}$$

Given $q \in \mathbb{Z}^+$ and integers $p_1, \ldots, p_n$, prime to q the matrix

$$\begin{pmatrix} R_{(p_1/q)} & & \\ & \ddots & 0 \\ 0 & & \ddots \\ & & R_{(p_n/q)} \end{pmatrix}$$

generates a cyclic subgroup of order q in $0(2n)$. The manifold $\Gamma\backslash S^{2n-1}$ with the canonical (constant curvature) metric is called a lens space and is denoted $L(q : p_1, \ldots, p_n)$.

Proposition. *The following are equivalent for*

$$L = L(q : p_1, \ldots, p_n) \quad and \quad L' = L(q : p_1', \ldots, p_n').$$

(1) L is isometric to L'.
(2) L is diffeomorphic to L'.
(3) L is homeomorphic to L'.
(4) There exists an integer ℓ and $e_i \in \{-1, 1\}$ such that $(p_1, \ldots, p_n)$ is a permutation of $(e_1 \ell p_1', \ldots, e_n \ell p_n')$.

Notice that eigenfunctions on the lens space $\Gamma\backslash S^{2n-1}$ lift to Γ-invariant eigenfunctions on the sphere S^{2n-1}. The eigenvalues of the sphere are precisely the numbers $k(k + 2n - 2)$, $k = 0, 1, 2 \ldots$. Thus to determine the spectrum of the lens space, one needs only to determine the multiplicity of the $k(k + 2n - 2)$-eigenspace, for each k. Consider the generating function

$$F_q(z : p_1, \ldots, p_n) = \sum_{n=0}^{\infty} a_k z^k$$

where $a_k = \dim k(k + 2n - 1)$-eigenspace.

Ikeda and Yamamoto showed that the generating function is rational:

$$F_q(z : p_1, \ldots, p_n) = \frac{1}{q} \sum_{\ell=1}^{q} \frac{1 - z^2}{\prod_{i=1}^{n}(z - \gamma^{p_i \ell})(z - \gamma^{-p_i \ell})} \qquad (*)$$

where γ is a primitive qth root of 1.

Now one has

$$\operatorname{spec} L(q : p_1, \ldots, p_n) = \operatorname{spec} L(q : p'_1, \ldots, p'_n)$$
$$\Leftrightarrow F_q(z : p_1, \ldots, p_n) = F_q(z : p'_1, \ldots, p'_n).$$

Using (*), Ikeda [I1] gives conditions for isospectrality and many examples, inclusing some which are not homotopically equivalent. In [I2], he studies the spectra on forms by similar methods, obtaining the theorem above.

REFERENCES

[Be1] P. Bérard, Spectral Geometry: Direct and Inverse Problems, *Springer Lecture Notes*, 1986.

[Be2] —, Variétés Riemanniennes isospectrales non isométriques, *Astérisque* 177–178, (1989), 127–154.

[Be3] —, Translplantation et isospectralité I, preprint.

[Be4] —, Transplantation et isospectralité II, preprint.

[BGM] M. Berger, P. Gauduchon, and E. Mazet, Le Spectre d'une variété Riemannienne, *Springer Lecture Notes* 194 (1971).

[Br1] R. Brooks, Constructing isospectral manifolds, *Amer. Math. Monthly* 95, (1988), 823–839.

[Br2] R. Brooks, On manifolds of negative curvature with isospectral potentials, *Topology* 26, (1987), 63–66.

[BrG] R. Brooks and C. S. Gordon, Isospectral families of conformally equivalent Riemannian metrics, *Bull. Amer. Math. Soc.* 23, no. 2 (1990), 433–436.

[BPY] R. Brooks, P. Perry, and P. Yang, Isospectral sets of conformally equivalent metrics, *Duke Math. J.* 58 (1989), 131–150.

[BrT] R. Brooks and R. Tse, Isospectral surfaces of small genus, *Nagoya Math. J.* 107 (1987), 13–24.

[Bu] P. Buser, Isospectral Riemann Surfaces, *Ann. Inst. Four., Grenoble* 36 (1986), 167–192.

[C] I. Chavel, Eigenvalues in Riemannian Geometry, *Academic Press*, 1984.

[D] D. M. DeTurck, Audible and inaudible geometric properties, preprint.

[DGGW1] D. M. DeTurck, C. S. Gordon, H. Gluck, and D. L. Webb, You cannot hear the mass of a homology class, *Comm. Math. Helv.* 64 (1989), 589–617.

[DGGW2] —, How can a drum change shape while sounding the same?, *to appear in DoCarmo Fetschrift.*

[DGGW3] —, How can a drum change shape while sounding the same II? *Mechanics, Analysis, and Geometry: 200 years after Lagrange* (North Holland).

[DGGW4] —, Inaudible geometry of nilmanifolds, preprint.

[DGGW5] —, Geometry of isospectral deformations, preprint.

[DGGW6] —, Conformal isospectral deformations, preprint.

[DG] D. M. DeTurck and C. S. Gordon, Isospectral deformations II: trace formulas, metrics and potentials, *Comm. Pure Appl. Math.* 42(1989), 1067–1095.

[Gi] P. Gilkey, On sperical space forms with meta-cyclic fundamental group which are isospectral but not equivariant cobordant, *Comp. Math.* 56 (1985), 171–200.

[G1] C. S. Gordon, When you can't hear the shape of a manifold, *Math. Intelligencer II*, no. 3 (1989), 39–47.

[G2] —, The Laplace spectra versus the length spectra of Riemannian manifolds, *Contemporary Mathematics* 51(1986), 63–79.

[G3] —, Riemannian manifolds isospectral on functions but not on 1-forms, *J. Differential Geometry* 24(1986), 79–96.

[GW1] C. S. Gordon and E. N. Wilson, Isospectral deformations of compact solvmanifolds, *Journal of Differential Geometry* 19(1984), 241–256.

[GW2] —, The spectrum of the Laplacian on Riemannian Heisenberg manifolds, *Mich. Math. J.* 33(1986), 253–271.

[I1] A. Ikeda, On lens spaces which are isospectral but not isometric, *Ann. Sci. Ecole. Norm. Sup.* 13(1980), 303–315.

I2] —, Riemannian manifolds p-isospectral but not $(p + 1)$-isospectral, preprint.

[I3] —, On sperical space forms which are isospectral but not isometric, *J. Math. Soc. Japan* 35(1983), 437–444.

M] J. Milnor, Eigenvalues of the Laplace Operator on certain manifolds, *Proc. Nat. Acad. Sci* USA 51(1964), 542.

Sp] R. J. Spatzier, On isospectral locally symmetric spaces and a theorem of Von Neumann, *Duke Math. J.* 59(1989), 289–294; Correction: *Duke Math. J.* 60 (1990), 561.

[S] T. Sunada, Riemannian coervings and isospectral manifolds, *Ann. Math.* 121 (1985), 169–186.

[V] M. F. Vignéras, Variétés Riemanniennes isospectrales et non isométriques, *Ann. math.* 112 (1980), 21–32.

[Z] S. Zelditch, On the generic spectrum of a Riemannian cover, *Ann. Inst. Fourier,* Grenoble 40, 2(1990), 407–442.

Department of Mathematics
Dartmouth College
Hanover, NH 03755

Kuznetsov Formulas

R. J. Miatello[*]

In [K] Kuznetsov proved a beautiful formula that gives a very explicit relationship between the Fourier coefficients of square integrable automorphic forms of weight 0 for the modular group, and Kloosterman sums (see also [Br1], [Br2]). In particular, Kuznetsov applied his formula to prove an important estimate on the average of Kloosterman sums (cf. [K], Theorem 3). The purpose of this note is to report on recent work (cf. [MW2], [MW3]) on generalizations of the formula to the case when Γ is a nonuniform, irreducible lattice in a product of semisimple **R**-rank one groups (see also [CPS]).

1. In this section we will assume that G is a real semisimple Lie group of **R**-rank one and Γ is a non uniform lattice satisfying Langlands assumptions ([L]). Let K be a maximal compact subgroup, and let $G = NAK$ be an Iwasawa decomposition of G. Let $\mathfrak{g} = \mathfrak{k} \oplus \mathfrak{a} \oplus \mathfrak{n}$ be the corresponding Lie algebra decomposition. If M is the centralizer of A in K, and $P = MAN$, let α be the simple root of (P, A) and let $\rho(H) = \frac{1}{2} tr\, ad(H)|\mathfrak{n}, H \in \mathfrak{a}$. Fix $H_o \in \mathfrak{a}$ such that $\alpha(H_o) = 1$. We will assume that P is Γ-percuspidal. Set $\Gamma_N = \Gamma \cap N$. Then $\Gamma_N \backslash N$ is compact and $\Gamma \cap P = \Gamma \cap MN$. If dg, dk are Haar measures on G, K with dk of mass one, we give to $X = G/K$ and $\Gamma \backslash X$ the canonical measures. Let Δ denote the Laplace-Beltrami operator. Then (cf. [L]) we have

$$L^2(\Gamma \backslash X) = L_d^2(\Gamma \backslash X) \oplus L_c^2(\Gamma \backslash X)$$

where the spectrum of Δ in $L_d^2(\Gamma \backslash X)$ (resp. $L_c^2(\Gamma \backslash X)$) is discrete (resp. continuous). Let ψ_j be an orthonormal basis of $L_d^2(\Gamma \backslash X)$ such that $\Delta \psi_j = \lambda(\nu_j)\psi_j$ where, if $\nu \in \mathbf{C}$, we set $\lambda(\nu) = (1 - \nu^2)\rho(H_o)^2$. Let $P_1, \ldots, P_n$ be a complete system of representatives for the Γ-conjugacy classes of Γ-cuspidal parabolic subgroups. Then $P_i = k_i P k_i^{-1}$, $k_i \in K$. Let $M_i = k_i M k_i^{-1}$, $A_i = k_i A k_i^{-1}$, $N_i = k_i N k_i^{-1}$ and let $\mathfrak{m}_i$, $\mathfrak{a}_i$, $\mathfrak{n}_i$ be the corresponding Lie algebras. If $a = exp(H)$, $H \in \mathfrak{a}_i$, and $\mu \in \mathfrak{a}_{i_c}^*$, set $a^\mu = e^{\mu(H)}$.

Let $E(P_i, \nu, g)$ denote the spherical Eisenstein series associated to P_i. That is, if $g = n_i(g)a_i(g)k_i(g)$ with $n_i(g) \in N, a_i(g) \in A, k_i(g) \in K$ and $\rho_i = \rho \circ Ad(k_i^{-1})$, then set, if $Re\,\nu > 1$,

$$E(P_i, \nu, g) = \sum_{\Gamma_{N_i} \backslash \Gamma} a_i(\gamma g)^{\nu \rho_i + \rho_i}.$$

If χ is a unitary character of N, trivial on $\Gamma \cap N$, we will write $\chi \in$

---
[*]Partially supported by CONICET and CONICOR, Argentina and ICTP, Italy.

$(\widehat{\Gamma_N \backslash N})$. We fix such a $\chi \neq 1$. Let ψ be an automorphic form on G/K, such that $\Delta \psi = \lambda(\nu)\psi$. Then (cf.[MW2])

$$\int_{\Gamma_N \backslash N} \chi(n_1)^{-1} \psi(n_1 nak) dn = \mathbf{c}_\chi(\psi) a^\rho K_\nu((aa_\chi)^\alpha)$$

where $K_\nu(x)$ denotes Mac Donald's K-Bessel function, and $a_\chi \in A$ depends only on χ.

The constant $\mathbf{c}_\chi(\psi)$ will be called the (P, χ)-Fourier coefficient of ψ. In the case when $\psi = E(Q, \nu, g)$, Q a Γ-cuspidal parabolic subgroup, we set $\mathbf{c}_\chi(E(Q, \nu, .)) = \mathbf{D}_\chi(P, Q, \nu)$.

Let $P = MAN$ be as above and let $N_G(A)$ denote the normalizer of A in G. Fix $s \in N_G(A) \cap K$, $s \notin M$. By the Bruhat lemma, $G = P \cup PsN$, a disjoint union, and any $g \in PsN$ can be written uniquely as a product $g = n_1(g) a_g m(g) s n_2(g)$ with $n_1(g), n_2(g) \in N$, $a_g \in A$ and $m(g) \in M$.

With this notation in place we now state the main result in [MW2] (see Thm.1.9).

1.1. Theorem. *Let G be a connected semisimple Lie group of real rank one and Γ a discrete subgroup of finite covolume. Fix $\chi, \chi' \in (\widehat{\Gamma_N \backslash N})$ and let $\varepsilon > 0, b > \rho(H_o)$. Let $f(\nu)$ be an even holomorphic function on the strip $S_{b+\varepsilon} = \{\nu : |Re(\nu)| < b + \varepsilon\}$ such that, if $\nu \in S_{b+\varepsilon}$, then*

$$|f(\nu)| \leq c \frac{e^{-\pi|Im(\nu)|}}{(1 + |\nu|)^{2+\delta}}$$

for some constants $c, \delta > 0$. We then have

$$\textbf{(K)} \quad \sum_{j=1}^{\infty} f(\nu_j) \overline{\mathbf{c}_\chi(\psi_j)} \mathbf{c}_{\chi'}(\psi_j) + \sum_{i=1}^{n} \int_{Re\nu=0} f(\nu) \overline{\mathbf{D}_\chi(P, P_i, \nu)} \mathbf{D}_{\chi'}(P, P_i, \nu) \, d\nu$$

$$= \frac{i}{\pi^2} \alpha_\chi \, \delta_{\chi,\chi'} \int_{Re\nu=0} f(\nu) \, \nu \sin\pi\nu \, d\nu +$$

$$\frac{1}{2\pi i} \sum_{\delta \in \Gamma_N \backslash \Gamma - \Gamma_P / \Gamma_N} \chi(n_1(\delta)) \chi(n_2(\delta)) \int_{Re\nu=0} f(\nu) b(\nu) \tau(\chi', \chi, m(\delta) a_\delta, \nu) \, d\nu$$

where $b(\nu) = 4 b_1^\nu \frac{c(\nu)}{\Gamma(\nu)^2}$, with $c(\nu)$ the Harish-Chandra c-function. The function $\tau(\chi', \chi, m(\delta) a_\delta, \nu)$ is a generalization of the classical Bessel functions. It is entire in ν and uniformly bounded in absolute value on closed vertical strips in $\{\nu \mid Re \nu > 0\}$. Furthermore b_1 is a positive real number depending only on χ, χ' and α_χ is a constant.

1.2. Remarks.

(i) Both sides of the formula converge absolutely for any f in the class. By choosing $f(\nu)$ appropriately, one obtains an estimate on the growth of $c_\chi(\psi_n)$, $n \in \mathbf{N}$ ([MW2], Corollary 1.10).

(ii) An entirely analogous formula holds if we use $\chi \in (\widehat{\Gamma_{N_Q} \backslash N_Q})$, $\chi' \in (\widehat{\Gamma_{N_{Q'}} \backslash N_{Q'}})$, Q, Q' two arbitrary Γ-cuspidal parabolic subgroups.

(iii) The function τ is computed explicitly in [MW] for $G = SO(n, 1), n \geq 2$ and for $G = SU(2, 1)$. As we shall see, when $G = SL(2, \mathbf{R})$ or $SL(2, \mathbf{C})$, it has a simple closed expression in terms of Bessel functions.

1.3. Example. We now indicate how to rewrite the second term in the right hand side of formula (**K**) in arithmetic terms, in the case when (G, Γ) is either $(SL(2, \mathbf{R}), SL(2, \mathbf{Z}))$ or $(SL(2, \mathbf{C}), SL(2, \mathbf{Z}[i]))$.

Let $G = SL(2, F), F = \mathbf{R}$ or $\mathbf{C}$; let $\mathcal{O}_F = \mathbf{Z}$, if $F = \mathbf{R}$, $\mathcal{O}_F = \mathbf{Z}[i]$ if $F = \mathbf{C}$.

If $x, z \in F, z \neq 0$, set $diag(z, z^{-1}) = \begin{bmatrix} z & 0 \\ 0 & z^{-1} \end{bmatrix}, n_x = \begin{bmatrix} 1 & x \\ 0 & 1 \end{bmatrix}$. Let

$$N = \{n_x : x \in F\}, A = \{diag(y, y^{-1}) : y > 0\}, M = \{diag(z, z^{-1}) : |z| = 1\}.$$

Let $P = MAN$ and $s = \begin{bmatrix} 0 & 1 \\ -1 & 0 \end{bmatrix}$. If $g \in G - P$, $g = \begin{bmatrix} a & b \\ c & d \end{bmatrix}, c \neq 0$, then $g = n_1(g) a_g m(g) s n_2(g)$ with $n_1(g) = n_{a/c}, n_2(g) = n_{d/c}, a_g = diag(|c|^{-1}, |c|)$, and $m(g) = diag(-(c/|c|)^{-1}, -c/|c|)$.

If $a = diag(y, y^{-1})$, then $a^\rho = y$. If $\chi \in \widehat{\Gamma_N \backslash N}$, then $\chi = \chi_r$ for some $r \in \mathcal{O}_F$, with $\chi_r(n_x) = e^{2\pi i d \, Re(rx)}$, where $d = \dim_\mathbf{R}(F)$.

A complete set of representatives for the double coset space $\Gamma_N \backslash \Gamma - \Gamma_P / \Gamma_N$ is given by

$$\left\{ \begin{bmatrix} a & b \\ c & d \end{bmatrix} : a, b, c, d \in \mathcal{O}_F, c \neq 0, (a, d) \in \mathcal{S}_c \right\}$$

where $\mathcal{S}_c = \{(a, d) \mid a, d \in \mathcal{O}_F \ a, d \ mod(c), ad \equiv 1 \ mod(c)\}$. If $c, r, r' \in \mathcal{O}_F, c \neq 0$, and $(\mathcal{O}_F / c\mathcal{O}_F)^*$ denotes the units in $\mathcal{O}_F / c\mathcal{O}_F$ define

$$S(r, r'; c) = \sum_{a \in (\mathcal{O}_F / c\mathcal{O}_F)^*} e^{2\pi i d \, Re(\frac{ra + r'a^{-1}}{c})}.$$

We note that when $F = \mathbf{R}$, $c > 0$ this is a classical Kloosterman sum. The computation of $b(\nu) \tau(\chi', \chi, x, \nu)$ in [MW2], Section 3, implies that the second summand in the right hand side of (**K**) is given by

$$\frac{2i}{\pi^2} \sum_{c>0} S(r, r'; c) \frac{1}{c} \int_{Re\nu=0} f(\nu) \nu \mathcal{J}_{2\nu}(4\pi (|rr'|)^{1/2}/c) \, d\nu$$

where $\mathcal{J}_{2\nu} = \begin{cases} I_{2\nu} & rr' < 0 \\ J_{2\nu} & rr' > 0 \end{cases}$. Here I_ν and J_ν are the classical I and J Bessel functions. It is now easy to see that (**K**) implies Kuznetsov's original formula (see [MW2] Corollary 3.1).

If $F = \mathbf{C}$, in light of the calculations above, and the computation of the function τ in [MW2], the second summand in the right hand side is given by

$$\frac{1}{2\pi i} \sum_{c \in \mathbf{Z}[\hat{\mathbf{i}}]^*} \frac{S(r, r'; c)}{|c|} \int_{\mathrm{Re}\nu=0} f(\nu)\nu\, \mathcal{I}_\nu(-4\pi(rr')^{1/2}/c)\, d\nu.$$

where

$$\mathcal{I}_\nu(z) = I_\nu(z)I_\nu(\bar{z}), \quad z \in \mathbf{C}.$$

We note here that $I_\nu(z)I_\nu(\bar{z})$ defines a continuous function on $\mathbf{C}$, though the individual factors are continuously defined only on $\mathbf{C} - \{t : t \leq 0\}$.

2. In [MW3] we proved that a formula analogous to (**K**) holds in the case when Γ is an irreducible, non uniform lattice in a product of semisimple **R**-rank one groups. In this section we will describe a very explicit version of a special case of this formula, in terms of generalizations of Kloosterman sums. Let F be a finite extension of $\mathbf{Q}$ and let $\mathcal{O}_F$ denote the ring of integers of F. Let $\mathbf{G} = R_{F/\mathbf{Q}}(\mathbf{SL(2)})$, the $\mathbf{Q}$-algebraic group defined by restriction of scalars ([Mo]).

Let $\mathcal{E}$ be the set of all field isomorphisms of F into $\mathbf{C}$ and set $\mathcal{E}_\mathbf{R} = \{\sigma \in \mathcal{E} \mid \sigma(F) \subset \mathbf{R}\}$. We write the infinite places of F out as the real places $\sigma_1, \ldots, \sigma_d$ and the complex places in complex conjugate pairs $\sigma_{d+1}, \overline{\sigma_{d+1}}, \ldots,$ $\sigma_{d+e}, \overline{\sigma_{d+e}}$. So the total number of infinite places is $d + 2e$. We have

$$\mathbf{G}_\mathbf{Q} = \{(x^{\sigma_1}, \ldots, x^{\sigma_d}, x^{\sigma_{d+1}}, \ldots, x^{\sigma_{d+e}}, x^{\overline{\sigma_{d+e}}}) \mid x \in SL(2, F)\}$$

$$\mathbf{G}_\mathbf{Z} = \{(x^{\sigma_1}, \ldots, x^{\sigma_d}, x^{\sigma_{d+1}}, \ldots, x^{\sigma_{d+e}}, x^{\overline{\sigma_{d+e}}}) \mid x \in SL(2, \mathcal{O}_F)\}.$$

We set $G = \mathbf{G}_\mathbf{R}^o$. Then if $r = d + e$ we have $G \cong \times_{i=1}^r G_i$, with $G_i = SL(2, \mathbf{R})$, if $1 \leq i \leq d$, and $G_i = SL(2, \mathbf{C})$, if $d+1 \leq i \leq r$. Now if $\Gamma = \mathbf{G}_\mathbf{Z} \cap G$, then Γ is an irreducible lattice in G and $\Gamma \backslash G$ is not compact. We note that if F is totally real ($e = 0$), then Γ is a Hilbert-Blumenthal group.

Choose

$$\mathbf{P}_1 = \left\{ \begin{bmatrix} a & b \\ 0 & a^{-1} \end{bmatrix} \mid a \in \mathbf{C}^\times, b \in \mathbf{C} \right\}.$$

Set $\mathbf{P} = R_{F/\mathbf{Q}}(\mathbf{P}_1)$ and $P = \mathbf{P}_\mathbf{R} \cap G$. Then $P = P_1 \times \cdots \times P_r$ with $P_i = \mathbf{P}_1 \cap SL(2, \mathbf{R})$ for $1 \leq i \leq d$ and $P_i = \{(x, \bar{x}) \mid x \in \mathbf{P}_1\}$ for $d+1 \leq i \leq r$. If $1 \leq i \leq d$, then let $M_i = \{\pm I\}$, $A_i = \{diag(a, a^{-1}) \mid a \in \mathbf{R}, a > 0\}$ and

$N_i = \mathbf{N}_1 \cap SL(2, \mathbf{R})$. If $d + 1 \leq i \leq r$, and if we set $\iota(x) = (x, \bar{x})$, then let $M_i = \iota\{diag(z, \bar{z}) | |z| = 1\}$, $A_i = \iota\{diag(a, a^{-1}) | a \in \mathbf{R}, a > 0\}$ and

$$N_i = \iota\left\{ \begin{bmatrix} 1 & b \\ 0 & 1 \end{bmatrix} \Big| b \in \mathbf{C} \right\}.$$

Then $P = MAN$ with $M = \times_i M_i$, $A = \times_i A_i$, $N = \times_i N_i$ is a Langlands decomposition of P.

We denote by α_i the simple root of (P_i, A_i). Let ρ_i be the "ρ" for (P_i, A_i). We identify $\mathfrak{a} = Lie(A)$ with $\times_{i=1}^r \mathfrak{a}_i$ and $\mathfrak{a}_C^*$ with $\times_{i=1}^r (\mathfrak{a}_i)_C^*$. An element in such a product of the form $(0, .., 0, \nu, 0, ...)$ will be denoted ν. Thus the set of simple roots of the pair P,A is $\{\alpha_1, ..., \alpha_r\}$ and $\rho_P = (\rho_1, ..., \rho_r) = \sum_i \rho_i$. We will often identify $\mathfrak{a}_C^*$ with $\mathbf{C}^r$ as follows. Let $H_i \in \mathfrak{a}_i$ be defined by $\alpha(H_i) = 1$. Then we identify ν with $(\nu(H_1), ..., \nu(H_r))$.

Furthermore we have a decomposition $A = {}^o A A_o$ with

$$^o A = \{(a_1, ..., a_d, \iota(a_1'), ..., \iota(a_e')) | \Pi_{i=1}^{i=d} a_i{}^\rho \, \Pi_{j=1}^{j=e} a_j'{}^{2\rho} = 1\}$$

$$A_o = \{(a, ..., a, \iota(a), ..., \iota(a)) \mid a \in A_1\}.$$

If we now set $^o P = M {}^o AN$ then we have $\Gamma \cap P = \Gamma \cap {}^o P$.

In general, if P' is a Γ-percuspidal parabolic subgroup of G with Langlands decomposition (over $\mathbf{R}$), $P' = M'A'N'$, then as above we have a splitting $A' = {}^o A' A_o'$ with A_o' a one dimensional vector subgroup. Furthermore, if $^o P' = M'({}^o A')N'$ then $(\Gamma \cap {}^o P') \backslash {}^o P$ is compact.

In the case at hand, one side of (K) will again be expressed in terms of the Fourier coefficients of eigenfunctions of the Laplacian on $\Gamma \backslash X$. Let $\{\psi_j\}$ be an orthonormal basis of square-integrable eigenfunctions of Δ in $L_d^2(\Gamma \backslash X)$ such that each ψ_j generates an irreducible $(\mathfrak{g}, K)$-submodule of $L_d^2(\Gamma \backslash X)$ which is a quotient of the space of K-finite vectors of the spherical principal P-series $H_K^{\overline{\mu}_j}$, where $\mu_j \in \mathfrak{a}_C^*$. If P' is a percuspidal subgroup of G, let $E(P', \nu, i\mu, g)$ denote the Eisenstein series as defined in [MW3]. Here $\mu \in L'$, a lattice in $^o \mathfrak{a}'^*$ defined as follows.

Let $\pi : {}^o P \to N' \backslash {}^o P'$ be the canonical projection. Notice that π is an isomorphism on $M'({}^o A')$ so we identify $N' \backslash {}^o P'$ with $M'({}^o A')$. Then $\pi(\Gamma \cap^o P')$ is a cocompact discrete subgroup of $M'({}^o A')$. Furthermore, $M' \pi(\Gamma \cap^o P')\backslash M'({}^o A') \cong \Lambda \backslash^o A'$ with Λ a lattice in $^o A'$. Then $L = log(\Lambda)$ is a lattice in $^o \mathfrak{a}'$. We set L' equal to the dual lattice. That is, $L' = \{\mu \in (^o \mathfrak{a}')^* | \mu(\mathbf{L}) \subset 2\pi \mathbf{Z}\}$. Hence L' is the character group of $M' \pi(\Gamma \cap^o P')\backslash M'({}^o A')$ under the map $ma \mapsto a^{i\mu}$.

The definition of (P, χ)-Fourier coefficient is entirely analogous to that in the rank one case (see [MW3],section 5). We will denote by $\mathbf{c}_\chi(\psi_j)$ (resp. $\mathbf{D}^\chi(P', \nu, i\mu)$) the (P, χ)-Fourier coefficient of ψ_j (resp. $E(P', \nu, i\mu, g)$).

Let $P^i, 1 \leq i \leq n$, be a complete system of representatives for the Γ-conjugacy classes of Γ-cuspidal parabolic subgroups. We will denote by L_i' the lattice L' defined for P^i in the same way as L' was above defined for P'.

Before we state our result, we will need to introduce some more notation. Each $\gamma \in \Gamma$ corresponds to

$$\begin{bmatrix} a(\gamma) & b(\gamma) \\ d(\gamma) & c(\gamma) \end{bmatrix} \in SL(2, F).$$

With our choice of N, if $\gamma \in \Gamma \cap N$ then $a(\gamma) = c(\gamma) = 1$ and $d(\gamma) = 0$. We consider characters of N of the form

$$\chi_r(n) = e^{2\pi i (\sum_{j=1}^{d} r^{\sigma_j} x_j + 2 \sum_{j=d+1}^{d+e} \operatorname{Re}(r^{\sigma_j} x_j))}$$

for $r \in \mathcal{O}_F$ and $n = (n_1, ..., n_{d+e})$ with $n_j = \begin{bmatrix} 1 & x_j \\ 0 & 1 \end{bmatrix}$ and $x_j \in \mathbf{R}$ if $j \le d$, $x_j \in \mathbf{C}$ if $j > d$. If $r, r', \in \mathcal{O}_F$, and $\delta \in \Gamma - \Gamma \cap P$ then a simple calculation yields

$$\chi_r(n_1(\delta))\chi_{r'}(n_2(\delta)) = \exp(2\pi i Tr_{F|\mathbf{Q}} \left(\frac{ra(\delta) + r'd(\delta)}{c(\delta)} \right)).$$

Also, if $r, r', c \in \mathcal{O}_F, c \ne 0$, define

$$S(r, r'; c) = \sum_{a \in (\mathcal{O}_F / c\mathcal{O}_F)^*} e^{2\pi i Tr_{F|\mathbf{Q}}(\frac{ra + r'a^{-1}}{c})}.$$

If $\delta \in \Gamma - \Gamma \cap P$ and $N(c) = Norm_{F|\mathbf{Q}}(c)$ for $c \in F$ then

$$a_\delta^\rho = \prod_{j \le d} |c(\delta)^{\sigma_j}|^{-1} \prod_{j > d} |c(\delta)^{\sigma_j}|^{-2} = |N(c(\delta))|^{-1}.$$

For each $j, 1 \le j \le d$, we set

$$\mathcal{J}_{2\nu_j} = \begin{cases} I_{2\nu_j} & (rr')^{\sigma_j} < 0 \\ J_{2\nu_j} & (rr')^{\sigma_j} > 0 \end{cases}.$$

Also, as in Section 1, let $\mathcal{I}_\nu(z) = I_\nu(z)I_\nu(\bar{z}), z \in \mathbf{C}$.

Finally we will describe the test functions to be used in the formula. Let $\mathbf{R}^+ = (0, \infty)$. If $\nu = (\nu_1, ..., \nu_t) \in \mathbf{C}^t$ and if $x \in (\mathbf{R}^+)^t$ then we set $K_\nu(x) = \prod_j K_{\nu_j}(x_j)$. Here, as above, K_ν is MacDonald's Bessel function. If $\phi \in C_c^\infty((\mathbf{R}^+)^t)$ then we define

$$K\phi(\nu) = \int_{(\mathbf{R}^+)^t} \phi(x)K_\nu(x)d^\times x \qquad (1)$$

with $d^\times x = (x_1 \cdots x_t)^{-1} dx_1 \cdots dx_t$. We will call $K\phi$ the Kantorovich-Lebedev transform of ϕ. If $\phi \in C_c^\infty((\mathbf{R}^+)^t)$ then $K\phi$ is holomorphic on

$\mathbf{C}^t$,even in each variable and for each $b > 0, n \in \mathbf{N}$ there exists a constant $C = C_{n,b}$ such that if $\nu \in S_b = \{\nu \in \mathbf{C}^t | \ |\nu_j| < b\}$ then

$$|K\phi(\nu)| \leq C \frac{e^{-\frac{\pi}{2}(\sum_{j=1}^{t} |Im\nu_j|)}}{(1 + \sum_{j=1}^{t} |\nu_j|)^n}. \tag{2}$$

We set $\mathcal{K} = \{K\phi | \phi \in C_c^\infty((\mathbf{R}^+)^t)\ \}$ and $\mathcal{K}^2 = \{\sum_{j=1}^{m} f_j g_j | f_j, g_j \in \mathcal{K}\}$.

We are now in a position to state a generalization of $(\mathbf{K})$ in the case at hand (cf. [MW3]).

2.1. Theorem. *Let* $\chi_r, \chi_{r'} \in (\widehat{\Gamma_N \backslash N}), r, r' \in \mathcal{O}_F - \{0\}$. *Let* $k \in \mathcal{K}^2$. *Then*

$$\sum_j k(\mu_j)\overline{\mathbf{c}_\chi(\psi_j)}\mathbf{c}_{\chi'}(\psi_j)$$

$$+ \sum_{i=1}^{n} \sum_{\mu \in L_j'} \int_{-\infty}^{\infty} k(i\nu + i\mu)\overline{\mathbf{D}_i^\chi(i\nu, i\mu)}\mathbf{D}_i^{\chi'}(i\nu, i\mu)d\nu$$

$$= \alpha(\chi, \chi') \int_{\mathbf{R}^r} k(i\nu) \prod_{j=1}^{r}(\nu_j \sinh \pi\nu_j)d\nu_1 \cdots d\nu_r + \frac{4^{d+e}}{\pi^d} \sum_{c \in \mathcal{O}_F^*} \frac{\overline{S(r, r'; c)}}{|N(c)|}$$

$$\times \int_{\mathbf{R}^r} k(i\nu) \prod_{j \leq d} \mathcal{J}_{2\nu}(-4\pi(rr')^{1/2}/c^{\sigma j}) \prod_{j > d} \mathcal{I}_\nu(-4\pi(rr')^{1/2}/c^{\sigma j})\, d\nu.$$

Furthermore if $\alpha(\chi, \chi') \neq 0$, *then* $r' = s^2 r$, *where* $s \in \mathcal{O}_F$ *is a unit.*

References

[Br1] Bruggeman, R., *Fourier coefficients of cusp forms*, Inventiones Math. 45, (1978), 1-18.

[Br2] Bruggeman,R.,*Fourier coefficients of automorphic forms*,Lecture Notes in Mathematics 865, Springer-Verlag,1981.

[CPS] Cogdell J., Piatetskii Shapiro I.,*The Arithmetic and Spectral Analysis of Poincaré Series*, Perspectives in Mathematics V.13, Ac. Press, 1990.

[GW] Goodman,R., Wallach,N.R.,*Whittaker vectors and conical vectors*, Jour.Funct.Anal. 39, (1980),199-279.

[K] Kuznetsov,N.V., *The conjecture of Petersson for forms of weight zero and the conjecture of Linnik*,Math.Sbornik 153(1980),334-389.

[L] Langlands,R.P.,*On the functional equations satisfied by Eisenstein series*,Lecture Notes in Mathematics 544,Springer-Verlag,1976.

[MW1] Miatello,R.,Wallach,N.R., *Automorphic forms constructed from Whittaker vectors*, Jour.Funct.Anal. 86(1989), 411-487.

[MW2] Miatello,R.,Wallach,N.R.,*Kuznetsov formulas for rank one groups,* Jour.Funct.Anal. 93, 1990, 171-207.

[MW3] Miatello, R., Wallach N.R., *Kuznetsov formulas for products of groups of* **R**-*rank one,* Israel Math. Conference Proceedings (volume in honor of I. Piatetskii-Shapiro), 305-321, 1990.

[Mo] Mostow,G., *Discrete subgroups of Lie groups,* Queen's Papers in Pure and Applied Mathematics 48,Proceedings of the Canadian Mathematical Congress, 1977.

[WW] Whittaker,E.T.,Watson,G.N., *Modern Analysis,* Cambridge University Press.

FAMAF, Univ.Nac. de Córdoba
C. Universitaria
5000 Córdoba, Argentina

Lefschetz Numbers and Cyclic Base Change for Purely Imaginary Extensions

by

Jurgen Rohlfs [1] and Birgit Speh [2]

Introduction: Let S be a locally symmetric space and V locally constant sheaf on S such that an automorphism σ of finite order of the underlying real Lie group G_∞ acts on S and V. Then a Lefschetz number $L(\sigma, S, V)$ of the induced σ-action on the cohomology $H^\cdot(S, V)$ is defined.

If S and V are given by sufficiently concrete local dae expects a closed formula for $L(\sigma, S, V)$ in terms of these local data. Moreover the Lefschetz number $L(\sigma, S, V)$ should have to do with the elliptic terms in a corresponding σ-twisted Arthur-Selberg trace formula. In addition there should exist a stabilisation of the formula for $L(\sigma, S, V)$ which would explain some examples (unpublished) where different $(S_i, V_i, \sigma_i), i = 1, 2$, lead to the same Lefschetz number.

We believe that all this can be established in full generalitt an example of general nature which supports this hope. The example is as follows:

Let l/k be totally imaginary and cyclic Galois extension of totally real number field k and $\sigma \in gal(l/k)$ a generator of the relative Galois group. Let G_0/k be a connected semi simple algebraic group defined over k and put $G = res_{l/\mathbb{Q}} G_0 \times l$ where $res_{l/\mathbb{Q}}$ denotes the Weil restriction. Then σ acts algebraically on G. Let $KK_f := K \Pi_p K_p \subset G(\mathbf{A})$ be a σ-stable compact subgroup where K is maximal compact in $G(\mathbb{R})$ and the K_p are open compact in the p-adic groups $G(\mathbb{Q}_p)$ and $\mathbf{A}$ denotes the adeles over $\mathbb{Q}$. We assume that KK_f acts freely on $G(\mathbf{A})/G(\mathbb{Q})$. Then $S(K_f) := KK_f \backslash G(\mathbf{A})/G(\mathbb{Q})$ is a locally symmetric manifold with σ-action. A finite dimensional complex representation of $G(\mathbb{R})$ with σ-action determines a sheaf again denoted by V on $S(K_f)$ and a Lefschetz number $L(\sigma, S(K_f), V)$ of the induced σ-action on the cohomology $H^\cdot(S(K_f), V)$ is defined. In this note we describe a formula $L(\sigma, S(K_f), V) = \varepsilon_0 S0(h_\infty, 1, \sigma) \Pi_p S0(h_p, 1, \sigma)$. Here ε_0 is zero or a sign and $S0(h_v, 1, \sigma)$ is the stable local σ-twisted orbital integral at the trivial σ-conjugacy class 1. The function h_p is the normalized characteristic funtion of K_p and h_∞ is the σ-discrete pseudo matrix coefficient introduced by Labesse [L]. This formula also is true without the assumption tha l is totally imaginary if G is simply connected. Generali-

[1]Supported by Deutsche Forschungsgemeinschaft
[2]Supported by N.S.F. grant Nr. DMS 880128

sations and details will appear in [R - S 2].

§1. A Lefschetz fixpoint formula and non abelian cohomology

We use the notation as introduced above and add some further explana-
tions. We assume that the finite dimensional complex representation V of
$G(\mathbb{R})$ is algebraic, i.e. we assume that V extends to a representation of
$G(\mathbb{C})$. The σ-action on V is given by a $\mathbb{C}$-linear map $\sigma : V \to V$ such that
$\sigma(gv) = \sigma(g)\,\sigma(v)$ for all $g \in G(\mathbb{C})$, $v \in V$. S $\dim_{\mathbb{C}} H^{\cdot}(S(K_f), V) < \infty$
the trace $tr(\sigma^i)$ of the map $\sigma^i : H^i(S(K_f), V) \to H^i(S(S_f), V)$ which is
induced by σ makes sense and by definition

$$L(\sigma, S(K_f), V) := \sum_{i=0}^{\infty} (-1)^i \, tr \, \sigma^i \, .$$

This number is called *the Lefschetz number of σ* on $S(K_f)$ and V. If G is
simply connected then $S(K_f) = K \backslash G(\mathbb{R})/\Gamma$ where $\Gamma = G(\mathbb{Q}) \cap K_f$ and
$S(K_f)$ is a connected locally symmetric space. In general $S(K_f)$ is a finite
disjoint union of such spaces.

Let $S(K_f)^\sigma$ be the set of fixpoints of σ in $S(K_f)$ and let $V|S(K_f)^\sigma$ be the
restriction of the sheaf V to the fixpoint set. Then σ acts on the fibres of
$V|S(K_f)^\sigma$ and the Lefschetz number $L(\sigma, S(K_f)^\sigma, V|S(K_f)^\sigma)$ of σ acting
on

$$H^{\cdot}(S(K_f)^\sigma, V|(K_f)^\sigma)$$

makes sense. One has, see [R 3]:

Proposition 1.1. *(Lefschetz fixpoint formula)*

$$L(\sigma, S(K_f), V) = L(\sigma, S(K_f)^\sigma, V|S(K_f)^\sigma) \, .$$

Next we describe $S(K_f)^\sigma$. For this we recall the notion of the non
abelian first cohomology $H^1(\sigma, H)$ of an action of σ on a group H, see [S].
By definition the set $H^1(\sigma, H)$ consists of classes of cocycles. A cocycle is
an element $a \in H$ such that $a \cdot \sigma(a) \ldots \sigma^i(a) = 1$, $i+1 = |<\sigma>|$. Cocycles
a and b are equivalent if there is a $c \in H$ such that $c^{-1}a\sigma(c) = b$. For a
cocycle a the a-twisted σ-action σ_a is given by $\sigma_a(h) = a\,\sigma(h)\,a^{-1}, h \in H$.

If $a \in G(\mathbf{A})$ represents a double coset in $S(K_f)^\sigma = (KK_p/G(Q))^\sigma$ then
$\sigma(a) = c^{-1}a\,\gamma$ for uniquely determined $c \in KK_f$ and $\gamma \in G(\mathbb{Q})$ and c rep.
γ determine classes in $H^1(\sigma, KK_f)$ resp. $H^1(\sigma, G(\mathbb{Q}))$ which are mapped
to the same class in $H^1(\sigma, G(\mathbf{A}))$ by the maps induced by the inclusions

$KK_f \subset G(A)$ resp. $G(\mathbb{Q}) \to G(A)$. Therefore there is a natural surjection ∂ to the fibre product

$$(KK_f \setminus G(A)/G(\mathbb{Q}))^\sigma \xrightarrow{\partial} H^1(\sigma, KK_f) \prod_{H^1(\sigma, G(A))} H^1(\sigma, G(\mathbb{Q})).$$

If b is an element of the fibre product we denote the fibre of ∂ over b by $F(b)$.

We describe $F(b)$. For this we assume that b is represented by (c, γ) and $\sigma(a) = c^{-1}a\gamma$. Let $G(\gamma)$ be the group such that $G(\gamma)(A) = \{g \in G(A) | g = \gamma\sigma(g)\gamma^{-1}\}$ for any $\mathbb{Q}$-algebra Similarly $KK_f(c) = \{x \in KK_f \setminus x = c\sigma(x)c^{-1}\}$. Then one has

$$F(b) \xrightarrow{\sim} a^{-1}(KK_f)(c)a \setminus G(\gamma)(A)$$

We introduce a γ-twisted σ-action σ_γ on V by $\sigma_\gamma(v) = \gamma\sigma(v), v \in V$. Then the trace $tr(\sigma_\gamma|V)$ of the σ_γ-action on V makes sense and this number depends only on b and is denoted by $tr(\sigma_b|V)$. We arrive at the following:

Proposition 1.2. *With the notation given above one has*

$$L(\sigma, S(K_f), V) = \sum_{\substack{b \in H^1(\sigma, KK_f) \prod H^1(\sigma, G(\mathbb{Q})) \\ H^1(\sigma, G(A))}} \chi(F(b)) \, tr(\sigma_b|V)$$

where $\chi(F(b))$ is the Euler-Poincaré characteristic of the space $F(b)$.

§2. The summation of the fixpoint contributions

We use an adelic version of Harder's Gaus-Bonnet formula [H] to compute $\chi(F(b))$. For this we need some notation.

Let $\omega = \prod \omega_v = \omega_\infty \prod_p p = \omega_\infty \omega_f$ be a fixed Tamagawa measure on $G(A)$. Take dk to be the normalized Haar measure on K and let dx be the corresponding measure on $X := K \setminus G(\mathbb{R})$. Let $\mathfrak{g} = \mathfrak{k} \oplus \mathfrak{p}$ be the Cartan decomposition of the Lie algebra $\mathfrak{g}$ of $G(\mathbb{R})$ which corresponds to K and let G_u be the compact connected subgroup of $G(\mathbb{C})$ corresponding to the Lie subalgebra $\mathfrak{k} \oplus i\mathfrak{p}$ of $\mathfrak{g} \otimes_{\mathbb{R}} \mathbb{C}$. Then there is an embedding $X \longrightarrow X_u$ of X into its compact dual $X_u = K^0 \setminus G_u$, where K^0 is the connected component of K. With respect to the natural Riemannian metrics of X and X_u this embedding is an isometry at the point x_0 which corresponds to K. Therefore dx induces a natural G_u left-invariant measure du on X_u by the requirement that dx and du coincide at $x_0 \in X \subset X_u$, where both measures now are concidered as differential forms. Hence $vol(G_u) := vol(X_u)$ is well defined and depends only on ω_∞.

If $G(\mathbb{R})$ contains a compact Cartan subgroup T we put $W_{\mathbb{R}} = N_{G(\mathbb{R})}(T)/T$ where $N_{G(\mathbb{R})}(T)$ is the normalizer of T in $G(\mathbb{R})$. By $W_{\mathbb{C}}$ we denote the complex Weyl group, i.e. the Weyl group of the Lie algebra $\mathfrak{g} \otimes \mathbb{C}$

2.1. Harder proved: If $G(\mathbb{R})$ has no compact Cartan subgroup then $\chi(KK_f \setminus G(A)/G(\mathbb{Q})) = 0$. If $G(\mathbb{R})$ has a compact Cartan subgroup then

$$\chi(KK_f \setminus G(A)/G(\mathbb{Q})) = (-1)^{dimX/2} \frac{|W_{\mathbb{C}}|}{|W_{\mathbb{R}}|} vol(G_u)^{-1} \tau(G) \, vol_{\omega_f}(K_f)^{-1}$$

where $\tau(G)$ is the Tamagawa number of G.

2.2. We recall that $H^1(\sigma, K) \xrightarrow{\sim} H^1(\sigma, G(\mathbb{R}))$, see [R 1]. Hence the fibre of the second projection

$$H^1(\sigma, KK_f) \prod_{H^1(\sigma, G(A))} H^1(\sigma, G(\mathbb{Q})) \longrightarrow H^1(\sigma, G, (\mathbb{Q}))$$

over a class represented by $\gamma \in G(\mathbb{Q})$ can be identified with a subset $H^1(\sigma, K_f)(\gamma)$ of $H^1(\sigma, K_f)$. The fixpoint contribution corresponding to such a subset can be expressed in terms of orbital integrals.

For a compactly supported continuous function h_p on $G(\mathbb{Q}_p)$ we put

$$0(h_p, \gamma, \sigma) = \int_{G(\gamma)(\mathbb{Q}_p)\setminus G(\mathbb{Q}_p)} h_p(y^{-1}\gamma\, ^\sigma y)dy$$

and call this the *σ-twisted orbital integral of h_p at the σ-conjugacy class γ*. Here we assume that dy is the quotient measure determined by the choosen local components of the Tamagawa measures on G and $G(\gamma)$.

We choose h_p to be such that $h_p(x) = vol(K_p)^{-1}$ if $x \in K_p$ and $h_p(x) = 0$ if $x \notin K_p$. Then we arrive at a formula

$$L(\sigma, S(K_f), V) = \sum_{\gamma \in H^1(\sigma, G(\mathbb{Q}))} a(\gamma, \infty) \, \tau(G(\gamma)) \prod_p 0(h_p, \gamma, \sigma)$$

where $a(\gamma, \infty) \in \mathbb{R}$ depends only on the image of γ in $H^1(\sigma, G(\mathbb{C}))$ and it is determined by Harder's Gaus-Bonnet formula.

2.3. Kottwitz [K 2] has explained how to stabilize the singular elliptic contributions to the trace formula. We follow his approach. There are simplifications because $H^1(\sigma, G(\bar{\mathbb{Q}})) = \{1\}$ where $\bar{\mathbb{Q}}$ is an algebraic closure of $\mathbb{Q}$. This holds since $G(\bar{\mathbb{Q}})$ is an induced group with σ-action.

For $c_p \in G(\mathbb{Q}_p)$ representing a class in $H^1(\sigma, G(\mathbb{Q}_p))$ we denote by $e(c_p)$ the Kottwitz sign, [K 1], of the group of fixpoints $G(c_p)/\mathbb{Q}_p$ of the c_p-twisted σ-action. We put

$$S0(h_p, 1, \sigma) := \sum_{c_p \in H^1(\sigma, G(\bar{\mathbb{Q}}_p))} e(c_p)\, 0(h_p, c_p, \sigma)$$

and call this the σ-twisted stable orbital integral of h_p at 1. For the occurring orbital integrals we have to use compatible measures, see [K 3]. To explain the notation we recall that for an e $\bar{\mathbb{Q}}_p$ of $\mathbb{Q}_p$ one has $H^1(\sigma, G(\bar{\mathbb{Q}}_p)) = \{1\}$.

In principle one has to expect stable κ-orbital in, in the formula for $L(\sigma, S(K_f), V)$. However a careful analysis of the contributions of the place ∞ of $\mathbb{Q}$ shows that only $\kappa = 1$ contributes. For this we use the assumption that l is totally imaginary. We point out that $L(\sigma, S(K_f), V) = 0$ if k is not totally real. So we have

$$L(\sigma, S(K_f), V) = \varepsilon_0\, tr(\sigma|V)\, 2^{l_0}\, vol(G_u(\sigma))^{-1} \prod_p S0(h_p, 1, \sigma)\,.$$

Here ε_0 is zero if $G(\sigma)(\mathbb{R})$ does not contain a compact Cartan subgroup and $\varepsilon_0 = (-1)^{dim\, X_0(\sigma)/2}$ otherwise where $X_0(\sigma)$ is the symmetric space associated to the quasi split inner $\mathbb{Q}$-form $G_0(\sigma)$ of $G(\sigma)$. Morever l_0 is the dimension of a compact Cartan subgroup of $G(\sigma)(\mathbb{R})$. Of course $G_u(\sigma)$ is the compact dual of $G(\sigma)(\mathbb{R})$.

2.4. It is possible to write the "real" factor in the formula for $L(\sigma, S(K_f), V)$ as a stable orbital integral. We thank Bouaziz, Labesse, Clozel and Delorme for helpful remarks on this.

We recall that Labesse [L] has constructed a compactly supported C^∞-function h_∞ of $G(\mathbb{R})$ such that

$$tr(\pi(h_\infty \times \sigma)) = L(\sigma, \pi \otimes V)$$

for all irreducible representations π of $G(\mathbb{R})$. Here delicate choices of equivalences $\pi \xrightarrow{\sim} \pi \circ \sigma$ are involved. On the left we have the Lefschetz number of the σ-action on the relative Lie algebra cohomology $H^{\cdot}(\mathfrak{g}, K, \pi \otimes V$ The stable σ-twisted orbital integrals of the function h_∞ can be determined using results of Bouaziz, [B]. We obtain

$$S0(h_\infty, 1, \sigma) = tr(\sigma|V) 2^{l_0}\, vol(G_u(\sigma))^{-1}$$

if we use compatible measures for the real groups. These are naturally determid by the infinite component of our fixed Tamagawa measures on G and $G(\sigma)$. So finally we can state our main result. For this we use the notation which has been introduced up to now, and put $h = h_\infty \prod_p h_p$ and

$$S0(h, 1, \sigma) = \prod_v S0(h_v, 1, \sigma)\,.$$

2.5. Theorem. *In the notation introduced above we have*

$$L(\sigma, S(K_f), V) = \varepsilon_0 \, S0(h, 1, \sigma) \, .$$

We indicate a special situation where this formula simplifies. To formulate this we introduce the notion K_f is *small enough*. This means by definition that the natural map

$$H^1(\sigma, KK_f) \prod_{H^1(\sigma, G(\mathbf{A}))} H^1(\sigma, G(\mathbb{Q})) \longrightarrow H^1(\sigma, G(\mathbf{A}))$$

has an image contained in $H^1(\sigma, G(\mathbb{R})) \times 1... \times 1... \subset H^1(\sigma, G(\mathbf{A}))$. It can be shown that K_f contains a congruence subgroup which is small enough, [R3].

2.6. Corollary. *If K_f is small enough then*

$$L(\sigma, S(K_f), V) = \varepsilon_0 2^{l_0} |H^1(\sigma, K_f)(1)| \, tr(\sigma|V) \, vol(G_u(\sigma))^{-1} \prod_p vol(K_p^\sigma)^{-1}$$

where K_p^σ are the fixpoints of σ acting on K_p.

We observe that by Kneser's Theorem $H^1(\sigma, G(\mathbb{Q}_p)) = \{1\}$ if G is simply connected. Hence if G is simply connected then all K_f are small enough. In the summation then instead of 2^{l_0} the Lefschetz number $L(\sigma, X_u) \geq 0$ of the σ-action on the cohomology $H^\cdot(X_u, \mathbb{C})$ of the compact dual X_u occurs. One obtains:

2.7. Proposition. *Assume that G is simply connected such that $G(\mathbb{R})$ is non compact and that $l|k$ is an arbitrary cyclic extension. Then*

$$L(\sigma, S(K_f), V)$$
$$= \varepsilon_0 \, L(\sigma, X_u) \, |H^1(\sigma, K_f)| \, tr(\sigma|V) \, vol(G_u(\sigma))^{-1} \prod_p vol(K_p^\sigma)^{-1} \, .$$

References

[B] A. Bouaziz, Formule d'inversion d'intégrals orbitales tordue, manuscript, Paris 1989.

[D] D. Delorme, Theoreme de Paley-Wiener invariant tordu pour le changement de base $\mathbb{C}|\mathbb{R}$, manuscript, Marseille, 1990.

[C] L. Clozel, Representations galoisiennes associées aux représentations auto-morphes autoduales de $GL(n)$, manuscript, Paris, 1990.

[H] G. Harder, A Gaus-Bonnet theorem for discrete arithmetically defined groups, Ann. Scient. Éc. Norm. Sup., 4e série, 1971, 409 - 455.

[K 1] R. Kottwitz, Sign changes in harmonic analysis on reductive groups, Transactions A.M.S. vol. 278, 1983, 289 - 297.

[K 2] R. Kottwitz, Stable trace formula, Elliptic singular terms, Math. Ann. vol. 275, 1986, 365 - 399.

[K 3] R. Kottwitz, Tamagawa numbers, Ann. of Math. vol. 127, 1988, 629 - 646.

[L] J.P. Labesse, Formule des traces tordue et représentations σ-discrètes, manuscript, Paris, 1989.

[R 1] J. Rohlfs, Arithmetisch definierte Gruppen mit Galois operation Invent. math. 48, 1978, 185 - 205.

[R 2] J. Rohlfs, The Lefschetz number of an involution on the space of classes of positive definite quadratic forms, Comment. Math. Helvetici 56 1981, 272 - 296.

[R 3] J. Rohlfs, Lefschetz numbers for arithmetic groups I, in preparation.

[R - Sp 1] J. Rohlfs and B. Speh, Automorphic representations and Lefschetz numbers, Ann. Scient. Éc. Norm. Sup. 4e série, t. 22, 1989, 473 - 499.

[R - Sp 2] J. Rohlfs and B. Speh, Lefschetz numbers for arithmetic groups II, in preparation.

[S] J.P. Serre, *Cohomologie Galoisienne,* Lecture Notes in Math. Vol. 5, 1965, Springer Verlag, Berlin, Heidelberg, New York.

J. Rohlfs
Katholische Universitat Eichstatt
Ostenstr. 26 - 28
8078 Eichstatt
Fed. Rep. of Germany

B. Speh
Department of Mathematics
Cornell University
Ithaca, N.Y. 14853
U.S.A.

Some Zeta Functions Attached to Γ\G/K

Floyd L. Williams *

1. Introduction

In an epochal paper [28] Selberg introduced his famous trace formula, and to a lattice Γ in $G = \mathrm{SL}(2, R)$ and a finite-dimensional unitary representation T of Γ he assigned a certain remarkable analytic function $Z_\Gamma(\cdot, T)$ (of one complex variable) whose zeros, for example, capture both topological and spectral properties of the space form $\Gamma \setminus G/K$ where $K = \mathrm{SO}(2)$. $Z_\Gamma(\cdot, T)$, now called the *Selberg zeta function*, satisfies a functional equation $s \to 1 - s$ (involving Harish–Chandra's c-function) and, up to finite exceptions involving the possible occurrence of *complementary series* representations of G in $L^2(\Gamma \setminus G)$, $Z_\Gamma(\cdot, T)$ satisfies a *Riemann hypothesis*: its "nontrivial" zeros have real part equal $\frac{1}{2}$.

Twenty years after [28] Gangolli, using the trace formula, showed that a Selberg zeta function $Z_\Gamma(\cdot, T)$ could be attached to an arbitrary compact space form $X_\Gamma = \Gamma \setminus G/K$ of a rank 1 symmetric space G/K. This work [11] was extended to the context of more general nonuniform lattices by Gangolli and Warner in [14].

One can also attach to X_Γ a certain Minakshisundaram–Pleijel type zeta function $s \to D_\Gamma(s; b)$, where $b > 0$ is a fixed parameter. One of the results presented in this paper is an explicit meromorphic continuation of $D_\Gamma(\; ; b)$; see Theorem 4.10 which extends the work of Randol [23] who focused on the case $G = \mathrm{SL}(2, R)$. In particular if Δ_Γ is the Laplace–Beltrami operator on X_Γ we can define, by "zeta regularization", the *determinant* $\det(-\Delta_\Gamma + b1)$. The latter, as a function of b, is in fact related to the Gangolli–Selberg zeta function $Z_\Gamma(\cdot, 1)$; cf. Theorem 5.6 which is a tentative result. Stronger results along this line abound in the case $G = \mathrm{SL}(2, R)$ and are useful in mathematical physics; cf. [1], [4], [5], [8], [18], [19], [24], [25], [26], [30], [31].[1]

Theorem 4.10 leads to an integral formula (formula (4.14)) for the multiplicity in $L^2(\Gamma \setminus G)$ of a class 1 unitary representation π of G.

We also consider a family $\{Z_\Gamma(\cdot; \alpha, b)\}_{\substack{\alpha > 0 \\ b \geq 0}}$ of "periodic" zeta functions

---

* This research is supported by NSF Grant No. DMS-8802597.

[1] Added in proof: a stronger and more explicit version of Theorem 5.6 is now available in [37]; also see [38].

attached to X_Γ. These functions are studied in [2] in joint work with Miss Barchini. In case $G = \mathrm{SL}(2, R)$, $\Gamma = \mathrm{SL}(2, Z)$, $b = 0$, $Z_\Gamma(\cdot; \alpha, 0)$ is the function $s \to \sum_{r_j > 0} \frac{\sin(\alpha r_j)}{r_j^s}$ (Re $s > 1$) introduced by Fujii [9], where $\lambda_j = \frac{1}{4} + r_j^2$ varies over the discrete spectrum of $-\Delta_\Gamma$ on $L^2(\Gamma \backslash$ upper $1/2$-plane). In Theorem 6.7 we formulate a general version of Fujii's *limit formula* for the special value $Z_\Gamma(0; \alpha, 0)$.

2. Some Notation

Throughout, G will denote a connected noncompact simple Lie group with finite center, and K a maximal compact subgroup of G. Fix a co-compact torsion free discrete subgroup Γ of G and let $X_\Gamma = \Gamma \backslash G / K$. We assume the rank of the symmetric space G/K is 1. Then in an Iwasawa decomposition KA_PN of G the Lie algebra n_0 of N is a sum of (real) restricted root spaces g_α:

$$(2.1) \qquad n_0 = \sum_{\alpha \in \Sigma^+} g_\alpha \text{ where } \Sigma^+ = \{\beta\} \text{ or } \{\beta, 2\beta\};$$

and we may normalize a choice of basis vector H_0 of the Lie algebra a_p of A_P (the abelian part of G) by

$$(2.2) \qquad \beta(H_0) = 1.$$

Let θ be the Cartan involution of G, let g_0, g denote the Lie algebra of G and its complexification $g_0^{\mathbb{C}}$, respectively, let $a \supset a_p$ be a θ-stable Cartan subalgebra of g_0, and let Φ^+ be an a_p-compatible system of positive roots in the set of nonzero roots Φ of $(g, a^{\mathbb{C}})$. Then we can take $\Sigma^+ = \{\alpha|_{a_p} \mid \alpha \in P^+\}$ where

$$(2.3) \qquad P^+ = \{\alpha \in \Phi^+ \mid \alpha \not\equiv 0 \text{ on } a_p\}.$$

We set

$$(2.4) \qquad \rho = \frac{1}{2} \sum_{\alpha \in P^+} \alpha, \qquad \rho_0 = \rho(H_0)$$

$$a_p^{*,\mathbb{C}} = \text{space of } R\text{-linear maps } a_p \to \mathbb{C}.$$

Let $H: G \to a_p$ be the smooth map defined by the Iwasawa decomposition of G: for each $x \in G$

$$(2.5) \qquad x = k(x) \exp H(x) n(x) \in KA_PN.$$

If

$$(2.6) \qquad \begin{aligned} p &= \left|\{\alpha \in P^+ |\alpha|_{a_p} = \beta\}\right| \\ q &= \left|\{\alpha \in P^+ |\alpha|_{a_p} \neq \beta\}\right| \end{aligned}$$

then $\Sigma^+ = \{\beta\}$ in (2.1) $\Leftrightarrow q = 0$. Also in (2.4), $\rho_0 = \frac{1}{2}(p+2q)$ and Harish–Chandra's c-function for the spherical Plancherel measure on G/K is given by

(2.7) $\qquad c(v)^{-1} = \dfrac{\Gamma\left(\frac{p+q}{2}\right)\Gamma\left(iv(H_0) + \frac{p}{2}\right)\Gamma\left(\frac{iv(H_0)}{2} + \frac{p}{4} + \frac{q}{2}\right)}{\Gamma(p+q)\Gamma(iv(H_0))\Gamma\left(\frac{iv(H_0)}{2} + \frac{p}{4}\right)}$

for $v \in a_p^{*,\mathbb{C}}$, for a suitable normalization of measures. Namely, dk always will denote normalized Haar measure on K; Haar measures dn, da, dx, dv on N, A_p, G, a_p^* will be normalized by

(2.8)
$$\int_N e^{-2\rho(H(\theta n))}\,dn = 1, \qquad \int_{A_p} f(a)\,da = \int_R f(\exp t H_0)\,dt$$
$$\int_G h(x)\,dx = \int_N \int_{A_p} \int_K h(kan)e^{2\rho(\log a)}\,dk\,da\,dn$$
$$\int_{a_p^*} \psi(v)\,dv = \frac{1}{2\pi}\int_R \psi(t\beta)\,dt \qquad (dt = \text{Lebesgue measure})$$

for $(f, h, \psi) \in C_c(A_p) \times C_c(G) \times C_c(a_p^*)$. Given these normalizations of Haar measures (which differs from that in [21]), Miatello's computation of the Plancherel density $|c(r)|^{-2} = c(r)^{-1}c(-r)^{-1}$ (for $c(r) = c(r\beta)$, $r \in \mathbb{C}$) takes the following form, where G is represented up to a local isomorphism.

(2.9)
$$|c(r)|^{-2} = C_G \pi r P(r)\tanh \pi r \quad \text{for } G = SO_1(2n, 1)$$
$$= C_G \pi r P(r)\begin{bmatrix} \tanh \frac{\pi r}{2} \\ \text{or} \\ \coth \frac{\pi r}{2} \end{bmatrix} \quad \text{for } G = SU(n, 1)$$
$$= C_G \pi r P(r)\tanh \frac{\pi r}{2} \quad \text{for } G = Sp(n, 1) \text{ or } F_{4(-20)}$$
$$= C_G \pi P(r) \quad \text{for } G = SO_1(2n + 1, 1)$$

where for $d = \dim G/K$, $G \neq SO_1(2n+1, 1)$, $P(r)$ is an even polynomial of degree $d - 2$; for $G = SO_1(2n + 1, 1)$, $P(r)$ is an even polynomial of degree $d - 1 = 2n$; C_G is a constant which depends only on G. C_G, $P(r)$ are given by

(2.10)

$$C_G = \frac{1}{2^{4n-4}\Gamma(n)^2}$$

(i)
$$P(r) = \prod_{j=2}^{n}\left(r^2 + \left(n - j + \frac{1}{2}\right)^2\right) \qquad \text{for } G = SO_1(2n, 1),\ (n \geq 2)$$

$$C_G = \frac{1}{2^{2n-2} \cdot 2\Gamma(n)^2}$$

(ii)

$$P(r) = \prod_{j=1}^{n-1} \left(\left(\frac{r}{2}\right)^2 + \frac{(n-2j)^2}{4} \right) \quad \text{for } G = \mathrm{SU}(n,1), (n \geq 2)$$

$$C_G = \frac{1}{2^{4n} \cdot 2\Gamma(2n)^2} \quad \text{for } G = \mathrm{SP}(n,1), (n \geq 2)$$

(iii)

$$P(r) = \prod_{j=3}^{n+1} \left[\left(\frac{r}{2}\right)^2 + \left(n-j+\frac{3}{2}\right)^2 \right] \cdot$$

$$\left[\left(\frac{r}{2}\right)^2 + \left(n-j+\frac{5}{2}\right)^2 \right] \left[\left(\frac{r}{2}\right)^2 + \left(\frac{1}{2}\right)^2 \right]$$

$$C_G = \frac{1}{2^{20} \cdot 2\Gamma(8)^2} \quad \text{for } G = F_{4(-20)}$$

(iv)

$$P(r) = \left[\left(\frac{r}{2}\right)^2 + \left(\frac{1}{2}\right)^2 \right]^2 \left[\left(\frac{r}{2}\right)^2 + \left(\frac{3}{2}\right)^2 \right]^2 \left[\left(\frac{r}{2}\right)^2 + \left(\frac{5}{2}\right)^2 \right] \cdot$$

$$\left[\left(\frac{r}{2}\right)^2 + \left(\frac{7}{2}\right)^2 \right] \left[\left(\frac{r}{2}\right)^2 + \left(\frac{9}{2}\right)^2 \right]$$

$$C_G = \frac{1}{2^{4n-2}\Gamma\left(n+\frac{1}{2}\right)^2} \quad \text{for } G = \mathrm{SO}_1(2n+1,1), (n \geq 1)$$

(v)

$$P(r) = \prod_{j=1}^{n} (r^2 + (n-j)^2);$$

see [20], [21]. We can now define the *Miatello coefficients* a_{2j} of G as follows:

$$P(r) = a_0 + a_2 r^2 + a_4 r^4 + \cdots + a_{2(\frac{d}{2}-1)} r^{2(\frac{d}{2}-1)}$$

(2.11)   for $G \neq \mathrm{SO}_1(2n+1,1)$, with $P(r)$ given by (2.10)((i)–(iv)), $d = \dim G/K$;

$$P(r) = a_0 + a_2 r^2 + a_4 r^4 + \cdots + a_{2n} r^{2n}$$

(2.11′)   for $G = \mathrm{SO}_1(2n+1,1)$, with $P(r)$ given by (2.10)(v); $2n = d - 1$.

Next we consider some notation on which the trace formula is based. As Γ is co-compact and torsion free, each $\gamma \in \Gamma - \{1\}$ is hyperbolic and can be expressed uniquely by an integer $j(\gamma) \geq 1$ and a primitive element $\delta \in \Gamma - \{1\}$: $\gamma = \delta^{j(\gamma)}$ where δ cannot be expressed as γ_1^j for some $\gamma_1 \in \Gamma$ $j > 1$; see [3], [12]. Moreover we can find $t_\gamma > 0$ such that γ is G-conjugate to $m_\gamma \exp t_\gamma H_0$ for some $m_\gamma \in M$, the centralizer of A_p in K; m_γ is unique up to conjugation in M. By [33] t_γ can be determined by

$$
\begin{aligned}
& e^{t_\gamma} = \max\{|c| \mid c = \text{an eigenvalue of } \mathrm{Ad}(\gamma) \colon g \to g\} \\
& \quad \text{if } \Sigma^+ = \{\beta\}; \\
& e^{t_\gamma} = \max\{|c|^{1/2} \mid c = \text{an eigenvalue of } \mathrm{Ad}(\gamma) \colon g \to g\} \\
& \quad \text{if } \Sigma^+ = \{\beta, 2\beta\}.
\end{aligned}
$$

(2.12)

Define the function C on $\Gamma - \{1\}$ by

$$
(2.13) \qquad C(\gamma)^{-1} = e^{t_\gamma \rho_0} \left| \det_{n_0}(\mathrm{Ad}(m_\gamma \exp t_\gamma H_0)^{-1} - 1) \right|.
$$

Finally, let C_Γ be a complete set of representatives in Γ of its conjugacy classes. If one wished, one could choose C_Γ such that

$$
(2.14) \qquad C_\Gamma - \{1\} = \bigcup_{\delta \in P_\Gamma} \{\delta^j \mid j \geq 1\}
$$

where P_Γ is a complete set of representatives in Γ for the conjugacy classes of primitive elements of $\Gamma - \{1\}$.

3. The Gangolli–Selberg Zeta Function

Given Haar measure dx on G let m_Γ be the unique G-invariant Radon measure on $\Gamma \backslash G$ such that for each $h \in C_c(G)$

$$
(3.1) \qquad \int_G h(x)dx = \int_{\Gamma \backslash G} \left(\sum_{\gamma \in \Gamma} h(\gamma x) \right) dm_\Gamma(\Gamma x).
$$

For $\pi \in \hat{G}$ (the set of equivalence classes of irreducible unitary representations of G) let $m_\pi(\Gamma)$ be the multiplicity of π in the right regular representation R_Γ of G on $L^2(\Gamma \backslash G, m_\Gamma)$, since m_Γ is G-invariant; $0 \leq m_\pi(\Gamma) < \infty$. $\pi \in \hat{G}$ is *spherical* (or class 1 with respect to K) if $\pi\big|_K$ contains the trivial representation of K. Let $\hat{G}_1$ be the subset of spherical classes in $\hat{G}$. Each $\pi \in \hat{G}_1$ is determined by its corresponding spherical function ϕ_π or, alternately, by the Harish–Chandra parameter $\nu_\pi \in a_p^{*,\mathbb{C}}$ of ϕ_π [15]. That is, for $x \in G$

$$
(3.2) \qquad \phi_\pi(x) = \langle v, \pi(x)v \rangle = \int_K e^{(i\nu_\pi - \rho)(H(xk))} dk
$$

where v is a $\pi(K)$-fixed unit vector in the representation space H_π of π; $\langle\,,\,\rangle$ is the inner product on H_π; cf. (2.4), (2.5). Let $\{\pi_j\}_{j\geq 0}$ represent all the classes of spherical representations which occur in R_Γ: $m_{\pi_j}(\Gamma) > 0$. For convenience we shall write $n_j = m_{\pi_j}(\Gamma)$ and $\nu_j = \nu_{\pi_j}$ for the spherical parameter of π_j, and we choose the labeling so that π_0 is the trivial representation 1 of G; thus in (3.2), $\nu_j = i\rho$. We may also set $\lambda_j \overset{\text{def}}{=} \nu_j(H_0) + \rho_0^2$ and arrange the order $0 = \lambda_0 < \lambda_1 < \lambda_2 < \cdots$. Then we remark that, for a canonical Riemannian metric on G/K pushed down to X_Γ, $\{\lambda_j\}_{j\geq 0}$ is the spectrum of the Laplacian $-\Delta_\Gamma$; and in fact n_j is the multiplicity of the eigenvalue λ_j on $C^\infty(X_\Gamma)$.[1]

With the preceding notation in place we can now define a θ-function θ_Γ,

(3.3)

$$\theta_\Gamma(t) = \sum_{j\geq 0} n_j e^{-\lambda_j t} - \frac{\text{vol}(\Gamma \setminus G)}{4\pi} \int_R e^{-(r^2+\rho_0^2)t}|c(r)|^{-2}dr \text{ for } t > 0,$$

$$\text{vol}(\Gamma \setminus G) = \int_{\Gamma \setminus G} 1\,dm_\Gamma,$$ and a complex-valued transform ψ_Γ of θ_Γ

(3.4)
$$\psi_\Gamma(s) = \kappa 2(s - \rho_0) \int_0^\infty e^{-s(s-2\rho_0)t}\theta_\Gamma(t)dt,$$

for Re s sufficiently large where κ is a constant.

Theorem 3.5. (Gangolli [11]). (i) *The integral in* (3.4) *converges for* $\text{Re}\,s > 2\rho_0$. ψ_Γ *defined thereby is holomorphic and nonzero.* (ii) ψ_Γ *admits a meromorphic continuation to the whole plane* $\mathbb{C}$ *and all of its poles are simple.* (iii) $\exists$ *a choice of the constant* κ, *a choice which depends only on* G, K, *such that every residue to* ψ_Γ *is an integer;* κ *is a positive integer.*

(iii), which involves the Gauss Bonnet theorem for X_Γ (say $G \neq SO_1(2n + 1, 1)$ so that $\dim X_\Gamma$ is even) is very delicate. Given (ii), (iii) one has by general principles

Corollary 3.6. *There exists a meromorphic function* Z_Γ *unique up to a constant such that* $Z'_\Gamma/Z_\Gamma = \psi_\Gamma$.

Z_Γ suitably normalized (for example we choose $Z_\Gamma(\rho_0) = 1$ if every $\nu_j \neq 0$) is the *Gangolli–Selberg zeta function* attached to X_Γ. In the notation of the introduction $Z_\Gamma = Z_\Gamma(\cdot, 1)$. Thus we have chosen the trivial representation 1 of Γ for our purposes rather than a general finite-dimensional

---

[1] $n_0 = 1$.

unitary representation T of Γ. Z_Γ satisfies the following functional equation $s \to 2\rho_0 - s$:

(3.7) $Z_\Gamma(2\rho_0 - s) = e^{\beta_\Gamma(s)} Z_\Gamma(s)$ (for $G \neq SO_1(2n+1, 1)$)

where

$$\beta_\Gamma(s) = \kappa \operatorname{vol}(\Gamma \backslash G) \int_{C_s} c(ir)^{-1} c(-ir)^{-1} dr$$

for any curve C_s from 0 to $s - \rho_0$ not passing through a pole of $z \to c(iz)^{-1}c(-iz)$; the latter function has integer residues and thus β_Γ is well-defined. A very beautiful property of the zeta function Z_Γ is that it provides for the *arithmetic interpretation* of the class 1 spectrum. Namely, Gangolli shows that, up to the factor κ, the multiplicity $n_j \overset{\text{def}}{=} m_{\pi_j}(\Gamma)$ coincides with the multiplicity of the order of a ("nontrivial") zero $s_j^+ = \rho_0 + i\nu_j(H_0)$ of Z_Γ. A much more thorough discussion of Z_Γ, including the Riemann hypothesis, is presented in [35] (where some trivial errors in [11] are corrected). Also see [6], [7], [14], [16], [25], [27], [28], [29], [30], [31], [32]. It is interesting to note in passing that even for $G = SL(2, k)$, k a locally compact field under a discrete valuation, an analogue of the Selberg zeta function Z_Γ is available—due to Ihara [17].

4. The Zeta Function

$D_\Gamma(\cdot, b)$. Using the notation of Section 3 we set

(4.1) $$D_\Gamma(s, b) = \sum_{j \geq 0}^{\infty} \frac{n_j}{(b + \lambda_j)^s}$$

for $b > 0$ and $\operatorname{Re} s$ sufficiently large.

Theorem 4.2. (Corollary of Gangolli–Warner [13], Wallach [34]). *The series in* (4.1) *converges absolutely for* $\operatorname{Re} s > \frac{d}{2}$, $d = \dim G/K$.

$D(s, b)$ is an analogue of the Minakshisundaram–Pleijel series [22]. The latter series, in our context, is

(4.3) $$D_0(s) = \sum_{j=1}^{\infty} \frac{n_j}{\lambda_j^s}$$

which, by Theorem 4.2, also converges absolutely for $\operatorname{Re} s > \frac{d}{2}$. One of the main points of this paper is a result on the *explicit* meromorphic continuation of $s \to D_\Gamma(s, b)$ to $\mathbb{C}$. Towards this end we consider the following definitions, recalling the notation of Section 2. K_s denotes the K-Bessel function:

(4.4) $$K_s(z) = \frac{1}{2} \int_0^{\infty} e^{-z/2(t+(1/t))} t^{s-1} dt$$

for $s, z \in \mathbb{C}$, $\text{Re}\, z > 0$. We define T_Γ by

(4.5)
$$T_\Gamma(s,b) = \frac{2}{\sqrt{4\pi}\left[2\sqrt{b+\rho_0^2}\right]^{s-1/2}} \sum_{\gamma \in C_\Gamma - \{1\}} j(\gamma)^{-1} \cdot$$
$$\cdot\, C(\gamma) t_\gamma^{s+(1/2)} K_{-s+(1/2)}\left(t_\gamma \sqrt{b+\rho_0^2}\right)$$

for $s \in \mathbb{C}$, $b > 0$. Similarly for $s \in \mathbb{C}$, $b > 0$, $a \in R$, $n = 0, 1, 2, 3, \ldots$ we set

(4.6)
$$K_n^\epsilon(s,b,a) = \left[\begin{array}{ll} \displaystyle\int_R \frac{t^{2n}\text{sech}^2 at\, dt}{(b+\rho_0^2+t)^s} & \text{for } \epsilon = 1 \\[2ex] \displaystyle\int_R \frac{t^{2n}\text{csch}^2 at\, dt}{(b+\rho_0^2+t^2)^s} & \text{for } \epsilon = -1 \end{array}\right].$$

Theorem 4.7. $s \to T_\Gamma(s,b)$ and $s \to K_n^\epsilon(s,b,a)$ are entire functions; see [36].

Next, for constants $\epsilon(G)$, $C(G)$, $a(G)$ which depend only on G, which we specify below, we set

(4.8)
$$\tilde{T}_\Gamma(s,b) = \frac{\epsilon(G)C(G)\text{vol}(\Gamma \backslash G)}{4\pi} \sum_{j=0}^{d/2-1} \sum_{\ell=0}^{j} \cdot$$
$$\cdot\, \frac{a_{2j}\, j!\, K_{j-\ell}^{\epsilon(G)}(s-\ell-1,b,a(G))}{(j-\ell)!(s-1)(s-2)\cdots(s-(\ell+1))}$$

for $s \in \mathbb{C}$, $b > 0$, where (for $G \neq SO_1(2n+1,1)$) the a_{2j} are the Miatello coefficients given by (2.11). $\epsilon(G)$, $C(G)$, $a(G)$, ρ_0 are given by the following table.

Table 1

G (up to local isomorphism)	$\epsilon(G)$	$a(G)$	$C(G)$	ρ_0
$SO_1(2n,1)$ $(n \geq 2)$	1	π	$a(G)\frac{\pi}{2}C_G$	$n - \frac{1}{2}$
$SU(n,1)$ $(n \geq 2)$	$(-1)^{n+1}$	$\frac{\pi}{2}$	$a(G)\frac{\pi}{2}C_G$	n
$Sp(n,1)$ $(n \geq 2)$	1	$\frac{\pi}{2}$	$a(G)\frac{\pi}{2}C_G$	$2n+1$
$F_{4(-20)}$	1	$\frac{\pi}{2}$	$a(G)\frac{\pi}{2}C_G$	11

where C_G is given by (2.10) ((i)–(iv)). By Theorem 4.7, $s \to \tilde{T}_\Gamma(s,b)$ is meromorphic with at most $\frac{d}{2}$ poles, each of which is simple. (4.5) and (4.8) provide for the following *main definition*:

(4.9)
$$\phi_\Gamma(s,b) = \tilde{T}_\Gamma(s,b) + \frac{T_\Gamma(s,b)}{\Gamma(s)}$$

for $s \in \mathbb{C}$, $b > 0$. By Theorem 4.7 and the preceding remark, $s \to \phi_\Gamma(s, b)$ is meromorphic with at most $\frac{d}{2}$ poles, each of which is simple and of the form $s = n$ an integer, $1 \leq n \leq \frac{d}{2}$. In particular we see that $s \to \phi_\Gamma(s, b)$ is *regular* at $s = 0$.

Theorem 4.10. *Assume G is not locally isomorphic to $SO_1(2n+1, 1)$ (we shall consider the latter case below).[1] Then for the series in (4.1) we have $D_\Gamma(s, b) = \phi_\Gamma(s, b)$ for $\operatorname{Re} s > \frac{d}{2}$, $b > 0$; $d = \dim G/K$.*

Thus Theorem 4.10 provides for the explicit meromorphic continuation of $s \to D_\Gamma(s, b)$ to the whole plane $\mathbb{C}$. This result is proved in [36], in case $b = 1$. The proof given there, based on the Selberg trace formula, clearly works for any $b > 0$. In [36] we similarly explicate the meromorphic continuation of $s \to D_0(s)$ in (4.3) and thus extend the result Randol obtained for $G = SL(2, R)$ in [23].

In case G is locally isomorphic to $SO_1(2n + 1, 1)$, the statement of Theorem 4.10 is modified as follows. In place of definition (4.8) we set

$$(4.11) \qquad \tilde{T}_\Gamma'(s, b) = \frac{C_G \operatorname{vol}(\Gamma \backslash G)}{4\pi (b + \rho_0^2)^s} \sum_{j=0}^{n} a_{2j} \frac{\Gamma\left(j + \frac{1}{2}\right) \Gamma\left(s - j - \frac{1}{2}\right)}{(b + \rho_0^2)^{-j-1/2}}$$

where

$$\prod_{j=1}^{n}(r^2 + (n - j)^2) = \sum_{j=0}^{n} a_{2j} r^{2n}$$

(as in (2.11)'), $\rho_0^2 = n^2$, and $C_G = \frac{\pi}{2^{4n-2}\Gamma(n+\frac{1}{2})^2}$.

Theorem 4.12. $D_\Gamma(s, b) = \tilde{T}_\Gamma'(s, b) + \frac{Tr(s,b)}{\Gamma(s)} \overset{\text{def}}{=} \phi_\Gamma(s, b)$ *for $b > 0$, $\operatorname{Re} s > \frac{d}{2} = n + \frac{1}{2}$, G locally isomorphic to $SO_1(2n + 1, 1)$. In particular $s \to \phi_\Gamma(s, b)$ is regular at $s = 0$.*

One immediate application of Theorems 4.10, and 4.12 yields an integral formula for class 1 multiplicities in $L^2(\Gamma \backslash G)$. Namely

Theorem 4.13. *Let $\pi \in \hat{G}_1$ with Harish–Chandra parameter $\nu_\pi \in a_p^{*,\mathbb{C}}$, as in (3.2). Then for $b > 0$, $\sigma > \frac{1}{2}\dim G/K$*

$$(4.14) \qquad m_\pi(\Gamma) = \lim_{T \to \infty} \frac{1}{2T} \int_{-T}^{T} [b + \rho_0^2 + \nu_\pi(H_0)^2]^{\sigma+it} \phi_\Gamma(\sigma + it, b) dt.$$

(4.14) is a simple consequence of a formula of Hadamard for the coefficients of a Dirichlet series—for example the series (4.1); see [36].

---

[1] Also if $G = SU(n, 1)$, assume (for technical reasons) n is odd.

5. Z_Γ and Φ_Γ Related

The meromorphic functions Z_Γ, ϕ_Γ of Sections 3 and 4 at first glance would appear to be unrelated. However, as we point out in this section, they are indeed related. To discover such a relationship we consider first a definition of determinant of the operator $-\Delta_\Gamma + b1$, say for $b > 0$. By the remarks preceding (3.3) we would want something like $\prod_{j=0}^{\infty}(\lambda_j + b)^{n_j}$ for the determinant. On the other hand if one formally differentiates with respect to s in (4.1) and sets $s = 0$ (which is invalid, as the series there diverges for $s = 0$) one obtains

$$e^{-\frac{\partial}{\partial s}D_\Gamma(s,b)}\Big|_{(0,b)} = b\prod_{j=1}^{\infty}(\lambda_j + b)^{n_j},$$

using $n_0 = 1$, $\lambda_0 = 0$. It is now clear how to arrive at a valid definition. Namely given Theorems 4.10, 4.12 we set (for $b > 0$)

$$(5.1) \qquad \det(-\Delta_\Gamma + b1) \stackrel{\text{def}}{=} e^{-\frac{\partial}{\partial s}\phi_\Gamma(s,b)}\Big|_{(0,b)},$$

having observed earlier that $s \to \phi_\Gamma(s,b)$ is regular at $s = 0$ (the "zeta regularization" of which we spoke in the introduction). Note that

$$\frac{\partial}{\partial s}\frac{T_\Gamma(s,b)}{\Gamma(s)}\Big|_{(0,b)} \stackrel{\text{(iv)}}{=} T_\Gamma(0,b),$$

using $\frac{1}{\Gamma(s)}\big|_{s=0} = 0$, $\frac{d}{ds}\frac{1}{\Gamma(s)}\big|_{s=0} = 1$, where

$$T_\Gamma(0,b) \stackrel{\text{(ii)}}{=} \sum_{\gamma \in C_\Gamma - \{1\}} j(\gamma)^{-1}C(\gamma)e^{-t_\gamma\sqrt{b+\rho_0^2}}$$

by (4.5), using $K_{1/2}(y) = e^{-y}\sqrt{\frac{\pi}{2y}}$ for $y > 0$. On the other hand by way of the trace formula we have the alternate description of ψ_Γ: For $\operatorname{Re} s > 2\rho_0$

$$(5.2) \qquad \psi_\Gamma(s) = \kappa \sum_{\gamma \in C_\Gamma - \{1\}} j(\gamma)^{-1}t_\gamma C(\gamma)e^{(\rho_0 - s)t_\gamma},$$

where by Proposition 2.7 of [11] the series here converges absolutely, and the convergence is uniform in s in any $\frac{1}{2}$-plane of the form $\operatorname{Re} s \geq 2\rho_0 + \delta$, $\delta > 0$. It follows that for x, $x_0 > 2\rho_0$

$$(5.3) \qquad \begin{aligned} \int_{x_0}^{x}\psi_\Gamma(t)dt &= \kappa \sum_{\gamma \in C_\Gamma - \{1\}} j(\gamma)^{-1}t_\gamma C(\gamma)\int_{x_0}^{x}e^{(\rho_0 - t)t_\gamma}dt \\ &= -\kappa \sum_{\gamma \in C_\Gamma - \{1\}} j(\gamma)^{-1}C(\gamma)\left[e^{(\rho_0 - x)t_\gamma} - e^{(\rho_0 - x_0)t_\gamma}\right] \\ &= -\kappa[T_\Gamma(0, x(x - 2\rho_0)) - T_\Gamma(0, x_0(x_0 - 2\rho_0))], \end{aligned}$$

by (ii). By the fundamental theorem of calculus, $x \to T_\Gamma(0, x(x - 2\rho_0))$ is differentiable for $x > 2\rho_0$ and

(5.4) $\qquad -\kappa \dfrac{d}{dx} T_\Gamma(0, x(x - 2\rho_0)) = \psi_\Gamma(x) \quad \text{for } x > 2\rho_0.$

By definition $Z_\Gamma'(x)/Z_\Gamma(x) = \psi_\Gamma(x)$ for $x > 2\rho_0$ so that we derive

(5.5) $\qquad Z_\Gamma(x) = Z_\Gamma(x_0) e^{-\kappa[T_\Gamma(0, x(x - 2\rho_0)) - T_\Gamma(0, x_0(x_0 - 2\rho_0))]}$

for $x > 2\rho_0$, where by (4.9) and (i) (say G is not locally isomorphic to $SO_1(2n + 1, 1)$) we have

$$-\kappa T_\Gamma(0, x(x - 2\rho_0)) = -\kappa \frac{\partial}{\partial s}[\phi_\Gamma(s, x(x - 2\rho_0))$$
$$- \tilde{T}_\Gamma(s, x(s - 2\rho_0))]|_{0, x(x - 2\rho_0)},$$

and where by (5.1)

$$e^{-\kappa \frac{\partial}{\partial s} \psi_\Gamma(s, x(x - 2\rho_0))} = [\det(-\Delta_\Gamma + x(x - 2\rho_0)1)]^\kappa.$$

We have therefore the following result which relates the Gangolli–Selberg zeta function Z_Γ and the "zeta" function ϕ_Γ.

Theorem 5.6. *Fix $x_0 > 2\rho_0$ arbitrary. Then for any $x > 2\rho_0$,*

$$Z_\Gamma(x) = Z_\Gamma(x_0) e^{-\kappa \frac{\partial \tilde{T}_\Gamma}{\partial s}(0, x_0(x_0 - 2\rho_0))} e^{\kappa \frac{\partial \phi_\Gamma}{\partial s}(0, x_0(x_0 - 2\rho_0))}.$$
$$\cdot e^{\kappa \frac{\partial \tilde{T}_\Gamma}{\partial s}(0, x(x - 2\rho_0))} e^{-\kappa \frac{\partial \phi_\Gamma}{\partial s}(0, x(x - 2\rho_0))}$$

for G non-locally isomorphic to $SO_1(2n+1, 1)$. [2] If G is locally isomorphic to $SO_1(2n+1, 1)$ then $\tilde{T}_\Gamma$ (defined in (4.8)) is replaced by $\tilde{T}_\Gamma'$ in (4.11) (with ϕ_Γ defined analogously as in Theorem 4.12). By (5.1),

$$e^{-\kappa \frac{\partial \phi_\Gamma}{\partial s}(0, x(x - 2\rho_0))} = [\det(-\Delta_\Gamma + x(x - 2\rho_0))]^\kappa.$$

One could improve Theorems 4.10 and 5.6 by evaluating the integrals in (4.6). So far attempts by the author to determine these integrals have been unsuccessful.[3]

---

[2] For $G = SO_1(2n, 1)$, $\kappa = (2n - 1)^n$; for $G = SO_1(2n + 1, 1)$ we may take $\kappa = 1$.

[3] As indicated in footnote 1 of the introduction, Theorem 5.6 is now successfully completed and explicated in [37].

6. The Family $\{Z_\Gamma(\ ;\alpha,b)\}_{\substack{\alpha>0 \\ b\geq 0}}$

In [2] the author with L. Barchini studied a family of "periodic" zeta functions attached to X_Γ given by[4]

$$(6.1) \qquad Z_\Gamma(\ ;a,b): s \to \sum_{\substack{j\theta \\ \nu_j(H_0)>0}} \frac{n_j \nu_j(H_0) \sin \alpha \nu_j(H_0)}{(b+\nu_j(H_0)^2)^{(s+1)/2}}$$

for $\operatorname{Re} s$ sufficiently large, where $\alpha > 0$, $b \geq 0$ are fixed. As pointed out in the introduction, we obtain Fujii's zeta function [9], [10] in case $b = 0$, $G = \operatorname{SL}(2,R)$ and Γ is the modular group (though $\Gamma \backslash G$ in this case is noncompact). Similar to Theorem 4.2 we have

Proposition 6.2. *The series in* (6.1) *converges absolutely for* $\operatorname{Re} s > \dim G/K$.

Theorem 6.3. $Z_\Gamma(\ ;\alpha,b)$, *defined initially in the* $\frac{1}{2}$-*plane* $\operatorname{Re} s > \dim G/K$, *extends to an entire function.*

The proof of Theorem 6.3 is a bit long and intricate [2] (even for $G = \operatorname{SL}(2,R)$). If one were only interested in the meromorphic continuation of $Z_\Gamma(\ ;\alpha,b)$ simpler arguments would suffice.

We define a *von Mangoldt function* $\tilde{\Lambda}$ on $\Gamma - \{1\}$ by

$$(6.4) \qquad \tilde{\Lambda}(\gamma) = e^{t_\gamma/2} j(\gamma)^{-1} t_\gamma C(\gamma).$$

In case $G = \operatorname{SL}(2,R)$, $\tilde{\Lambda}$ reduces to the usual von Mangoldt function for the Selberg zeta function:

$$(6.5) \qquad \tilde{\Lambda}(\gamma) = \frac{\log N(\delta)}{1 - N(\gamma)^{-1}} \quad \text{for } \gamma = \delta^{j(\gamma)},$$

δ primitive, $N(\gamma) = \max |c|^2$; $c = $ an eigenvalue of γ; we appeal to the notation of Section 2.

In general we set

$$(6.6) \qquad N(\gamma) = e^{t_\gamma}, \quad \gamma \in \Gamma - \{1\}.$$

Given definitions (6.4), and (6.6) we have the following higher dimensional version of Fujii's limit formula; cf. [10]

Theorem 6.7. *For any* $\gamma_1 \in \Gamma - \{1\}$

$$\lim_{\alpha \to \log N(\gamma_1)} (\alpha - \log N(\gamma_1)) Z_\Gamma(0;\alpha,0) = \frac{1}{2} \sum_{\substack{\gamma \in C_\Gamma - \{1\} \\ N(\gamma)=N(\gamma_1)}} \tilde{\Lambda}(\gamma)/\sqrt{N(\gamma)}.$$

The proof of Theorem 6.7 has not been written up yet.

---

[4] Recall the notation of Section 3.

The results of this paper (and much of its notation) are based on the Selberg trace formula. The reader may consult [6], [12], [13], [16], [33], [35]. This paper and [36] are dedicated to the memory of Marshall H. Stone.

REFERENCES

1. L. Alvarez-Gaume', G. Moore, C. Vafa, *Theta functions, modular invariance, and strings*, Comm. Math. Physics **106** (1986), 1–40.

2. L. Barchini, F. Williams, *Periodic zeta functions for rank 1 space forms of symmetric spaces*, 1989, Hiroshima Math. J. **22**(1992), 26–48.

3. D. De George, *Length spectrum for compact locally symmetric spaces of strictly negative curvature*, Ann. École Norm. Sup. **10** (1977), 133–152.

4. E. D'Hoker, D. Phong, *Multiloop amplitudes for the bosonic Polyakov string*, Nuclear Physics **B269** (1986), 205–234.

5. I. Efrat, *Determinant of Laplacians on surfaces of finite volume*, Comm. Math. Physics **119** (1988), 443–451.

6. J. Fischer, *An approach to the Selberg trace formula via the Selberg zeta function*, Lecture Notes in Math. **1253**, Springer-Verlag, 1987.

7. D. Fried, *The zeta function of Ruelle and Selberg I*, Ann. École Norm. Sup. **19** (1986), 491–517.

8. D. Fried, *Analytic torsion and closed geodesics on hyperbolic manifolds*, Inven. Math. **84** (1986), 523–540.

9. A. Fujii, *A zeta function connected with the eigenvalues of the Laplace–Beltrami operator on the fundamental domain of the modular group*, Nagoya Math. J. **96** (1984), 167–174.

10. A. Fujii, *Arithmetic of some zeta function connected with the eigenvalues of the Laplace–Beltrami operator*, in *Investigations in Number Theory*, Advanced Studies in Pure Math. 13, Academic Press, 1988, 237–260.

11. R. Gangolli, *Zeta functions of Selberg's type for compact space forms of symmetric spaces of rank one*, Ill. J. Math. **21** (1977), 1–42.

12. R. Gangolli, G. Warner, *On Selberg's trace formula*, J. Math. Soc. Japan **27** (1975), 328–343.

14. R. Gangolli, G. Warner, *Zeta functions of Selberg's type for some noncompact quotients of symmetric spaces of rank one*, Nagoya Math. J. **78** (1980), 1–44.

15. Harish-Chandra, *Representations of semisimple Lie groups II*, Trans. Am. Math. Soc. **76** (1954), 26–65.

16. D. Hejhal, *The Selberg trace formula for* SL(2, R), I, Lecture Notes in Math. **548**, Springer-Verlag, 1976.

17. Y. Ihara, *Discrete subgroups of* PL(2, k_ρ), in *Algebraic Groups and Discontinuous Subgroups*, Proc. Sympos. Pure Math., Boulder, Colorado, 1965, 272–278, Am. Math. Soc. Pub., 1966.

18. S. Koyama, *Determinant expression of Selberg zeta functions I*, Trans. A.M.S. **324** (1991), 149–168.

19. S. Koyama, *A note on the determinant of the Laplacian*, preprint, 1989.

20. R. Miatello, *The Minakshisundaram–Pleijel coefficients for the vector-valued heat kernel on compact locally symmetric spaces of negative curvature*, Thesis, Appendix 2, Rutgers Univ., 1976; also see Trans. Am. Math. Soc. **260** (1980), 1–33.

21. R. Miatello, *On the Plancherel measure for linear Lie groups of rank one*, Manuscripta Math. **29** (1979), 249–276.

22. S. Minakshisundaram, A. Pleijel, *Some properties of the eigenfunctions of the Laplace operator on Riemannian manifolds*, Canadian J. Math. **1** (1949), 242–256.

23. B. Randol, *On the analytic continuation of the Minakshisundaram–Pleijel zeta function for compact Riemann surfaces*, Trans. Am. Math. Soc. **201** (1975), 241–246.

24. P. Sarnak, *Determinants of Laplacians*, Comm. Math. Physics **110** (1987), 113–120.

25. P. Sarnak, *Special values of Selberg's zeta function*, in *Number Theory, Trace Formulas and Discrete Groups*, Oslo Symposium in honor of Atle Selberg, Academic Press, 1987, 457–465.

26. P. Sarnak, *Determinant of Laplacians; heights and finiteness*, preprint, 1989.

27. D. Scott, *Selberg type zeta functions for the group of complex two by two matrices of determinant one*, Math. Ann. **253** (1980), 177–194.

28. A. Selberg, *Harmonic analysis on discontinuous groups in weakly symmetric Riemannian spaces with applications to Dirichlet series*, J. Indian Math. Soc. **20** (1956), 47–87.

29. A. Sitaram, *A Selberg type zeta function and the representations of* SL(2, R), Thesis, Univ. Washington, 1975.

30. M-F. Vignéras, *L'equation fonctionnelle de la fonction zêta de Selberg du groupe modulaire* PSL(2, Z), in Journées Arithmétiques de Luminy, Soc. Math. de France, Astérisque **61** (1979), 235–249.

31. A. Voros, *Spectral functions, special functions and the Selberg zeta function*, Comm. Math. Physics **110** (1987), 439–465.

32. M. Wakayama, *Zeta functions of Selberg's type associated with homogeneous vector bundles*, Hiroshima Math. J. **15** (1985), 235–295.

33. N. Wallach, *On the Selberg trace formula in the case of compact quotient*, Bull. Am. Math. Soc. **82** (1976), 171–195.

34. N. Wallach, *On an asymptotic formula of Gelfand and Gangolli for the spectrum of* $\Gamma \setminus G$, J. Diff. Geometry **11** (1976), 91–101.

35. F. Williams, *Lectures on the spectrum of* $L^2(\Gamma \setminus G)$, Trabajos de Math., Instituto de Math., Astronomía y Física, Univ. Nacional de Córboda, Argentina, 1989.

36. F. Williams, *Formula for the class 1 spectrum and the meromorphic continuation of Minakshisundaram–Pleijel series*, preprint, 1989.

37. F. Williams, *A factorization of the Selberg zeta function attached to a rank 1 space form*, preprint 1990.
38. F. Williams, *On certain definite integrals which arise in automorphic Lie theory*, preprint 1990.
39. F. Williams, *Lectures on the spectrum of $L^2(\Gamma \setminus G)$*, Pitman Research Notes in Mathematics Series, Vol. 242, Longman House Pub., 1990.

Floyd L. Williams
Department of Mathematics and Statistics
University of Massachusetts
Amherst, Massachusetts 01003

ON THE CENTRALIZER OF K IN THE UNIVERSAL ENVELOPING ALGEBRA OF $SO(n,1)$ AND $SU(n,1)$

JUAN A. TIRAO

INTRODUCTION

Let G_o be a non compact real semisimple Lie group with finite center, and let K_o denote a maximal compact subgroup of G_o. If $\mathfrak{k} \subset \mathfrak{g}$ denote the respective complexified Lie algebras, let $U(\mathfrak{g})$ be the universal enveloping algebra of $\mathfrak{g}$ and let $U(\mathfrak{g})^K$ denote the centralizer of K_o in $U(\mathfrak{g})$.

To study $U(\mathfrak{g})^K$, B. Kostant suggested to consider the projection map $P : U(\mathfrak{g}) \to U(\mathfrak{k}) \otimes U(\mathfrak{a})$, corresponding to the direct sum

$$U(\mathfrak{g}) = \big(U(\mathfrak{k}) \otimes U(\mathfrak{a})\big) \oplus U(\mathfrak{g})\mathfrak{n},$$

associated to an Iwasawa decomposition $G_o = K_o A_o N_o$ of G_o, adapted to K_o. In Lepowsky [12] it is shown that $P : U(\mathfrak{g})^K \to U(\mathfrak{k})^M \otimes U(\mathfrak{a})$ is an injective anti-homomorphism of algebras. Here $U(\mathfrak{k})^M$ denotes the centralizer of M_o in $U(\mathfrak{k})$, M_o being the centralizer of A_o in K_o. To pursue this idea further it is necessary to have a good characterization of the image f $U(\mathfrak{g})^K$ in $U(\mathfrak{k})^M \otimes U(\mathfrak{a})$. In this opportunity, we shall describe such image when $G_o = SO(n,1)_e$ or $SU(n,1)$.

Let $\mathfrak{g} = \mathfrak{k} \oplus \mathfrak{p}$ be the complexified Cartan decomposition, corresponding to K_o, and let θ denote the corresponding Cartan involution, either of G_o or $\mathfrak{g}$. Also let M'_o denote the normalizer of A_o in K_o.

Given $w \in M'_o$ let $\bar{N}_w = \bar{N}_o \cap w^{-1} N_o w$, where $\bar{N}_o = \theta(N_o)$. If $\lambda \in \mathfrak{a}^*$ we consider

$$T(w, \lambda) : f \to \int_{\bar{N}_w} e^{-(\lambda+\rho)H(v)} f(w\kappa(v))dv, \quad f \in C^\infty(K_o).$$

For λ in an open cone $S(w) \subset \mathfrak{a}^*$, $T(w,\lambda)$ defines a distribution on K_0 (see Schiffmann [13]). Moreover from the fact that $T(w,\lambda)$ is closely related to the Kunze-Stein intertwining operators, it was shown in Kostant, Tirao [11,Theorem 3.2] that

$$T(w, \lambda) * P(u)(-\lambda - \rho) = P(u)(-w(\lambda) - \rho) * T(w, \lambda)$$

Supported in part by CONICET and CONICOR grants..

holds for all $u \in U(\mathfrak{g})^K$. Now let $\left(U(\mathfrak{k})^M \otimes U(\mathfrak{a})\right)^{\bar{W}}$ denote the set of al $b \in U(\mathfrak{k})^M \otimes U(\mathfrak{a})$ such that

$$T(w, \lambda) * b(-\lambda - \rho) = b(-w(\lambda) - \rho) * T(w, \lambda), \quad \lambda \in S(w).$$

This is an equation in the convolution algebra $D'(K_\circ) \supset U(\mathfrak{k})$ of al distributions on $K_\circ$.

If $U(\mathfrak{a}) = \bigoplus_{j \geq 0} U_j(\mathfrak{a})$ is the cannonical grading of $U(\mathfrak{a})$ and $b \in U(\mathfrak{k})^M \otimes U(\mathfrak{a})$, we write $b = b_d + b_{d-1} + \cdots + b_0$ where $b_j \in U(\mathfrak{k})^M \otimes U_j(\mathfrak{a})$. Then in Kostant, Tirao [11, Theorem 4.5] it was also proved that if $b \in \left(U(\mathfrak{k})^M \otimes U(\mathfrak{a})\right)^{\bar{W}}$ then b_d is W-invariant, under the cannonical action of the Weyl group on $U(\mathfrak{k})^M \otimes U(\mathfrak{a})$. From this it was shown that a suitable completion of a localization of $U(\mathfrak{g})^K$ becomes isomorphic to a corresponding completion of a localization of $\left(U(\mathfrak{k})^M \otimes U(\mathfrak{a})\right)^{\bar{W}}$ (cf. Kostant, Tirao [11, Theorem 7.4]).

THE ALGEBRA B

To characterize $P(U(\mathfrak{g})^K)$ we need more relations satisfied by is elements. In order to get them we shall consider some imbeddings $M(\mu_1) \subset M(\mu_2)$ among Verma modules.

Let $\mathfrak{b} = \mathfrak{h} \oplus \mathfrak{g}^+$ be a Borel subalgebra of $\mathfrak{g}$ and let Δ^+ be the corresponding set of positive roots. Given $\mu \in \mathfrak{h}^*$ let

$$M(\mu) = U(\mathfrak{g}) \otimes_{U(\mathfrak{b})} \mathbf{C}_{\mu-\rho}$$

where $\mathbf{C}_{\mu-\rho}$ denotes the 1-dimensional $\mathfrak{b}$-module where $\mathfrak{h}$ acts by $\mu - \rho$ and $\mathfrak{g}^+$ acts trivially. Then $M(\mu)$ is a $U(\mathfrak{g})$-module by left multiplication in the first factor. All the pairs (μ_1, μ_2) for which $\dim \operatorname{Hom}_{U(\mathfrak{g})}(M(\mu_1), M(\mu_2)) = 1$ (that is, the embedding $M(\mu_1) \subset M(\mu_2)$ exists) are described in the so called B-G-G theorem (see Bernstein, Gelfand, Gelfand [3]). In particular it was shown that if $(\mu, \alpha) = 2\langle \mu, \alpha \rangle / \langle \alpha, \alpha \rangle = n \in \mathbf{N}$ for some $\alpha \in \Delta^+$ then $M(\mu - n\alpha) \subset M(\mu)$. Moreover, every embedding $M(\mu_1) \subset M(\mu_2)$ is a composition of embeddings of this kind. The following proposition, due to Shapovalov [14], is a refinement of results in Bernstein, Gelfand, Gelfand [3].

Proposition 1. For every $\alpha \in \Delta^+$ and $n \in \mathbf{N}$ there exists an element $\theta_{\alpha,n} \in U(\mathfrak{g}^- + \mathfrak{h})$ of weight $-n\alpha$ with the following properties:

(i) $[X_\gamma, \theta_{\alpha,n}] \in U(\mathfrak{g})(H_\alpha + \rho(H_\alpha) - n) + U(\mathfrak{g})\mathfrak{g}^+$ for all $\gamma \in \Delta^+$,

where $H_\alpha \in \mathfrak{h}$ is defined by $\mu(H_\alpha) = (\mu, \alpha), \mu \in \mathfrak{h}^*$.
(ii) If $\{\alpha_1, \ldots, \alpha_r\}$ is the set of all simple roots and $\alpha = \sum_i l_i \alpha_i$, then

$$\theta_{\alpha,n} = \prod_i X_{-\alpha_i}^{n l_i} + \sum_j a_j b_j,$$

where $a_j \in U(\mathfrak{g}^-)$ is of weight $-n\alpha$, $b_j \in U(\mathfrak{h})$ and the degree of a_j is less than $n \sum_i l_i$.

Moreover $\theta_{\alpha,n}$ is uniquely determined by properties (i) and (ii) modulo the left ideal of $U(\mathfrak{g}^- \oplus \mathfrak{h})$ generated by $(H_\alpha + \rho(H_\alpha) - n)$.

Remarks. (i) If α is a simple root then we may choose $\theta_{\alpha,n} = X_{-\alpha}^n$.
(ii) If $1_\mu = 1 \otimes 1$ is the cannonical generator of $M(\mu)$ then $\theta_{\alpha,n} \cdot 1_\mu$ can be identified with the cannonical generator $1_{\mu-n\alpha}$ of $M(\mu - n\alpha) \subset M(\mu)$.

In our case we take a Cartan subalgebra $\mathfrak{t}_o$ of $\mathfrak{m}_o$, set $\mathfrak{h}_o = \mathfrak{t}_o + \mathfrak{a}_o$ and let $\mathfrak{h} = \mathfrak{t} \oplus \mathfrak{a}$ be the corresponding complexification. Then $\mathfrak{h}_o$ and $\mathfrak{h}$ are Cartan subalgebras of $\mathfrak{g}_o$ and $\mathfrak{g}$, respectively. Now we choose a Borel subalgebra $\mathfrak{t} + \mathfrak{m}^+$ of $\mathfrak{m}$ and take $\mathfrak{b} = \mathfrak{h} + \mathfrak{m}^+ + \mathfrak{n}$ as a Borel subalgebra of $\mathfrak{g}$.

If $\mu \in \mathfrak{h}^*$ let $A_\mu = \{X \in U(\mathfrak{t}) : X \cdot 1_\mu = 0\}$. Then we have

Proposition 2. (i) If $\alpha \in \Delta^+$ and $(\mu, \alpha) = n \in \mathbf{N}$ then

$$P(\theta_{\alpha,n})(\mu - \rho)P(u)(\mu - \rho) \equiv P(u)(\mu - n\alpha - \rho)P(\theta_{\alpha,n})(\mu - \rho) \qquad (mod\, A_\mu)$$

for all $u \in U(\mathfrak{g})^K$.

(ii)
$$A_\mu = U(\mathfrak{t})\mathfrak{m}^+ + \sum_{H \in \mathfrak{t}} U(\mathfrak{t})\big(H - \mu(H) + \rho(H)\big).$$

Proof. We shall only prove (i). Given $u \in U(\mathfrak{g})^K$ we have

$$u \cdot (\theta_{\alpha,n} \cdot 1_\mu) = P(u) \cdot (\theta_{\alpha,n} \cdot 1_\mu) = P(u)(\mu - n\alpha - \rho) \cdot (\theta_{\alpha,n} \cdot 1_\mu)$$
$$= P(u)(\mu - n\alpha - \rho)P(\theta_{\alpha,n})(\mu - \rho) \cdot 1_\mu.$$

On the other hand

$$u \cdot (\theta_{\alpha,n} \cdot 1_\mu)$$
$$= u \cdot (P(\theta_{\alpha,n})(\mu - \rho) \cdot 1_\mu) = P(\theta_{\alpha,n})(\mu - \rho)$$
$$= P(\theta_{\alpha,n})(\mu - \rho)P(u)(\mu - \rho) \cdot 1_\mu.$$

From this the assertion follows.

From now on we shall assume that we are in the split rank one case, i.e. $\dim \mathfrak{a} = 1$. If $\alpha \in \Delta^+$ let $H_\alpha = Y_\alpha + Z_\alpha$ with $Y_\alpha \in \mathfrak{t}$ and $Z_\alpha \in \mathfrak{a}$. Let $P^+ = \{\alpha \in \Delta^+ : Z_\alpha \neq 0\}$. If $\alpha \in P^+$ we shall consider the elements in $U(\mathfrak{t}) \otimes U(\mathfrak{a})$ as polynomials in Z_α with coefficients in $U(\mathfrak{t})$. When $\alpha \in P^+$ Proposition 2 (i) is equivalent to

Proposition 3. If $\alpha \in P^+$ and $n \in \mathbf{N}$, for all $u \in U(\mathfrak{g})^K$ we have

$$P(\theta_{\alpha,n})(n - Y_\alpha - \rho(H_\alpha))P(u)(n - Y_\alpha - \rho(H_\alpha))$$
$$\equiv P(u)(-n - Y_\alpha - \rho(H_\alpha))P(\theta_{\alpha,n})(n - Y_\alpha - \rho(H_\alpha))$$

$\mathrm{mod}\,(U(\mathfrak{k})\mathfrak{m}^+)$.

If $\alpha \in P^+$ is a simple root, taking $\theta_{\alpha,n} = X^n_{-\alpha}$ we have $P(\theta_{\alpha,n}) = (X_{-\alpha} + \theta X_{-\alpha})^n$ because $Y_\alpha \neq 0$. Let $E_\alpha = X_{-\alpha} + \theta X_{-\alpha} \in \mathfrak{k}$. Then from Proposition 3 we obtain

$$(1) \qquad E^n_\alpha b(n - Y_\alpha - 1) \equiv b(-n - Y_\alpha - 1)E^n_\alpha \qquad \mathrm{mod}\,(U(\mathfrak{k})\mathfrak{m}^+)$$

for all $b = P(u)$, $u \in U(\mathfrak{g})^K$, and all $n \in \mathbf{N}$.

Equations (1) define a subalgebra B of $U(\mathfrak{k})^M \otimes U(\mathfrak{a})$. More precisely let

$$B = \{b \in U(\mathfrak{k})^M \otimes U(\mathfrak{a}) : (1) \text{ holds for all } \alpha \in P^+ \text{ simple, } n \in \mathbf{N}\}$$

and

$$B^{\tilde{W}} = B \cap \left(U(\mathfrak{k})^M \otimes U(\mathfrak{a})\right)^{\tilde{W}}.$$

Therefore we have proved that

$$(2) \qquad\qquad P(U(\mathfrak{g})^K) \subset B^{\tilde{W}}.$$

THE IMAGE OF $U(\mathfrak{g})^K$

We have checked that equality holds in (2) when $G_o = SL(2, \mathbf{R})$, $SL(2, \mathbf{C})$ (see Tirao [16]), $SO(4,1)_e$ (see Brega, Tirao [4]), $SU(2,1)$ and $SU(3,1)$. In all these cases $\dim \mathfrak{m}^+ \leq 1$. In this paper we want to give an idea of the proof of the following theorem

Theorem 4. If $G_o = SO(n,1)_e$ or $SU(n,1)$ we have $P(U(\mathfrak{g})^K) = B^{\tilde{W}}$.

Remark. If $G_o = SO(n,1)_e$ or $SU(n,1)$ then

$$B^{\tilde{W}} = \{b \in B : \delta_w * b(\lambda - \rho) = b(w(\lambda) - \rho) * \delta_w \text{ for all } w \in M'_o, \lambda \in \mathfrak{a}^*\}.$$

(See Kostant, Tirao [11, Corollary 3.3].)

Let G be the adjoint group of $\mathfrak{g}$ and let K be the connected Lie subgroup of G with Lie algebra $ad_{\mathfrak{g}}(\mathfrak{k})$. Also let $M = \mathrm{Centr}_K(\mathfrak{a})$, $M' = \mathrm{Norm}_K(\mathfrak{a})$ and $W = M'/M$. If H is a group and V a finite dimensional H-module over $\mathbf{C}$, let $S'(V)$ denote the ring of all polynomial functions on V, and let $S'(V)^H$ denote the subring of all H-invariants. We shall need to know the image of the homomorphism $\pi : S'(\mathfrak{g})^K \to S'(\mathfrak{k}+\mathfrak{a}) = S'(\mathfrak{k})\otimes S'(\mathfrak{a})$ induced by restriction from $\mathfrak{g}$ to $\mathfrak{k} \oplus \mathfrak{a}$.

Let Γ denote the set of all equivalence classes of irreducible holomorphic finite dimensional K-modules V_γ such that $V^M_\gamma \neq 0$. Any $\gamma \in \Gamma$ can be realized as a submodule of all harmonic polynomial functions on $\mathfrak{p}$, homogeneous of degree d for a uniquely determined $d = d(\gamma)$ (Kostant, see Kostant, Rallis [10]).

Let $C = S'(\mathfrak{k})^M$ and let $C_d = \bigoplus S'(\mathfrak{k})^M_\gamma$, the sum over all $\gamma \in \Gamma$ such that $d(\gamma) \leq d$. Then $C = \bigcup_{d \geq 0} C_d$ is a nice ascending filtration of C. Now

$$D = \bigoplus_{d \geq 0}(C_d \otimes S'_d(\mathfrak{a}))$$

is an algebra, precisely de Rees algebra associated to the filtration $C = \bigcup_{d \geq 0} C_d$ (we are in the case $\dim \mathfrak{a} = 1$).

Theorem 5. (See Tirao [15] and Andruskiewitsch, Tirao [1].) *The operation of restriction from $\mathfrak{g}$ to $\mathfrak{k} + \mathfrak{a}$ induces an isomorphism of $S'(\mathfrak{g})^K$ onto D^W.*

If $\mathfrak{l}$ is any complex reductive Lie algebra let $\lambda : S(\mathfrak{l}) \to U(\mathfrak{l})$ denote the symmetrization mapping and let $\delta : S'(\mathfrak{l}) \to S(\mathfrak{l})$ be the algebra isomorphism defind by any $Ad(\mathfrak{l})$-invariant non-degenerate symmetric bilinear form of $\mathfrak{l}$.

If $b = b_m \otimes Z_\alpha^m + \cdots + b_0 \in B^{\tilde{W}}$ then we know that $b_m \otimes Z_\alpha^m \in \left(U(\mathfrak{k})^M \otimes U(\mathfrak{a})\right)^W$ (Kostant, Tirao [11, Theorem 4.5]). Thus to prove Theorem 4 it is enough to establish the following proposition.

Proposition 6. *If $b = b_m \otimes Z_\alpha^m + \cdots + b_0 \in B$ and $c_m = \delta^{-1}(\lambda^{-1}(b_m))$ then $c_m \in C_m$.*

In fact, in such a case Theorem 5 implies that there exists $f \in S'(\mathfrak{g})^K$ such that $c_m \otimes Z_\alpha^m = \pi(f)$. Let $u = \lambda(\delta(f)) \in U(\mathfrak{g})^K$. Then the leading term of $P(u)$ is $b_m \otimes Z_\alpha^m$ (see Kostant, Tirao [11,Proposition 7.2]). Now $b - P(u) \in B^{\tilde{W}}$ and its degree is strictly smaller than m. Thus an inductive argument on m completes the proof of Theorem 4 as soon as we realize that if $b = b_0 \in B^{\tilde{W}}$ then $c_0 \in C_0 = S'(\mathfrak{k})^K$, thus $b_0 \in U(\mathfrak{k})^K \subset P(U(\mathfrak{g})^K)$.

We move on to establish the main steps in the proof of Proposition 6. When $G_\circ = SO(n,1)_e$ there is only one simple root $\alpha_1 \in P^+$ and when $G_\circ = SU(n,1)$ there are exactly two simple roots α_1, α_n in P^+. Set $E_1 = E_{\alpha_1}$ in the first case, and $E_2 = E_{\alpha_1}, E_3 = E_{\alpha_n}$ in the second case. We shall also use E to designate any one of the vectors E_1, E_2 or E_3 and α for α_1, α_1 or α_n, respectively.

Since we are only interested in those $\gamma \in \Gamma$ that appear as subrepresentations of $U(\mathfrak{k})$ we set $\Gamma_1 = \{\gamma \in \Gamma : \gamma$ is trivial in the center of $K\}$. If $\gamma \in \Gamma$ then $\dim V_\gamma^M = 1$ (Kostant [9]). Also if $G_\circ = SU(n,1)$ and $\gamma \in \Gamma_1$ then

$$\max\{q \in \mathbf{N}_\circ : E_2^q(V_\gamma^M) \neq 0\} = \max\{q \in \mathbf{N}_\circ : E_3^q(V_\gamma^M) \neq 0\}.$$

Therefore, $q(\gamma) = max\{q \in \mathbf{N}_\circ : E^q(V_\gamma^M) \neq 0\}$ is well defined in either case. Moreover for all $\gamma \in \Gamma_1$ $q(\gamma) = d(\gamma)$ if $G_\circ = SO(n,1)_e$ and $q(\gamma) = d(\gamma)/2$ if $G_\circ = SU(n,1)$ (see Brega, Tirao [5]). Thus Proposition 6 is equivalent to the following proposition. In what follows $\dot{E}$ denotes the adjoint action of   on $U(\mathfrak{k})$ and $\equiv$ will always stand for congruence modulo $U(\mathfrak{k})\mathfrak{m}^+$.

Proposition 7. *If $b = b_m \otimes Z_\alpha^m + \cdots + b_0 \in B$ then $\dot{E}^{m+1}(b_m) = 0$ when $G_\circ = SO(n,1)_e$, and $\dot{E}^{[m/2]+1}(b_m) = 0$ when $G_\circ = SU(n,1)$.*

Lemma 8. *(i) If $b = b_m \otimes Z_\alpha^m + \cdots + b_0 \in U(\mathfrak{k})^M \otimes U(\mathfrak{a})$ then*

$$\sum_{t=m}^{2m}(-1)^t \binom{m}{t-m} \dot{E}^{2m-t}\left[\sum_{i=0}^{t+1}(-1)^i \binom{t+1}{i} E^{t+1-i}b(t-Y_\alpha-i)E^i\right]$$

$$= m!E^m(\dot{E}^{m+1}b_m).$$

(ii) If $b = b_m \otimes Z_\alpha^m + \cdots + b_0 \in B$ then

$$\sum_{i=0}^{t+1} (-1)^i \binom{t+1}{i} E^{t+1-i} b(t - Y_\alpha - i) E^i \equiv 0.$$

From this and because $[E, m^+] = 0$ we obtain that $\dot{E}^{m+1} b_m \equiv 0$. Another fundamental step in the proof of Proposition, and thus in the proof of Theorem 4, is the following fact established in Brega, Tirao [5, Theorem 2.11].

Proposition 9. *The infinite sum* $\sum_{j\geq 0} \dot{E}^j (U(\mathfrak{k})^M)$ *is a direct sum. More-over*

$$\left(\sum_{j\geq 0} \dot{E}^j (U(\mathfrak{k})^M) \right) \cap U(\mathfrak{k}) m^+ = 0.$$

This completes the proof of Proposition 7 in the case $G_o = SO(n, 1)$. To finish the proof when $G_o = SU(n, 1)$ we have to work a bit more. First of all, we generalize Lemma 8 in the following way.

Lemma 10. *Let* $w \in N_o$, $0 \leq w \leq m$. *If* $b = b_m \otimes Z_\alpha^m + \cdots + b_0 \in B$ *and* $\dot{E}^{2m-j+1} b_j \equiv 0$ *for all* j *such that* $w < j \leq m$, *then*

$$\sum_{t=m}^{m+w} (-1)^t \binom{w}{t-m} \dot{E}^{2m-t} \left[\sum_{i=0}^{t+1} (-1)^i \binom{t+1}{i} E^{t+1-i} b(t - Y_\alpha - i) E^i \right]$$

$$\equiv (-1)^{m+w} w! E^w (\dot{E}^{2m-w+1} b_w).$$

By decreasing induction on w and using Lemma 8 (ii) we obtain that $\dot{E}^{2m-j+1} b_j \equiv 0$ for all $0 \leq j \leq m$. Also one proves

Lemma 11. *If* $w, \alpha \in N_o, 0 \leq w \leq m, 0 \leq \alpha \leq m, \alpha + w \geq m + 1$ *and* $\dot{E}^{m+\alpha+1-j} b_j \equiv 0$ *for all* $0 \leq j \leq m$, *then*

$$\sum_{j=m-w}^{m} (-2)^{-j} j! \binom{\alpha+w}{j+w-m} (\dot{E}^{m+\alpha-j} b_j) E^j \equiv 0.$$

Lemma 12. *Assume that for* $E = E_2, E_3$, $\sum_{j=0}^{m} (\dot{E}^{m+\alpha-j} b_j) E^j \equiv 0$, $b_j \in U(k)^M$ *and* $\dot{E}^{m+\alpha+1-j} b_j \equiv 0$ *for for all* $0 \leq j \leq m$, *then*

$$\sum_{\substack{0 \leq j \leq m \\ j \text{ even}}} (\dot{E}^{m+\alpha-j} b_j) E^j \equiv 0 \equiv \sum_{\substack{0 \leq j \leq m \\ j \text{ odd}}} (\dot{E}^{m+\alpha-j} b_j) E^j.$$

Thus if $0 \leq w \leq m, 0 \leq \alpha \leq m, \alpha + w \geq m + 1$ and $\dot{E}^{m+\alpha+1-j} b_j \equiv 0$ for all $0 \leq j \leq m$, then

$$(3) \qquad \sum_{i=0}^{[w/2]} (-2)^{m-2i} (m - 2i)! \binom{\alpha+w}{w-2i} (\dot{E}^{\alpha+2i} b_{m-2i}) E^{m-2i} \equiv 0.$$

Now we start with $\alpha = m$. From (3) we obtain inductively on w in the range $1 \leq w \leq m$, that $\dot{E}^{m+2i}b_{m-2i} \equiv 0$ for all $0 \leq i \leq [m/2]$. Then we can take $\alpha = m - 1$ and look at (3) as a system of $m - 1$ linear equations in $[m/2] + 1$ unknowns. Since the coefficient matrix is of rank $[m/2] + 1$ we obtain that $\dot{E}^{m-1+2i}b_{m-2i} \equiv 0$ for all $0 \leq i \leq [m/2]$. In this way we can continue up to $\alpha = [m/2] + 1$ where the number of equations equals the number of unknowns. At all these steps the coefficient matrix has rank $[m/2] + 1$. Hence $\dot{E}^{[m/2]+1+2i}b_{m-2i} \equiv 0$ for all $0 \leq i \leq [m/2]$. Proposition 7 now follows from Proposition 9.

Let $Z(\mathfrak{g}) = U(\mathfrak{g})^G$ be the center of $U(\mathfrak{g})$ and let $Z(\mathfrak{k}) = U(\mathfrak{k})^K$ be the center of $U(\mathfrak{k})$. Then it is well known that the canonical homomorphism $\mu : Z(\mathfrak{g}) \otimes Z(\mathfrak{k}) \to U(\mathfrak{g})^K$ is injective. Now as a corollary of Theorem 4 we obtain

Theorem 13. *If $G_o = SO(n, 1)_e$ or $SU(n, 1)$ we have*

(i) $P(U(\mathfrak{g})^G) = \left(U(\mathfrak{k})^M \otimes U(\mathfrak{a})\right) \cap B^{\tilde{W}}$ and $P(U(\mathfrak{k})^K) = U(\mathfrak{k})^K \otimes 1$.

(ii) $B^{\tilde{W}} = P(U(\mathfrak{g})^G)P(U(\mathfrak{k})^K)$.

(iii) $\mu : Z(\mathfrak{g}) \otimes Z(\mathfrak{k}) \to U(\mathfrak{g})^K$ is a surjective isomorphism.

The last statement was first established by Cooper [6]. The same result was obtain by Benabdallah [2] when $G_o = SO(n, 1)_e$ and by Johnson [7] when $G_o = SU(n, 1)$. Of course for μ to be surjective $U(\mathfrak{g})^K$ has to be commutative and this happens precisely in these two cases (see Kostant, Tirao [11, Corollary 3.3]). See also Knop [8] for an exposition of this.

REFERENCES

1. N. Andruskiewitsch and J. Tirao, *A restriction theorem for modules having a spherical submodule*, Trans. Amer. Math. Soc. (to appear).

2. A. Benabdallah, *Generateurs de l'algebre $U(G)^K$ avec $G = SO(n)$ ou SO_o $(1, m-1)$ et $K = SO(m - 1)$*, Bull. Soc. Math. France **111** (1983), 303-326.

3. J. Bernstein, I.M. Gelfand and S.I. Gelfand, *The structure of representations generated by vectors of the highest weight*, Funktsional Analiz ego Prilozhen **5, Nr.1** (1971), 1-9.

4. A. Brega and J. Tirao, *K-invariants in the universal enveloping algebra of the De-Sitter group*, Manuscripta Math. **58** (1987), 1-36.

5. A. Brega and J. Tirao, *A transversality property of a derivation of the universal enveloping algebra $U(k)$, for $G = SO(n, 1), SU(n, 1)$*, Manuscripta Math. (to appear).

6. A. Cooper, *The classifying ring of groups whose classifying ring is commutative*, Ph. D. Thesis, MIT,1975.

7. K. Johnson, *The centralizer of a Lie algebra in an enveloping algebra*, J. reine angew. Math. **395** (1989), 196-201.

8. F. Knop, *Der zentralisator einer Liealgebra in einer einhüllenden Algebra*, J. reine angew. Math. **406** (1990), 5-9.

9. B. Kostant, *On the existence and irreducibility of certain series of representations*, Lie groups and their representations, 1971 Summer School in Math., Wiley, New York (1975), 231-330.

10. B. Kostant and S. Rallis, *Orbits and representations associated with symmetric spaces*, Amer. J. Math. **93** (1971), 753-809.

11. B. Kostant and J. Tirao, *On the structure of certain subalgebras of a universal enveloping algebra*, Trans. Amer. Math. Soc. **218** (1976), 133-154.

12. J. Lepowsky, *Algebraic results on representations of semisimple Lie groups*, Trans. Amer. Math. Soc. **176** (1973), 1-44.

13. G. Schiffmann, *Integrales d'entrelacement et fonctions de Whittaker*, Bull. Soc. Math. France **99** (1971), 3-72.

14. N. Shapovalov, *On a bilinear form on the universal enveloping algebra of a complex semisimple Lie algebra*, Funct. Anal. Appl. **6** (1972), 307-312.

15. J. Tirao, *A restriction theorem for semisimple Lie groups of rank one*, Trans. Amer. Math. Soc. **279, Nr. 2** (1983), 651-660.

16. J. Tirao, *The classifying ring of $SL(2, \mathbf{C})$*, Lecture Notes in Math., 597, Springer-Verlag (1977), 641-678.

FAMAF
Universidad N. de Córdoba
Córdoba Argentina

ON SPHERICAL MODULES

Nicolás Andruskiewitsch

1. Introduction

Let L be a group, V a finite dimensional complex vector space and $\varphi : L \to GL(V)$ a representation. Let us denote by $S'(V)$ the ring of all polynomial functions on V; clearly L acts on $S'(V)$. The main problem of the classical invariant theory can be phrased as follows: To find explicitly all the L-invariant polynomial functions on V.

More generally, the invariant theory is concerned with the understanding of the action of L on V.

From now on, let us assume that L is a reductive complex linear algebraic group. Under this hypothesis, the Hilbert-Nagata theorem says that $S'(V)^L$ (the C-algebra of L-invariant polynomial functions on V) is finitely generated. Thus we can consider V/L , the affine variety associated to $S'(V)^L$ and $V \to V/L$ the projection corresponding to the inclusion. It turns out that V/L parametrizes the closed orbits of L in V. Let $\mathcal{N} = \mathcal{N}(V, L)$ be the fiber $\pi^{-1}(\pi(0))$; $\mathcal{N}$ is called the nilpotent cone and plays an important role, since it is, roughly speaking, the worst fiber.

Now concerning the classification of the L-orbits in V, there are some criteria of "non-complicatedness". (See [Kc] or [M, p. 160]). The criteria are:

A. $\mathcal{N}$ is a finite union of orbits. ((V, L) is visible).

B. All the fibres of π are of the same dimension.

C. $S'(V)^L$ is a polynomial ring. ((V, L) is coregular).

D. $S'(V)$ is a $S'(V)^L$-free module. ((V, L) is cofree). In such a case, $S'(V)^L$ is a polynomial ring (i.e., (V, L) is coregular) and

$$S'(V) \simeq S'(V)^L \otimes H$$

where $\otimes$ is given by multiplication and H is a homogenous L-submodule of $S'(V)$.

E. The isotropy subgroup L^x is non trivial for every $x \in V$.

This work was supported by CONICET, CONICOR and FAMAF

In fact, for a fixed L, there are only few modules V satisfying any of the above criteria; if L is simple, it is known precisely what are, see the references in [Kc] or [M].

In order to discuss some examples, let us introduce some notation. Let $g_{\mathbf{R}} = k_{\mathbf{R}} \oplus p_{\mathbf{R}}$ be a Cartan decomposition of a real semisimple Lie algebra $g_{\mathbf{R}}$ and let $g = k \oplus p$ be the corresponding complexification. Let θ be the associated Cartan involution. Also let $a_{\mathbf{R}}$ be a maximal abelian subspace of $p_{\mathbf{R}}$ and let a be its complexification. Now let G be the adjoint group of g and let K be the analytic subgroup of G with Lie algebra $\mathrm{ad}_g(k)$. Also let M be the centralizer of a in K. The particular case that we shall consider in this note is the action of K in g given by the restriction of the Adjoint representation. But let us say a few words about the cases (g, G) and (p, K). In both cases, the criterion C is satisfied. Indeed, the well known Chevalley Restriction Theorem states that the restriction homomorphism $S'(p) \to S'(a)$ maps $S'(p)^K$ isomorphically onto $S'(a)^W$. (Here W denotes the Weyl group $N_K(a)/M$). Since W is generated by pseudoreflections, another theorem of Chevalley guarantees that $S'(p)^K$ is regular.

The Restriction Theorem was generalized by Luna and Richardson [L R]:

Let V, L be as above and let us assume for simplicity that (V, L) has generically closed orbits; i.e., the union of all closed orbits contains a non-empty Zariski open subset of V. Then there exists a non empty (Zariski) open subset U of V such that for every $x, y \in U$, the orbits Lx and Ly are isomorphic, and closed. Pick any $x \in U$. The conjugacy class of the isotropy subgroup $M = L^x$ is called a principal isotropy class. The generalization of the Chevalley Restriction Theorem given in [L R] states that the restriction map $S'(V) \to S'(V^M)$ maps $S'(V)^L$ isomorphically onto $S'(V^M)^W$, where $W = N_L(M)/M$.

(g, G), as well as (p, K), also satisfies the other criteria, as was shown by Kostant [K] and Kostant-Rallis [K R]. But, unfortunately, (g, K) never satisfies criteria B nor E; it satisfies criteria A, C, D if and only if $g_{\mathbf{R}}$ is a product of copies of $so(p,1)$ or $su(p,1)$. (See [A 1]). In other words, the invariant theory of the pair (g, K) seems to be rather complicated. The purpose of this note is to present a technique (developed in [T], [A 2], [A 3], [A T], see also [B T]) which seems to be useful to make some explicit computations under the additional hypothesis dim $a = 1$. It applies in fact in the more general setting of the spherical modules and consists of two steps. The first is to characterize the image of $S'(g)^K$ in $S'(k \oplus a)$ via te restriction map, see section 2. The second is to interpret certain quotient of $S'(g)^K$ as the graded ring of a filtration in $S'(k)^M$; this statement can be extended to representations wich are

no longer spherical as we shall see in Section 3, extending the results of
[A T].

This note is an expanded version of a talk given at the Third Workshop
on Representation Theory of Lie Groups and its Applications.

2. SPHERICAL MODULES

As usual, $\hat{L}$ denotes the set of equivalence classes of finite dimensional irreducible representations of an algebraic reductive complex linear group L. We will identify $\tau \in \hat{L}$ with the space on which L acts. If
E is any L-module, $\tau \in \hat{L}$, let us denote by E_τ the isotypic component
of type τ.

Let us recall briefly the notion of a spherical pair. Let G be a reductive
connected algebraic group and $\mathcal{H}$ a closed subgroup, over an algebraically
closed field of characteristic zero. $(G, \mathcal{H})$ is a spherical pair if satisfies
one of the following equivalent conditions:

i) $\mathcal{H}$ has an open orbit in the flag variety of G.

ii) $\mathcal{H}$ has a finite number of orbits in the flag variety of G.

iii) Let Z be an algebraic G-variety and let $z \in Z^{\mathcal{H}}$; then G has a
finite number of orbits in the closure of Gz.

iv) Let χ be a one dimensional representation of $\mathcal{H}$ and let C be the
representation of G induced by χ (i.e., the space of global sections of
the associated line bundle over $G/\mathcal{H}$). Then for every $\gamma \in \hat{G}$, dim
$\mathrm{Hom}_G(\gamma, C) \leq 1$.

(See [B L V] for the history of the paternity of this result). It follows
from Frobenius reciprocity that iv) can be also stated as follows: for
every $\gamma \in \hat{G}$, dim $\mathrm{Hom}_{\mathcal{H}}(\chi, \gamma) \leq 1$, viewing γ as an $\mathcal{H}$-module. In
particular, taking χ trivial, dim $\gamma^{\mathcal{H}} \leq 1$ for all $\gamma \in \hat{G}$. On the other
hand this implies the above conditions, whenever $\mathcal{H}$ is reductive. (See
[V K].)

DEFINITION. We will say that (V, L) is a *spherical representation of
rank one* if it has generically closed orbits, dim $A = 1$ and for all $\rho \in \hat{L}$,
dim $\rho^M \leq 1$, where M is a principal isotropy group of (V, L) and $A =
V^M$. In particular, (V, L) is irreducible.

REMARKS: i) M is reductive thanks to a theorem by Matsushima.

ii) Kramer has listed all the spherical pairs (G, H) with G simple and
Elashvili gave a list of all the representations of simple groups satisfying
criteria E. (see [E], [V K]). In fact, the intersection of Elashvili's and
Kramer's tables gives us all the rank one spherical representations of
simple groups.

Using standard techniques we can prove:

PROPOSITION 1. *Let (V, L) be a spherical representation of rank one. Then it satisfies criteria A, B, C, D, E. Let H be a graded submodule of $S'(V)$ such that*

$$S'(V) = S'(V)^L \otimes H$$

Then the multiplicity of ρ in H is $\leq 1 \forall \rho \in \hat{L}$.

DEFINITION. We will say that (V, L) is a *spherical representation* (of rank s) if

$$V = V_1 \oplus \cdots \oplus V_s, \quad L = L_1 \times \cdots \times L_s,$$

each V_i is a L_i-spherical module of rank one and L acts on V via

$$(k_1, ..., k_s)(v_1, ..., v_s) = (k_1 v_1, ..., k_s v_s).$$

Note that it has generically closed orbits and that Proposition 1 extends easily to spherical representations of arbitrary rank.

Now let us assume that (V, L) is a spherical representation of rank one, $S'(V) = S'(V)^L \otimes H$ and let

$$\Gamma = \{\rho \in \hat{L} : \rho^M \neq 0\}.$$

Then $H = \oplus_{\rho \in \Gamma} H_\rho$. Moreover, H_ρ is irreducible and homogeneous.
Clearly, $\rho \in \Gamma \Rightarrow \rho^* \in \Gamma$. So we put

$$m(\gamma) = \text{degree of homogeneity of } H_{\gamma^*}, \quad \gamma \in \Gamma.$$

Now let

$$(V, L) = (V_1, L_1) \oplus \cdots \oplus (V_s, L_s)$$

be a spherical representation, where (V_i, L_i) are spherical representations of rank one. We introduce the following notation:

$$A = A_1 \oplus \cdots \oplus A_s, \quad \Gamma = \Gamma_1 \times \cdots \times \Gamma_s,$$
$$H = H_1 \otimes \cdots \otimes H_s, \quad M = M_1 \times \cdots \times M_s,$$

where M_i, Γ_i, A_i, H_i correspond to (V_i, L_i).

Recall now that $\hat{L}$ can be identified with $\hat{L} \times \cdots \times \hat{L}_s$. Thus $\gamma \in \Gamma$ if and only if γ appears in H; and in such a case, it does with multiplicity one. Moreover, let us consider the N_0^s-grading in $S'(V)$ given by the decomposition $V = V_1 \oplus \cdots \oplus V_s$. Therefore, H is an N_0^s-graded L-submodule of $S'(V)$ and for $\gamma = \gamma_1 \otimes \cdots \otimes \gamma_s \in \Gamma$, γ^* appears only in $H_{m(\gamma)}$, where

$$m(\gamma) = (m(\gamma_1), ..., m(\gamma_s)) \in N_0^s$$

and $m(\gamma_i)$ corresponds to $(V_i, L_i), \gamma_i$.

Clearly, M is a principal isotropy group of (V, L) and $A = V^M$. Let $W = N_L(M)/M$. We consider the order in $\mathbf{N}_0^s$ given by

$$(a_1, ..., a_s) \le (b_1, ..., b_s) \iff a_i \le b_i, \quad i = 1, ..., s$$

and we set

$$\Gamma_r = \{\gamma \in \Gamma : m(\gamma) \le r\}$$

for every $r \in \mathbf{N}_0^s$.

THEOREM 1. *Keep the notations and the hypothesis as above. Let U be a finite dimensional L-module; then the restriction from $U \oplus V$ to $U \oplus A$ induces an isomorphism σ of $S'(U \oplus V)^L$ onto*

$$(\oplus_{r \in N^s} (\oplus_{\gamma \in \Gamma_r} S'(U)_\gamma^M \otimes S_r'(A)))^W$$

PROOF: The injectivity follows from the generic closedness of the orbits. The characterization of the image is straightforward, taking into account the well known Schur Lemma: For $\tau, \lambda \in \hat{L}$, $\dim(\tau \otimes \lambda)^L = 1$ if $\tau = \lambda^*$, 0 otherwise.

REMARK: The above proof is given essentially in [A 2]. The original proof in [T] uses geometric considerations which rassembles some recent work of Ranee Brylinski; and is based in a Kostant's local Restriction Theorem (see [K T]). On the other hand, Kostant's theorem was generalized in [A 2], using the tools of [L R].

3. REPRESENTATIONS CONTAINING THE ADJOINT

Clearly, the adjoint representation is spherical only for $SL(2, \mathbf{C})$; and it is not evident how to generalize Theorem 1 to other representations. In [A T], the characterization of the image in Theorem 1 was interpreted as the Rees ring of an ascending filtration of $S'(U)^M$ and it was shown that the associated graded algebra is isomorphic to $S'(U)^P$ for a suitable subgroup P of L. In this section we shall show that this last results can be generalized to arbitrary adjoint representations.

Let us first recall that the adjoint representation (g, G) enjoys the following properties:

(1) There exists $x \in \mathcal{N}$ such that the closure of the orbit Gx is $\mathcal{N}$. Moreover, $\mathcal{N}$ is normal, the ideal of $S'(g)$ generated by the homogeneous elements of positive degree in $S'(g)^G$ is prime and $\text{codim}(\mathcal{N}, \mathcal{N} - Gx) \ge 2$.

(2) Pick any $y \in g$ such that the orbit Gy is closed and G^z is conjugate to G^y for all z in an open neighborhood of y. Then $\mathcal{N}$ is contained in the Zariski closure of the union of all the orbits $G.ty$, over all the non-zero complex numbers t.

As the adjoint representation is cofree, we can use the notation H, M, Γ. Now let $P = G^x$. By (1), we have $\Gamma = \{\rho \in G^\wedge : \rho^P \neq 0\}$.

There exist non-negative integers $d_1^\rho \leq \cdots d_{m(\rho)}^\rho$ we will omit the superscript ρ in the following) and homogeneous submodules of degree d_j, $H_{\rho^*,d_j} \simeq \rho^*$, such that

$$H_{\rho^*} = \oplus_{1 \leq j \leq m(\rho)} H_{\rho^*,d_j}$$

Let $F(\rho) = Hom_G(\rho, H)$, for $\rho \in \Gamma$; we have $m(\rho) = \dim F(\rho^*)$.

Now, for any $z \in g$, let $\beta_z : F(\rho^*) \to \rho^{G^x}$ be given by

$$< \beta_z(\sigma), \alpha >= \sigma(\alpha)(z)$$

$\forall \alpha \in \rho^*$, $\sigma \in F(\rho^*)$. It is known that β_y and β_z are isomorphisms ([K, Corollary 4]). We define $\phi = \phi(\rho)$ by making commutative the following diagram:

$$\rho^M \xrightarrow{\phi} \rho^P$$
$$\beta_y \nwarrow \quad \nearrow \beta_x$$
$$F(\rho^*)$$

Now let $\rho^M[d]$ be the image via β_y of the linear span of the isomorphisms $(\rho^* \to H_{\rho^*,d_j})$, $d_j = d$. Clearly, if V is a G-module isomorphic to ρ, then $V^M[d]$ is well defined and does not depend on the choosen isomorphism with ρ. Thus we can define $V^M[d]$ for any isotypic G-module V of type ρ or even to any G-module which is a sum of finite dimensional submodules. More precisely, we obtain via β_y that $(V \otimes H)^G \simeq Hom_G(V^*, H) \simeq V^M$; hence we have an isomorphism of vector spaces from $(V \otimes H)^G$ onto V^M given explicitly by

$$\sum_j v_j \otimes h_j \to \sum_j h_j(y)v_j$$

Moreover, the image of $(V \otimes H_d)^G$ (the homogeneous component of degree d) is precisely $V^M[d]$. For example, if U is a finite dimensional representation of G, let $C = S'(U)^M$ and let us introduce

$$C[e] = \oplus_{\rho \in \Gamma}(S'(U)_\rho)^M[e]$$

and

$$C_{(d)} = \oplus_{e \leq d} C[e]$$

We are ready to state:

THEOREM 2.

i) $C_{(0)} \subseteq C_{(1)} \subseteq ... \subseteq C_{(j)} \subseteq ...$ is an ascending filtration of C.

ii) The isomorphisms $\phi(\rho) : \rho \in \Gamma$ give rise to an isomorphism of algebras between $\mathrm{gr}C$ and $S'(U)^P$. (Here $\mathrm{gr}C$ is the associated graded ring of the filtration in i).

iii) The quotient of $S'(g \oplus U)^G$ by the ideal $\mathcal{J}$ generated by the homogeneous polynomials of positive degree in $S'(g)^G$ is isomorphic to $\mathrm{gr}C$.

Before proceeding with the proof, we shall make some remarks.

Let λ, $\rho \in \Gamma$, and $f \in \lambda^M[c] - 0$, $g \in \rho^M[e] - 0$. We have

$$\lambda \otimes \rho = \delta_1 \oplus ... \oplus \delta_t, \quad \delta_k \in G^\wedge.$$

Let us decompose

$$f \otimes g = h_1 + ... + h_s, \quad s \leq t, \quad h_k \in \delta_k - 0,$$

reordering the index set if necessary. We have the following well known isomorphism of G-modules:

$$\mathbf{C}[G/M] \simeq \oplus_{\rho \in G^\wedge} \rho^M \otimes \rho^*.$$

From the inclusions given by f and g

$$\lambda^* \hookrightarrow \lambda^M \otimes \lambda^* \quad \text{and} \quad \rho^* \hookrightarrow \rho^M \otimes \rho^*$$

we have a morphism of G-modules

$$\lambda^* \otimes \rho^* \to (f \otimes \lambda^*)(g \otimes \rho^*) \subseteq \oplus_{k=1,...,t} \delta_k^M \otimes \delta_k^*.$$

Now we claim that

$$(f \otimes \lambda^*)(g \otimes \rho^*) \supseteq \oplus_{k=1,...,s} h_k \otimes \delta_k^*.$$

Indeed, if $c_k \in \delta_k^*$, there exist $\eta_q \in \lambda^*$, $\zeta_q \in \rho^*$ such that $\sum_k c_k = \sum_q \eta_q \otimes \zeta_q$ in $\lambda^* \otimes \rho^*$. Thus for every $t \in G$:

$$< \sum t c_k, \sum h_k > = < \sum_q t\eta_q \otimes t\zeta_q, f \otimes g > = \sum_q < t\eta_q, f > . < t\zeta_q, g >$$

That is,

$$\sum h_k \otimes c_k = \sum_q (f \otimes \eta_q).(g \otimes \zeta_q).$$

Now, we have the following diagram of G-modules, where the first horizontal arrow in the first row is the multiplication composed with $\beta_y^{-1}(f)$ and $\beta_y^{-1}(g)$:

$$\lambda^* \otimes \rho^* \to H_{\lambda^\bullet,c}.H_{\rho^\bullet,e} \hookrightarrow S'(g)$$
$$\searrow \qquad\qquad \nearrow \text{ Restriction}$$
$$\mathbf{C}[G/M] \longrightarrow \mathbf{C}[Gy]$$

This diagram is commutative: we need to check that $\forall t \in G$, $\eta \in \lambda^*$, $\zeta \in \rho^*$:

$$\beta_y^{-1}(f)(\eta)(ty)\beta_y^{-1}(g)(\zeta)(ty) = <tf,\eta><tg,\zeta>$$

But for example

$$\beta_y^{-1}(f)(\eta)(ty) = t^{-1}\beta_y^{-1}(f)(\eta)(y) = \beta_y^{-1}(f)(t^{-1}\eta)(y) = <f,t^{-1}\eta>$$

and the claim follows. Hence the homogeneous module, of degree $c+e$, $H_{\lambda^\bullet,c}.H_{\rho^\bullet,e}$ contains an irreducible G-module of type δ_k^* for each $k = 1,...,s$.

PROOF OF i): Let us fix k, $1 \leq k \leq s$, and set $\delta = \delta_k$, in order to avoid cumbersome notations. Let $\sigma_1,...,\sigma_{m(\delta)}$ be a basis of $F(\delta^*)$ such that $\text{Im}(\sigma_\ell) = H_{\delta^\bullet,d_\ell}$. As

$$S'(g)_{\delta^\bullet} = S'(g)^G \otimes H_{\delta^\bullet} = S'(g)^G \otimes (\oplus_{1\leq\ell\leq m(\delta)} H_{\delta^\bullet,d_\ell})$$

we deduce that

$$Hom_G(\delta^*, S'(g)) \simeq S'(g)^G \otimes F(\delta^*).$$

Indeed, the morphism from right to left is obvious. Let $w \in \delta^*$ be a highest weight vector (with respect to a fixed Borel subgroup B with unipotent radical N). Let $\sigma \in Hom_G(\delta^*, S'(g))$. We have

$$S'(g)_{\delta^\bullet}^N = S'(g)^G \otimes H_{\delta^\bullet}^N = S'(g)^G \otimes (\oplus_{1\leq\ell\leq m(\delta)} H_{\delta^\bullet,d_\ell}^N)$$

and hence there exist unique $P_1,...,P_{m(\delta)} \in S'(g)^G$ satisfying $\sigma(w) = P_1\sigma_1(w) + ... + P_{m(\delta)}\sigma_{m(\delta)}(w)$; it follows easily that $\sigma = P_1\sigma_1 + ... + P_{m(\delta)}\sigma_{m(\delta)}$.

In particular, let us assume that $\sigma = P_1\sigma_1 + ... + P_{m(\delta)}\sigma_{m(\delta)}$ corresponds to the morphism

$$h_k \otimes \delta_k^* \to H_{\lambda^\bullet,c}.H_{\rho^\bullet,e}$$

Then $P_1, ..., P_{m(\delta)}$ must be homogeneous. We claim that

$$\beta_y^{-1}(h_k) = P_1(y)\sigma_1 + ... + P_{m(\delta)}(y)\sigma_{m(\delta)}.$$

Let $c_k \in \delta_k^*$, and $\eta_q \in \lambda^*$, $\zeta_q \in \rho^*$ such that $c_k = \sum_q \eta_q \otimes \zeta_q$. Then

$$< \beta_y(P_1(y)\sigma_1 + ... + P_{m(\delta)}(y)\sigma_{m(\delta)}), c_k > =$$
$$(P_1(y)\sigma_1 + ... + P_{m(\delta)}(y)\sigma_{m(\delta)})(c_k)(y) =$$
$$(P_1\sigma_1 + ... + P_{m(\delta)}\sigma_{m(\delta)})(c_k)(y) =$$
$$\sum_q < f, \eta_q >< g, \zeta_q > = < f \otimes g, \sum_q \eta_q \otimes \zeta_q > = < h_k, c_k >$$

because we already showed that $\sigma(c_k)(ty) = \sum_q < tf, \eta_q >< tg, \zeta_q >$ for all $t \in G$.

But now $P_\ell(y) \neq 0$ implies of course $P_\ell \neq 0$ and $d_\ell \leq d_i + d_j$. In other words, $h_k \in \oplus_{d \leq c+e} \delta_k^M[d]$; from this $i)$ follows.

PROOF OF $ii)$: Let us retain the notation above. Let $f \in \lambda^M[c] - 0$, $g \in \rho^M[e] - 0$; as above, we have

$$f \otimes g = h_1 + ... + h_s, \quad s \leq t, \quad h_k \in \delta_k - 0.$$

Let $d = c + e$. For $1 \leq k \leq s$, let us decompose

$$h_k = \sum_{1 \leq r \leq d} h_{k,[r]}, \quad h_{k,[r]} \in \delta_k[r].$$

Let us check that

$$\phi(f) \otimes \phi(g) = \phi(h_{1,[d]}) + ... + \phi(h_{s,[d]})$$

Let $A \in F(\rho^*)$, $B \in F(\lambda^*)$, $C_{k,[r]} \in F(\delta_k^*)$ corresponding to f, g, $h_{k,[r]}$. This means, for example, that if $u \in \rho^*$:

$$< f, u > = A(u)(y), \qquad < \phi(f), u > = A(u)(x)$$

Let $\sigma_{k,1}, ..., \sigma_{k,m(\delta_k)}$ be a basis of $F(\delta_k^*)$ such that $Im\sigma_{k,\ell} = H_{\delta^*, d_\ell}$. From the proof of $i)$, we know the existence of $P_{k,1}, ..., P_{k,m(\delta_k)} \in S'(g)^G$ such that

$$h_{k,[r]} = \sum_{\ell:d_\ell = r} P_{k,\ell}(y)\beta_y(\sigma_{k,\ell})$$

or even

$$\mathcal{C}_{k,[r]} = \sum_{\ell:d_\ell=r} P_{k,\ell}(y)\sigma_{k,\ell}$$

In other words, if $w_k \in \delta_k^*$,

$$< h_{k,[r]}, w_k >= \sum_{\ell:d_\ell=r} P_{k,\ell}(y)\sigma_{k,\ell}(w_k)(y)$$

Note that $P_{k,\ell}$ is homogeneous of degree $d - d_\ell$. So let $u \in \rho^*$, $v \in \lambda^*$; with the above identification, there exist $w_k \in \delta_k^*$ such that

$$u \otimes v = w_1 + \ldots + w_t$$

Hence

$$< f \otimes g, u \otimes v >=< h_1, w_1 > +\ldots+ < h_s, w_s >= \sum_{k,r} < h_{k,[r]}, w_k >$$

i.e.

$$\mathcal{A}(u)(y)\mathcal{B}(v)(y) = \sum_{k,r} \sum_{\ell:d_\ell=r} P_{k,\ell}(y)\sigma_{k,\ell}(w_k)(y)$$

But it is easy to see that for all $t \in G$:

$$\mathcal{A}(u)(ty)\mathcal{B}(v)(ty) = \sum_{k,r} \sum_{\ell:d_\ell=r} P_{k,\ell}(ty)\sigma_{k,\ell}(w_k)(ty)$$

Thus $\mathcal{A}(u)\mathcal{B}(v)$ and $\sum_{k,r} \sum_{\ell:d_\ell=r} P_{k,\ell}\sigma_{k,\ell}(w_k)$ agree on $\overline{\bigcup_{c\in\mathbb{C}^\times} Gcy}$. By (2), as $x \in N = \{z \in U : J(z) = 0 \forall J \in S'(g)^G$ homogeneous of positive degree $\}$, we have

$$\mathcal{A}(u)(x)\mathcal{B}(v)(x) = \sum_{k} \sum_{\ell:d_\ell=d} P_{k,\ell}(x)\sigma_{k,\ell}(w_k)(x)$$

$$= \sum_{k} \sum_{\ell:d_\ell=d} P_{k,\ell}(y)\sigma_{k,\ell}(w_k)(x) = \sum_{k} \mathcal{C}_{k,[d]}(w_k)(x)$$

(because if $d_\ell = d$, $P_{k,\ell} \in \mathbb{C}$); i.e.

$$< \phi(f) \otimes \phi(g), u \otimes v >= \sum_{k} < \phi(h_{k,[d]}), w_k >$$

Now, we are essentially done, because if V is a G-module isomorphic to ρ, then $\phi : V^M \to V^P$ does not depend on the choosen isomorphism with ρ.

PROOF OF *iii*): We can deduce easily:

$$S'(g \oplus U)^G \simeq S'(g)^G \otimes (H \otimes S'(U))^G$$

LEMMA. *Let $\mathcal{A} = \oplus_{j \geq 0} \mathcal{A}_j$ be a graded $\mathbf{C}$-algebra, $\mathcal{B} = \oplus_{j \geq 0} \mathcal{B}_j$ a graded subalgebra, $H = \oplus_{j \geq 0} H_j$ a graded subspace ($\mathcal{A}_0 = \mathcal{B}_0 = \mathbf{C}$) such that the multiplication induces an isomorphism $\mathcal{A} \simeq \mathcal{B} \otimes H$. Let $\mathcal{I}$ be the ideal of $\mathcal{A}$ generated by $\oplus_{j > 0} \mathcal{B}_j$. Then $\mathcal{A} = \mathcal{I} \oplus H$.*

Considering the natural gradation of $S'(g \oplus U)^G$, we obtain from the preceding Lemma an isomorphism of vector spaces

$$S'(g \oplus U)^G / \mathcal{J} \simeq \oplus_{\rho \in \Gamma}(H_{\rho^*} \otimes S'(U)_\rho)^G.$$

On the other hand, let us recall the existence of an isomorphism of vector spaces from $(H \otimes S'(U))^G$ onto $C = S'(U)^M$ given explicitly by

$$\sum_j h_j \otimes v_j \mapsto \sum_j h_j(y)v_j.$$

Moreover, the image of $(S'(U) \otimes H_d)^G$ (the homogeneous component of degree d) is precisely $C[d]$. Composing with the obvious isomorphism $C \simeq \mathrm{gr}C$, we obtain an isomorphism of vector spaces $S'(g \oplus U)^G / \mathcal{J} \to \mathrm{gr}C$. Our last task is to show that it preserves the multiplication. So let $\sum_j v_j \otimes p_j \in (S'(U) \otimes H_d)^G$, $\sum_i w_i \otimes q_i \in (S'(U) \otimes H_e)^G$. There exist

$$h_r^{i,j} \in H_{d+e-r}, \qquad T_r^{i,j} \in S'(g)_r^G$$

such that

$$p_j q_i = \sum_{0 \leq r \leq d+e} T_r^{i,j} h_r^{i,j}$$

Hence, with the identification $S'(g \oplus U)^G / \mathcal{J} \simeq (H \otimes S'(U))^G$, we have

$$(\sum_j v_j \otimes p_j)(\sum_i w_i \otimes q_i) = \sum_{i,j} v_j w_i T_0^{i,j} h_0^{i,j}$$

On the other hand, we have in C:

$$(\sum_j p_j(y)v_j)(\sum_i q_i(y)w_i) = \sum_{i,j,r} T_r^{i,j}(y)h_r^{i,j}(y)v_j w_i$$

But as

$$\sum_{i,j} T_r^{i,j}(y)h_r^{i,j}(y)v_j w_i \in C[d+e-r]$$

the product $\mathrm{gr}C$ is

$$(\sum_j p_j(y)v_j)(\sum_i q_i(y)w_i) = \sum_{i,j} T_r^{i,j} h_r^{i,j}(y)v_j w_i$$

This concludes the proof of *iii*) .

REMARK: *iii*) is well-known and easy to prove in another way. We think however that this approach could be useful for explicit computations.

REFERENCES

[A 1] ANDRUSKIEWITSCH, N., *On the complicatedness of the pair (g,K)*, Rev. Mat. Univ. Complut. Madrid **2** 1 (1989), 12–28.

[A 2] *A new proof of Tirao's Restriction Theorem*, Revista de la UMA (To appear).

[A 3] *Computing some rings of invariants*, Proceedings of the IX ELAM (To appear).

[A T] ANDRUSKIEWITSCH, N., TIRAO, J., *A Restriction Theorem for modules having a spherical submodule.*, Trans. AMS (To appear).

[B T] BREGA, O.A. and TIRAO, J. A., *A property of a distinguished class of K-modules associated to the classical rank one semisimple Lie algebras.*, Proc. oof the IX ELAM (To appear).

[B L V] BRION, M., LUNA, D. and VUST, Th, *Espaces homogenes spheriques*, Inv. Math. **84** (1986), 617–632.

[E] ELASHVILI, A.G., *Canonical form and stationary subalgebras of points of general position for simple linear Lie groups*, Funct. Anal. Appl. **6** (1972).

[Kc] KAC, V.G., *Some remarks on nilpotent orbits*, Journal of Algebra **64** (1980), 190–213.

[K] KOSTANT, B., *Lie group representations on polynomial rings.*, Amer. J. of Math. **85** (1963), 327–404.

[K R] KOSTANT, B. and RALLIS, S., *Orbits and representations associated with symmetric spaces*, Amer. J. of Math. **93** (1971), 753–809.

[K T] KOSTANT, B. and TIRAO, J.A., *On the structure of certain subalgebras of a universal enveloping algebra*, Trans. AMS **218** (1976), 133–154.

[L R] LUNA, D. and RICHARDSON R.W., *A generalization of the Chevalley Restriction Theorem*, Duke J. of Math. **46** (1979), 487–497.

[M] MUMFORD, D., "Geometric Invariant Theory," Springer-Verlag, 1980. 2nd. edition.

[T] TIRAO, J., *A Restriction Theorem for Semisimple Lie Groups of Rank One*, Trans. A.M.S. **279** (1983), 651–660.

[V K] VINBERG, E.B. and KIMEL'FELD,B.N., *Homogeneous domains on flag manifolds and spherical subgroups*, Funct. Anal. Appl. **12** (1978), 168–174.

Valparaíso y R. Martínez. Ciudad Universitaria. CP: 5000. Córdoba. R. Argentina.

Generalized Weil Representations for SL(n,k), n odd, k a Finite Field

José Pantoja[1] and Jorge Soto-Andrade[2]

Abstract. Generalized Weil representations for the group $G = \mathrm{SL}(n, k)$, k a finite field, are constructed by contraction of a suitable complex G-vector bundle with the help of an appropriate connection. This extends a previous construction by the authors for the case where n is even, obtained with the help of "Grassmann–Heisenberg" groups.

0. Introduction

In a previous paper [P-SA], we constructed generalized Weil representations for the group $G = \mathrm{SL}(n, k)$, for n even, and k a finite field, by extending Lion–Vergne's version [L-V] of A. Weil's original method [W], applied to $\mathrm{SL}(2, k)$, to suitable higher order analogues of the Heisenberg group in one degree of freedom (which we call "Grassmann–Heisenberg groups"). The construction was restricted to the case where n is even, because only in this case there exist reasonable analogues of the Schrödinger representation for our Grassmann–Heisenberg groups, whose isomorphy type will be fixed by the natural action of the whole group G (in the case where n is odd one obtains only Schrödinger-like representations whose isomorphy type is fixed by very small subgroups of G).

We present in this paper a more geometric approach to this construction, using the method of contraction of a complex G-vector bundle over a base point with the help of a suitable G-equivariant connection [SA]. A first specialization of this method allows us to recover the representations constructed in [P-SA]. A second specialization allows us to obtain generalized Weil representations also for $\mathrm{SL}(n, k)$, n odd.

[1] Partially supported by FONDECYT (Project 89-1073) and DICYT, Univ. Valparaiso (Project UV 2/89).

[2] Partially supported by FONDECYT (Project 89-1073) and DTI. U. Chile (Project E-2587).

It should be noticed that our method could be applied to a local base field instead of a finite base field, a case to which we have reduced ourselves here for technical simplicity.

1. Construction of Representations by Contraction

We recall here the results in [SA]. Let G be a finite group and $\xi = (E, B, p, \eta)$ a complex G-vector bundle, where as usual E denotes the total space, B the base space, p the projection from E to B, and η the action of G on E and B. We have then

$$p(\eta_g(v)) = \eta_g(p(v)) \quad (g \in G, \ v \in E).$$

We denote by E_b the fiber $p^{-1}(b)$ over $b \in B$.

To obtain a linear representation of G out of ξ, by contraction over a base point $b_0 \in B$, one needs an equivariant connection on ξ, in the following sense.

Definition 1. An equivariant connection on the complex G-vector bundle $\xi = (E, B, P, \eta)$ is a family of linear isomorphisms $\Gamma = \{\gamma_{b,a}\}_{a,b \in B}$ such that

(i) each $\gamma_{b,a}$ is a linear isomorphism from the fiber E_a onto the fiber E_b and every $\gamma_{a,a}$ is the identity;

(ii) we have

$$\gamma_{\eta_g(b),\eta_g(a)} \circ \eta_g = \eta_g \circ \gamma_{b,a}$$

for all $a, b \in B$, $g \in G$;

(iii) we have $\gamma_{c,b} \circ \gamma_{b,a} = \mu_\Gamma(c, b, a)\gamma_{c,a}$, for all $a, b, c \in B$, for a suitable function $\mu_\Gamma \colon B \times B \times B \to \mathbb{C}^\times$ called the multiplier of Γ.

We say that the connection Γ is flat iff its multiplier μ_Γ is the constant function 1.

An easy calculation shows that given an equivariant connection Γ on ξ we can contract ξ to a projective representation of G.

Proposition 1. *Let us fix a base point* $b \in B$ *and suppose that an equivariant connection G is given on the complex G-vector bundle ξ with base B. A projective representation (V, σ) of G, with multiplier $\mu_\sigma(g, h) = \mu_\Gamma(b, \eta_g(b), \eta_{gh}(b))$ for $g, h \in G$, may be constructed as follows. Let $V = E_b$ and define σ by*

$$\sigma_g(v) = \gamma_{b,\eta_g(b)}(\eta_g(v)) \quad (g \in G, \ v \in V).$$

We call (V, σ) the representation of G obtained by contraction of ξ over b according to Γ.

2. A Geometric Approach to the Construction of Generalized Weil Representations for $G = \mathrm{SL}(n, k)$, n even and k a Finite Field

2.1. Generalized Weil representations for $G = \mathrm{SL}(n, k)$, k a finite field and n even, were constructed in [P-SA] in the following way.

Let V be an n-dimensional vector space over the field k and let $G = \mathrm{SL}(V) \cong \mathrm{SL}(n, k)$. The Grassmann–Heisenberg group of V is the subgroup $U(V) = 1 \oplus \left(\bigoplus_{i \geq 1} \Lambda^i(V) \right)$ of the multiplicative group $(\Lambda V)^\times$ of the Grassmann (or exterior) algebra ΛV of V. Notice that $U(V)$ coincides with the usual Heisenberg group in one degree of freedom when V is a plane. Only for n even has the group $U(V)$ a Schrödinger-like irreducible representation $\sigma^{\ell,c}$ (where ℓ is a 1-dimensional subspace of V and $c \in k^\times$) whose isomorphy type is fixed by the natural action of G by automorphisms of the group $U(V)$. Therefore for each $g \in G$ there exists an intertwining operator ρ_g^c of $\sigma^{\ell,c}$, defining in this way a projective representation ρ^c of G. This representation turns out to be a bona fide representation for $n \geq 4$, for the specific choice of the operators ρ^c which we recall below.

Let ψ be a fixed nontrivial character of the additive group k^+ of the base field k. Elements $\omega \in \Lambda V$ will be written in the form $\omega = \sum_{0 \leq i \leq n} \omega_i$, with $\omega_i \in \Lambda^i V$. Fix a volume element $e_V \in \Lambda^n V$ and let ψ_V be the character of the additive group $(\Lambda V)^+$ of the exterior algebra ΛV defined by $\psi_V(\omega) = \psi(\omega_n / e_V)$ for $\omega \in \Lambda V$. Introduce the subgroup

$$U^\ell = 1 \oplus (\ell \wedge \Lambda V)$$

and call Z' the center of $U(V)$. For each $c \in k^\times$ define a character $\psi_V^c \otimes 1$ of the subgroup $U^\ell Z'$ of $U(V)$ by

$$(\psi_V^c \otimes 1)((1 + \xi) \wedge (1 + \omega)) = \psi_V(c\xi) \quad (1 + \xi \in U^\ell, \ 1 + \omega \in Z').$$

Then the generalized Schrödinger representation $\sigma^{\ell,c}$ of $U(V)$ associated to the parameters $\ell < V$, $\dim \ell = 1$, $c \in k^\times$, is just the induced representation from the character $\psi_V^c \otimes 1$ of $U^\ell Z'$ up to $U(V)$.

Now for any two distinct vectorial lines ℓ, $\ell' < V$, we put

$$(1) \qquad (T_{\ell,\ell'} f)(h) = q^{-d(n)/2} \sum_{u \in U^{\ell'} Z' / U^{\ell'} Z' \cap U^\ell Z'} (\psi_V^c \otimes 1)(u^{-1}) f(uh)$$

for f in the space $E(\ell, c)$ of $\sigma^{\ell,c}$, $h \in U(V)$, where $d(n) = 2^{n-3}$ (resp. $d(n) = 1$) if $n \geq 4$ (resp. $n = 2$) and q is the number of elements in k.

The generalized Weil operators ρ_g^c ($g \in G$) are given by $\rho_g^c = T_{\ell,g(\ell)} \circ \tau_g$ in the space $E(\ell, c)$, where $\tau_g = ? \circ g^{-1}$ intertwines $\sigma^{\ell,c} \circ g^{-1}$ and $\sigma^{g(\ell),c}$. We recover in this way the classical projective Weil representations for $\mathrm{SL}(2, k)$ and we obtain a true representation of dimension $q^{2^{n-2}}$ for $\mathrm{SL}(V) \cong \mathrm{SL}(n, k)$ for $n \geq 4$, n even.

2.2. It turns out, however, that this construction may be thought of in a more geometric way as follows.

Proposition 2. *The generalized Weil representation* $(E(\ell, c), \rho)$ *of* $G = SL(V) \cong SL(n, k)$, *for* n *even may be obtained by the contraction procedure recalled above according to the following data:*

(i) *the complex* G-*vector bundle* $\xi = (E, B, p, \eta)$ *is defined as follows.* Let

$$W = \Lambda^1 V \oplus \Lambda^3 V \oplus \cdots \oplus \Lambda^{n-1} V,$$

$$\langle \omega, \zeta \rangle = (\omega \wedge \zeta)_n \qquad\qquad (\omega, \zeta \in W),$$

$$B = \{\ell < V \mid \dim \ell = 1\}.$$

We put for all $\ell \in B$,

$$W_\ell = \ell \oplus \ell \wedge \Lambda^2 V \oplus \cdots \oplus \ell \wedge \Lambda^{n-2} V,$$

and

$$E_\ell = \{f \colon W \to \mathbb{C} \mid f(\omega + \zeta) = \psi(c\langle \omega, \zeta \rangle) f(\omega) \quad \forall \omega \in W, \ \forall \zeta \in W_\ell\},$$

$$E = \bigcup_{\ell \in B} E_\ell,$$

p *is the obvious projection and* η *is given by*

$$(\eta_g f)(\omega) = f(\Lambda g^{-1} \omega) \qquad (g \in G, \ \omega \in W),$$

$$\eta_g(\ell) = g(\ell) \qquad\qquad (g \in G, \ \ell \in B);$$

(ii) *the connection* $\Gamma = \{\gamma_{\ell', \ell}\}_{\ell, \ell' \in B}$ *is defined by* $\gamma_{\ell, \ell} = Id_{E_\ell}$ *for all* $\ell \in B$ *and*

$$(2) \qquad (\gamma_{\ell', \ell} f)(\omega) = q^{-d(n)/2} \sum_{\zeta \in W_{\ell'} / W_\ell \cap W_{\ell'}} \psi(-c\langle \omega, \zeta \rangle) f(\omega + \zeta)$$

for all $f \in W_\ell$, $\omega \in W$, $\ell \in B$, $\ell \neq \ell'$.

Proof. This follows easily by associating to each function f in the space of the (right) Mackey model for the induced representation, the function f' on W defined by

$$f'(\omega_1 + \omega_3 + \cdots + \omega_{n-1}) = f(1 + \omega) \quad (\omega \in W). \qquad \square$$

3. Generalized Weil Representations for $G = SL(n, k)$, n Odd, k a Finite Field

3.1. Let V be a k-vector space of odd dimension $n = 2m + 1$. Put $W = \Lambda^1 V \oplus \Lambda^2 V \oplus \cdots \oplus \Lambda^{n-1} V$ and define an alternating form $\langle \, , \, \rangle$ on W by

$$(3) \qquad\qquad\qquad \langle \omega, \zeta \rangle = (\omega^\vee \wedge \zeta)_n$$

for $\omega, \zeta \in W$, where $\omega^{\vee} = \sum_i (-1)^i \omega_i$. Then $(W, \langle \, , \, \rangle)$ is a $2(2^{2m} - 1)$-dimensional nondegenerate symplectic space.

Let us denote by B the set of all one-dimensional subspaces ℓ of V. For each $\ell \in B$ let

$$(4) \qquad W_\ell = \ell \oplus (\ell \wedge \Lambda^1 V) \oplus \cdots \oplus (\ell \wedge \Lambda^{n-2} V).$$

Then W_ℓ is a Lagrangian subspace of $(W, \langle \, , \, \rangle)$, of dimension $2^{2m} - 1$. For $\ell, \ell' \in B$, put

$$(5) \qquad W_{\ell, \ell'} = W_\ell / W_\ell \cap W_{\ell'}.$$

3.2. We will need later the following lemmas, which can be easily checked.

Lemma 1. *For $\ell, \ell' \in B$, $\ell \neq \ell'$ and $V = \ell \oplus \ell' \oplus M$ we have $W_{\ell, \ell'} = \ell \wedge (\Lambda M)$, and so $\# W_{\ell, \ell'} = \# \Lambda M = q^{2^{n-2}}$.*

Lemma 2. (i) *For $\ell, \ell', \ell'' \in B$ with $\dim \langle \ell, \ell', \ell'' \rangle = 3$ and $V = \ell \oplus \ell' \oplus \ell'' \oplus N$, we have*

$$\left(W_{\ell, \ell''} \right)^{\perp} \cap \left(W_{\ell', \ell} + W_{\ell'', \ell'} \right) = \ell' + \ell \wedge \ell'' \wedge (\Lambda N).$$

(ii) *For $\ell, \ell', \ell'' \in B$ with $\ell < \ell' \oplus \ell''$ and $V = \ell' \oplus \ell'' \oplus M$ we have*

$$\left(W_{\ell, \ell''} \right)^{\perp} \cap \left(W_{\ell', \ell} + W_{\ell'', \ell'} \right) = \ell \wedge (\Lambda M).$$

Lemma 3. *Let F be an r-dimensional vector space over k, with r odd. Fix a nontrivial character ψ of k^+ and let $\# k = q$. Then*

$$\sum_{\eta \in \Lambda V} \psi_F((\eta \wedge \eta)_r) = q^{2^{r-1}}.$$

Proof. Since r is odd, say $r = 2m + 1$, we have

$$(\eta \wedge \eta)_r = 2 \sum_{0 \leq i \leq m} \eta_i \wedge \eta_{r-i}.$$

The lemma follows easily if we notice that $\eta_i + \eta_{r-i} \mapsto 2\eta_i \wedge \eta_{r-i}$ defines a nondegenerate hyperbolic quadratic form on the $2\binom{r}{i}$-dimensional space $\Lambda^i F \oplus \Lambda^{r-i} F$ and we recall that the Gauss sum $\sum_{x \in L} \psi(Q(x))$ associated to any nondegenerate hyperbolic quadratic form Q on a $2s$-dimensional k-vector space L has the value q^s. $\qquad \square$

3.3. Let us define the complex G-vector bundle $\eta = (E, B, p, \tau)$ as follows:

B stands as above for the set of all one-dimensional subspaces ℓ of W;

$$E_\ell = \{f \colon W \to \mathbb{C} \mid f(\omega + \zeta) = \psi(\langle \omega, \zeta \rangle) f(w) \ \forall \omega \in W, \ \forall \zeta \in W_\ell\};$$

$$E = \bigcup_{\ell \in B} E_\ell;$$

and p is the canonical projection from E to B.

The group $G = \mathrm{SL}(V)$ acts in a natural way on η by

$$\begin{aligned} (\tau_g f)(\omega) &= f(\Lambda g^{-1} \omega) && (g \in G, \ \omega \in W, \ f \in E), \\ \tau_g(\ell) &= g(\ell) && (g \in G, \ \ell \in B). \end{aligned}$$

3.4. We define the family of linear transformations $\Gamma = \{\gamma_{\ell', \ell}\}_{\ell, \ell' \in B}$ on the fibers of η by

$$(6) \qquad (\gamma_{\ell', \ell} f)(\omega) = q^{-2^{n-3}} \sum_{\zeta \in W_{\ell', \ell'}} \psi(-\langle \omega, \zeta \rangle) f(\omega + \zeta)$$

for $f \in E_\ell$, $\omega \in W$, $\ell, \ell' \in B$, $\ell \neq \ell'$ and $\gamma_{\ell, \ell} = \mathrm{Id}_{E_\ell}$ for all $\ell \in B$.

It is clear that the transformations $\gamma_{\ell', \ell}$ satisfy properties (i) and (ii) of Definition 1. Property (iii) will follow from the next two propositions.

Proposition 3. *We have* $\gamma_{\ell, \ell'} \circ \gamma_{\ell', \ell} = \mathrm{Id}_{E_\ell}$ *for* $\ell, \ell' \in B$.

Proof. In the nontrivial case $\ell \neq \ell'$, using Lemma 1 we obtain for $f \in E_\ell$, $\omega \in W$,

$$\begin{aligned} (\gamma_{\ell, \ell'} \circ \gamma_{\ell', \ell} f)(\omega) \\ &= q^{-2^{n-2}} \sum_{\zeta \in \ell \wedge \Lambda M} \sum_{\xi \in \ell' \wedge \Lambda M} \psi(-\langle \omega, \zeta \rangle - \langle \omega + \zeta, \xi \rangle) f(\omega + \zeta + \xi) \\ &= q^{-2^{n-2}} \sum_{\xi \in \ell' \wedge \Lambda M} \left(\sum_{\zeta \in \ell \wedge \Lambda M} \psi(2\langle \xi, \zeta \rangle) \right) \psi(-\langle \omega, \xi \rangle) f(\omega + \xi) \\ &= q^{-2^{n-2}} q^{2^{n-2}} f(\omega). \end{aligned}$$

Proposition 4. *We have*

$$\gamma_{\ell, \ell''} \circ \gamma_{\ell'', \ell'} \circ \gamma_{\ell', \ell} = \mathrm{Id}_{E_\ell}$$

for $\ell, \ell', \ell'' \in B$.

Proof. If $\#\{\ell,\ell',\ell''\} \le 2$ then Prop. 4 follows from Prop. 3 and the fact that $\gamma_{\ell,\ell} = \mathrm{Id}_{E_\ell}$ for all $\ell \in B$. Assume then that $\#\{\ell,\ell',\ell''\} = 3$ and put $S = (\gamma_{\ell,\ell''} \circ \gamma_{\ell'',\ell'} \circ \gamma_{\ell',\ell}f)(\omega)$ for $f \in E_\ell$ and $\omega \in W$. Then we have

$$
S = q^{-2^{n-2}-2^{n-3}} \sum_{\zeta,\zeta',\zeta''} \psi(-\langle\omega,\zeta\rangle - \langle\omega+\zeta,\zeta''\rangle
$$
$$
- \langle\omega+\zeta+\zeta'',\zeta'\rangle)f(\omega+\zeta+\zeta'+\zeta'')
$$
$$
= q^{-2^{n-2}-2^{n-3}} \sum_{\zeta',\zeta''} \left(\sum_\zeta \psi(2\langle\zeta'+\zeta'',\zeta\rangle)\right)\psi(\langle\zeta',\zeta''\rangle)
$$
$$
- \langle\omega,\zeta'+\zeta''\rangle)f(\omega+\zeta'+\zeta'')
$$

where we sum over $\zeta \in W_{\ell,\ell''}$, $\zeta' \in W_{\ell',\ell}$, and $\zeta'' \in W_{\ell'',\ell'}$ (see (5)). Therefore

$$
S = q^{-2^{n-3}} \sum_{\zeta',\zeta''} \psi(\langle\zeta',\zeta''\rangle - \langle\omega,\zeta'+\zeta''\rangle)f(\omega+\zeta'+\zeta'')
$$

where we sum now over all $\zeta' \in W_{\ell',\ell}$, $\zeta'' \in W_{\ell'',\ell'}$ such that $\zeta'+\zeta'' \in W_{\ell,\ell''}^\perp$. We have two cases:

(i) The lines ℓ, ℓ', ℓ'' span a 3-dimensional subspace of V. Write then $V = \ell \oplus \ell' \oplus \ell'' \oplus N$ where N is an $(n-3)$-dimensional subspace of V. According to Lemma 2.(i) we get, since $\zeta'' \in W_\ell$ and $f \in E_\ell$,

$$
S = q^{-2^{n-3}} \sum_{\zeta' \in \ell} \left(\sum_{\zeta'' \in \ell'' \wedge \ell \wedge N} \psi(\langle\zeta',\zeta''\rangle)\right)\psi(-\langle\omega,\zeta'\rangle)f(\omega+\zeta')
$$
$$
= \sum_{\zeta' \in \ell' \cap (\ell'' \wedge \ell \wedge N)^\perp = \{0\}} \psi(-\langle\omega,\zeta'\rangle)f(\omega+\zeta')
$$
$$
= f(\omega)
$$

as it was to be shown.

(ii) The line ℓ lies in the plane $\ell' \oplus \ell''$. In this case we write $V = \ell' \oplus \ell'' \oplus M$ where M is as $(n-2)$-dimensional subspace of V. We have now, according to Lemma 2.(ii),

$$
S = q^{-2^{n-3}} \sum_{\zeta',\zeta''} \psi(\langle\zeta',\zeta''\rangle - \langle\omega,\zeta'+\zeta''\rangle)f(\omega+\zeta'+\zeta'')
$$

where we sum over all $\zeta' \in \ell' \wedge \wedge M$, $\zeta'' \in \ell'' \wedge \wedge M$ such that $\zeta'+\zeta'' \in \ell \wedge \wedge M$. But then $\zeta'+\zeta'' \in W_\ell$ and since $f \in E_\ell$ we obtain

$$
S = q^{-2^{n-3}} \left(\sum_{\zeta \in \ell \wedge \wedge M} \psi(\langle\zeta',\zeta''\rangle)f(\omega)\right),
$$

where now ζ' (resp. ζ'') denotes the component of $\zeta \in \ell \wedge \Lambda M \subset (\ell' \oplus \ell'') \wedge \Lambda M$ in $\ell' \wedge \Lambda M$ (resp. $\ell'' \wedge \Lambda M$). Let us choose a generator v of ℓ and call v' (resp. v'') its component in ℓ' (resp. ℓ''). Then necessarily $\zeta \in \ell \wedge \Lambda M$ implies $\zeta = v \wedge \eta$, $\zeta' = v' \wedge \eta$ and $\zeta'' = v'' \wedge \eta$ for some $\eta \in \Lambda M$. Hence

$$\langle \zeta', \zeta'' \rangle = \langle v' \wedge \eta, v'' \wedge \eta \rangle = (v'' \wedge v') \wedge (\eta \wedge \eta)_{n-2}$$

and therefore, by Lemma 3, we get

$$S = q^{-2^{n-3}} \sum_{\eta \in M} \psi_M((\eta \wedge \eta)_{n-2}) f(\omega) = q^{-2^{n-3}} q^{2^{n-3}} f(\omega) = f(\omega). \quad \square$$

Theorem. *The family* $\Gamma = \left\{ \gamma_{\ell', \ell} \right\}_{\ell, \ell' \in B}$ *of linear transformations defined by* (6) *is a G-equivariant flat connection on the complex G-vector bundle* η.

Proof. It follows immediately from Props. 3 and 4. $\quad \square$

Corollary. *The bundle* η *defined above* (cf. 3.3) *gives raise by contraction* (cf. 1) *according to the flat equivariant connection* Γ *over any line* $\ell < V$ *to a* $q^{2^{n-1}} - 1$-*dimensional representation* (E_ℓ, ρ) *of* $G = \mathrm{SL}(V) \cong \mathrm{SL}(n, k)$, *which we call generalized Weil representation of* G.

REFERENCES

[L-V] G. Lion and M. Vergne, *The Weil Representation, Maslov Index and Theta Series*, Progress in Math., vol. 6, Birkhäuser Verlag, Basel-Boston, 1980.

[P-SA] J. Pantoja and J. Soto-Andrade, *Groupes de Grassmann–Heisenberg et représentations de Weil généralisées pour* SL_n, *n pair*, Astérisque **168** (1988), 167–189.

[SA] J. Soto-Andrade, *Geometrical Gel'fand models, tensor quotients and Weil representations*, Proc. Sympos. Pure Math., vol. 47, Amer. Math. Soc., Providence, RI, 1987, 305–316.

[W] A. Weil, *Sur certains groupes d'opérateurs unitaires*, Acta Math. **150** (1964), 143–211.

José Pantoja
Instituto de Matemáticas y Física
Universidad de Valparaíso
Casilla 1470, Valparaíso, CHILE

Jorge Soto-Andrade
Departamento de Matemáticas
Facultad de Ciencias
Universidad de Chile
Casilla 653, Santiago, CHILE

Local Multiplicity of the Intersection
of Lagrangian Cycles and the Index
# of Holonomic Modules

Alberto S. Dubson

Introduction

Let M be an oriented manaifold. Cycles on M, of complementary dimensions, have a well defined *multiplicity of intersection* at a given point, whenever they are in general position near it; if they are not, the multiplicity of intersection is not defined. The same applies to two chains, their *degree of intersection* is defined whenever their boundaries have disjoint support. Of course the chains can be deformed so that their boundaries be disjoint, but then, different deformations will give different degrees of intersection and in general there is no preferred deformation.

But if M is real analytic and symplectic and the chains or cycles are supported on subanalytic Lagrangian subsets a good definition of *multiplicity* or *degree of intersection* can be given, using deformations by Hamiltonian flows.

Let A and B be two subanalytic Lagrangian subsets of M (i.e. both contain a dense Lagrangian submanifold). Let x_0 be a point not in the boundaries of A and B, and let $H(x) = \text{distance}(x, x_0)$ be the distance function derived from an arbitrary analytic Riemann metric near x_0. We prove that there exists $\delta(x_0) > 0$ such that $\forall \varepsilon,\ 0 < \varepsilon \leq \delta(x_0)$ the Hamiltonian flow f_t associated to H deforms $A \cap H^{-1}(\varepsilon)$ away from B that is $f_t(A \cap H^{-1}(\varepsilon)) \cap B = \phi$ for $|t|$ sufficiently small. This result is false without analyticity assumptions. Let now A and B be two PL Lagrangian subanalytic chains in M (i.e. carried by subanalytic Lagrangian subsets). The preceding result, together with Lefschetz's theory, allows us to give a good definition of the local multiplicity of intersection $m(A, B, x_0; M)$ of A and B at any point x_0 not on their boundaries, even if they are not in general position near x_0. Using similar arguments we can define the degree of intersection of two subanalytic Lagrangian chains of complementary dimension, even when their boundaries are not disjoint. In fact we define two degrees (or multiplicities) of intersection: $\deg_+$ and $\deg_-$, depending on whether t is taken positive or negative. They are not necessarily equal in the real case but they coincide when M, A and B are complex analytic and the symplectic form on M is holomorphic. For any constructible complex of sheaves $K^\bullet$ or Holonomic Module $\mathcal{M}$ on a complex manifold M, we express their local indexes $\chi_{K^\bullet}(x)$, $\chi_{\mathcal{M}}(x)$ at any $x \in M$, as the local multiplicity of intersection at $(x, 0) \in T^*M$ of suitable Lagrangian cycles in T^*M.

The proofs depend on the theory of subanalytic sets and standard properties of PL chains. All the necessary definitions and results will be recalled (together with some new technical refinements) in Sections 1 and 2. In Section 3 we state the main theorems, after some results on Hamiltonian flows. The index formulas for Holonomic Modules in terms of multiplicity of intersection of Lagrangian cycles are detailed in Section 4. Section 5 is devoted to a brief sketch of the proof of the main deformation theorem.

A complete version will be published elsewhere. Although the preliminaries make this exposition somewhat superficial, I hope they will make it also more accessible.

1. Subanalytic Sets [H]

Throughout this paper, *proper map* means proper real analytic map between real analytic manifolds. A *proper image* is the image of a set under a proper map.

The *semianalytic* subsets of an analytic manifold M are those locally defined by a finite number of analytic inequalities. Their finite unions and intersections, complements and closures, in particular differences and boundaries $bA = \overline{A} \setminus A$, are also semianalytic. But their images under proper maps are usually not.

The family of *subanalytic* subsets of M is the smallest one containing the proper images of semianalytic sets and closed under the operations of (locally finite) union and intersection, closure and complement. Subanalytic subsets admit (subanalytic) *stratifications* that is, they can be decomposed into the disjoint union $\bigcup_{\alpha \in I} A_\alpha$ of a *locally finite* family $\{A_\alpha\}_{\alpha \in I}$ of connected real analytic submanifolds, each A_α a subanalytic subset of M, such that its closure $\overline{A}_\alpha$ (and so its boundary) is the union of some members of the family. Also any locally finite family $\{S_i\}_{i \in \Lambda}$ of subanalytic subsets of M can be refined by a stratification of its union $\bigcup_{i \in \Lambda} S_i$.

Any stratification of a subanalytic set can be refined to one which verifies conditions "a" and "b" of Whitney: they are used many times in the proofs of the main results, but as they are not necessary for the understanding of this brief presentation, they are not recalled.

A point x of a subanalytic subset A is *smooth* of dimension p if there exists a neighborhood U of x in M such that $A \cap U$ is a real analytic submanifold of dimension p of M.

The dimension $\dim_x A$ of A at a point $x \in A$ is equal to p if A contains smooth points of dimension p, but not of higher dimension, arbitrarily near x; $\dim A = \sup\{\dim_x A, \ x \in A\}$ is finite if A is compact; $\dim \overline{A} = \dim A$, $\dim bA \leq \dim A - 1$. Let $f: M \to N$ be a real analytic map, $A \subset M$, $B \subset N$ subanalytic subsets. If $\dim B = p$ and $\forall x \in B$, $\dim A \cap f^{-1}(x) \leq q$ then $\dim(A \cap f^{-1}(B)) \leq p + q$.

The connected components of a subanalytic sets are subanalytic and form a locally finite family. In particular a 0-dimensional subanalytic subset is locally finite, a fact which will often be used as follows: if $A \subset M$ is subanalytic, $K \subset M$ is compact and $A \cap K$ contains an infinite number of points, then $\dim A \geq 1$. This argument will be essential in many proofs: that explains, in part, why an "analyticity" condition is required in the statement of the main theorems.

Let $f : M \to \mathbb{R}$ be a real analytic map. For any compact K of M, the critical values taken by f over K are finite; of course this fact is false without analyticity conditions. More generally, any subanalytic subset A of M admit (Whitney) stratifications $\{A_\alpha\}_{\alpha \in I}$ such that for any compact K of M and any $\alpha in I$, the critical values taken by the function $f|_{A_\alpha} : A_\alpha \to \mathbb{R}$ over $K \cap A_\alpha$ are finite. Except for this last property (easily deduced from the existence of stratifications), all the other ones are explicitly stated and proved in the literature. We will need some other delicate properties, which are not very difficult to prove; in order to keep this exposition brief, they will not be proved here.

Theorem 1. *Let $f : X \to Y$ be a real analytic map and $\{B_i\}_{i \in I}$ a locally finite family of subanalytic subsets of X. There exist stratifications $\{S_\alpha\}_{\alpha \in \Lambda}$ of X which refine $\{B_i\}_{i \in I}$ such that*

i. $\forall \alpha \in \Lambda$ *the rank of the differential map $d(f \mid S_\alpha) : TS_\alpha \to TY$ associated to $f \mid_{S_\alpha} : S_\alpha \to Y$, a constant.*

ii. *If $\dim S_\alpha = \operatorname{rank} d(f \mid_{S_\alpha})$ then $f(S_\alpha)$ is smooth and $f \mid_{S_\alpha} : S_\alpha \to Y$ is proper.*

Proposition 1. *Let $A \subset X$ be a subanalytic subset and $f : X \to Y$ as before. If the restriction $f \mid_A : A \to Y$ is proper then $f(A)$ and $f(bA)$ are disjoint $(bA = \overline{A} \setminus A)$.*

Corollary 1. *There exist stratifications of X which verify conditions (i) and (ii) in Theorem 1 and such that if $\dim S_\alpha = \operatorname{rank} d(f \mid S_\alpha)$, then $f(S_\alpha)$ is smooth and $f(S_\alpha) \cap f(bS_\alpha) = \phi$. They are called stratifications adapted to f.*

Subanalytic subsets admit subanalytic triangulations. More generally given any locally finite family of subanalytic subsets of a manifold M there exist subanalytic triangulations of M which refine the family. Recall that the geometric realization $|K|$ of a locally finite simplicial complex K is a subanalytic subset of $\mathbb{R}^N$ for some N and that a subanalytic triangulation of M is a subanalytic homeomorphism of $|K|$ onto M (a map is subanalytic if its graph is subanalytic). With the help of Theorem 1 and Corollary 1 we obtain

Theorem 2. *Let $f : X \to Y$ be a real analytic map, $\{S_\alpha\}_{\alpha \in \Lambda}$ a stratification adapted to f and subordinated to the pair (A, bA) where $A \subset X$ is subanalytic and relatively compact. There exist subanalytic triangulations $T_{\bar{A}}$ of $\bar{A}$, subordinated to the family $\{S_\alpha\}_{\alpha \in \Lambda}$ such that for each open simplex $\sigma \in T_{\bar{A}}$, the restriction*

$$f \mid_{\bar{\sigma}} : \bar{\sigma} \to Y \quad \text{is injective}$$

if the following two conditions are satisfied

   i. *If S_σ is the stratum which contains σ then $\operatorname{rank} d(f \mid_{S_\sigma}) = \dim S_\sigma$.*

  ii. *Let $b\sigma = \bigcup_{i=1}^{k} \sigma_i$, $\sigma_i \in T_{\bar{A}}$, and for each i, $1 \le i \le k$, let S_{α_i} be the stratum which contains σ_i. Then $\operatorname{rank} d(f \mid_{S_{\alpha_i}}) = \dim S_{\alpha_i}$.*

2. Subanalytic Chains and Cycles.
Multiplicity of Intersection. Integration

2.1. *Borel–Moore Homology* [B-H]

Let X be a topological space and ϕ be a family of supports. We denote by $H_*^\phi(X, Z)$ (resp. $H_\phi^*(X, Z)$) the Borel–Moore homology (resp. cohomology) groups of X with supports in ϕ and coefficients in Z. The family of closed subsets of a closed subset F is also denoted by F, if $F = X$ we will write $H_*(X, Z)$ instead of $H_*^X(X, Z)$.

  a. For every closed subset $F \subset X$ we have an exact sequence

$$H_p(F, Z) \xrightarrow{i_{F,X}} H_p(X, Z) \xrightarrow{r_{X,U}} H_p(U, Z) \xrightarrow{\partial_{U,F}} H_{p-1}(F, Z) \longrightarrow$$

where $U = X - F$. For any $c \in H_p(X, Z)$ we will usually write $c \mid_U$ instead of $r_{X,U}(c)$.

  b. For any pair of supports ϕ and ψ, we denote by

$$\cup : H_p^\phi(X, Z) \times H_\psi^q(X, Z) \to H_{p-q}^{\phi \cap \psi}(X, Z) \quad \text{the cup-product.}$$

  c. For any closed subset $F \subset X$ there exists a canonical isomorphism

$$H_p(F, Z) \xrightarrow{\sim} H_p^F(X, Z).$$

Both groups will be identified.

  d. If X is compact, the singular homology of X coincides with the Borel–Moore homology with closed supports. If X admits a (locally finite) triangulation then its Borel–Moore homology is isomorphic with the homology of the complex of locally finite (but eventually infinite) PL chains.

e. If X is an orientable and connected manifold of dimension r then $H_r(X, Z) \simeq Z$.

f. For each $x \in X$ there exists a canonical generator $c_x \in H_0(\{x\}, Z) \simeq Z$.

g. (Poincaré duality). Let X be an orientable connected manifold of dimension n and $\theta_x \in H_n(X, Z)$ a generator. The map $\Delta_p : H_\phi^{n-p}(X, Z) \rightarrow H_p^\phi(X, Z)$ defined by $\Delta_p(c) = \bigcup(\theta_x, c)$ is an isomorphism.

h. If $\dim A = p$, $H_{p+i}(A, Z) = 0$ $\forall i > 0$.

2.2. Subanalytic Chains [D_1, D_2]

Let M be a real analytic manifold. A *subanalytic chain* of dimension p is a pair (A, c_A) where $A \subset M$ is a locally closed subanalytic subset of dimension p and $c_A \in H_p(A, Z)$. Its *boundary* $b(A, c_A)$ is the $(p-1)$-chain $(bA, \partial_{A, bA}(c_A))$, $bA = \overline{A} \backslash A$ is closed because A is locally closed. A cycle is a chain (A, c_A) such that $\partial(c_A) \in H(bA, Z)$ is zero. The support $\mathrm{supp}(A, c_A)$ is the set of all points of A which have a filter of neighborhoods $\{U_\alpha\}$ such that $c_A \mid_{U_\alpha} \neq 0$; it is a subanalytic subset of M, closed in A (but not necessarily in M). It is either empty or of the same dimension as (A, c_A). A chain (A, c_A) is *faithful* if $\mathrm{supp}(c_A) = A$ and $\mathrm{supp}(\partial c_A) = bA$.

Let $A' = \mathrm{supp}(c_A)$. There exists a unique class $c_{A'} \in H_p(A', Z)$ such that $i_{A', A}(c_{A'}) = c_A$, because $c_A \mid_{A-A'} = 0$ and the sequence

$$0 = H_{p+1}(A - A', Z) \rightarrow H_p(A', Z) \rightarrow H_p(A, Z) \rightarrow H_p(A - A', Z) \text{ is exact.}$$

We denote this uniquely defined class by c_A^f.

Two subanalytic chains (A, c_A) and (B, c_B) are *equivalent* and we will write $(A, c_A) \sim (B, c_B)$ iff $\mathrm{supp}\, c_A = \mathrm{supp}\, c_B$ and $c_A^f = c_B^f$. Each equivalence class has a *unique* faithful representation.

Two subanalytic chains (A, c_A) and (B, c_B) in M, of dimensions respectively p and q are said to be in *general position* or to *intersect properly* if

$$\dim(\mathrm{supp}(A, c_A) \cap \mathrm{supp}(B, c_B)) \le p + q - n$$
$$\dim(\mathrm{supp}\, \partial(A, c_A) \cap \mathrm{supp}(B, c_B)) \le p + q - n - 1$$
$$\dim(\mathrm{supp}(A, c_A) \cap \mathrm{supp}\, \partial(B, c_B)) \le p + q - n - 1 \qquad n = \dim M$$
$$\dim(\mathrm{supp}\, \partial(A, c_A) \cap \mathrm{supp}\, \partial(B, c_B)) \le p + q - n - 2.$$

Assume from now on that M is oriented and connected. For any two classes $c_1 \in H_p^{\phi_1}(M, Z)$ and $c_2 \in H_q^{\phi_2}(X, Z)$ we define their intersection product $c_1 \cdot c_2$ by

$$\Delta_{p+q-n}[\Delta_p^{-1}(c_1) \cap \Delta_q^{-1}(c_2)] \quad \text{where } \cap \text{ is the cup-product in cohomology.}$$

For two subanalytic chains (A, c_A) and (B, c_B) of dimensions p and q, in general position, we define their intersection $(A, c_A) \cdot (B, c_B)$ by

$$(A, c_A) \cdot (B, c_B) = (A \cap B, c'_A \cdot c'_B)$$

where c'_A and c'_B are the images of c_A and c_B respectively by the composition of morphisms in the sequences

$$H_p(A, Z) \to H_p(A - bB, Z) \to H_p^{A - bB}(X - (bA \cup bB), Z)$$
$$H_q(B, Z) \to H_q(B - bA, Z) \to H_q^{B - bA}(X - (bA \cup bB), Z).$$

Obviously $c'_A \cdot c'_B \in H_{p+q-n}(A \cap B, Z)$. The chain $(A, c_A) \cdot (B, c_B)$ obtained by intersection is subanalytic, of dimension $p + q - n$. We have

$$(A, c_A) \cdot (B, c_B) = (-1)^{p \cdot q}(B, c_B) \cdot (A, c_A).$$

. Let (D, c_D) be a 0-cycle. Then $D = \{x_i\}_{i \in I}$ is a discrete subset of M and $H_0(D, Z)$ is canonically isomorphic to $\bigoplus_{i \in I} H_0(\{x_i\}, Z)$. So $c_D = \sum e_i$, $e_i \in H_0(\{x_i\}, Z)$ and $e_i = m_i \cdot c_i$ where c_i is the canonical generator of $H_0(\{x_i\}, Z)$ and $m_i \in Z$ are uniquely determined.

We define the *multiplicity* of (D, c_D) at $x_i \in D$, to be the integer m_i, the *degree* $\deg(D, c_D)$ of (D, c_D) is defined by

$$\deg(D, c_D) = \sum_{x_i \in D} m_i.$$

If D is compact then $\deg(D, c_D)$ is finite.

Let (A, c_A) and (B, c_B) be two subanalytic chains, of complementary dimensions and in general position. Their intersection $(A, c_A) \cdot (B, c_B) = (D, c_D)$ is a zero cycle, $D = A \cap B = \{x_i\}$ is a discrete subset.

Definition. The *multiplicity of intersection* at $x_i \in A \cap B$ is equal to $\text{mult}(D, c_D)(x_i)$. The degree of the intersection $\text{def}[(A, c_A) \cdot (B, c_B)]$ is the degree of (D, c_D), that is

$$\deg(A, c_A) \cdot (B, c_B) = \sum_{x_i \in D} \text{mult}(D, c_D)(x_i) = \sum m_i.$$

If A or B are relatively compact, then $\deg[(A, c_A) \cdot (B, c_B)] < \infty$. The degree and multiplicity of intersection only depend on the *equivalence* class of the chains.

The sum of two subanalytic chains (A, c_A) and (B, c_B) of equal dimension p is the chain $(C, c'_A + c'_B)$ where $C = A \cup B \setminus (bA \cup bB)$ and c'_A is the image of c_A under the composition of the morphisms in the sequence

$$H_p(A, Z) \xrightarrow{r} H_p(A - bB, Z) \xrightarrow{i} H_p(A \cup B - (bA \cup bB), Z)$$

and c'_B is defined in an analogous way.

Addition of chains is associative and commutative and compatible to the equivalence relation $\sim$.

2.3. *PL Chains and Deformation* [L]

The invariance of the degree of intersection under deformation is a consequence, via the triangulation theorem, of Lefschetz's theory of PL chains [L].

Let K be a simplicial complex, $|K|$ its geometric realization; for each simplex $\sigma \in K$ let $|\sigma| \subset |K|$ be its geometric realization. We denote by $|\dot\sigma|$ the interior of $|\sigma|$. Fix an orientation θ_σ for each $\sigma \in K$. It determines a generator $c_\sigma \in H_{\dim \sigma}(|\dot\sigma|, Z)$. The PL chains of $|K|$ of dimension p ar the linear combinations $\sum_{\sigma \in C} k_\sigma \cdot (|\dot\sigma|, \theta_\sigma)$, $k_\sigma \in Z$, where C is the set of all p-simplexes in $\bigcup_n K^{(n)} = $ the union of all baricentric subdivisions of K. The boundary of each chain is defined as usual.

Let $\eta : |K| \to M$ be a subanalytic triangulation, subordinated to a subanalytic chain (A, c_A) that is a subanalytic homeomorphism such that A, $\operatorname{supp} c_A$, bA and $\operatorname{supp} \partial(c_A)$ are unions of images of open simplexes in $|K|$. Let $\mathcal{C}_A = \{\sigma : \sigma \in K, \dim \sigma = \dim A, \eta(|\dot\sigma|) \subset A\}$ and let $c_\sigma \in H_{\dim \sigma}(\eta(|\dot\sigma|), Z)$ be the generator uniquely determined by θ_σ. Then $c_A \mid_{|\dot\sigma|} = k_\sigma \cdot c_\sigma$, $k_\sigma \in Z$ and the subanalytic chain $\sum_{\sigma \in A_A} k_\sigma \cdot (|\dot\sigma|, c_\sigma)$ is equivalent to (A, c_A). So to (A, c_A) we associate the chain $\sum_{\sigma \in C_A} k_\sigma \cdot (|\dot\sigma|, \theta_\sigma)$ of $|K|$.

Conversely to each PL-chain $\sum a_\sigma \cdot (|\dot\sigma|, \theta_\sigma)$ we associate the subanalytic chain $\sum(\eta(|\dot\sigma|), a_\sigma \cdot c_\sigma)$. Lefschetz defined *multiplicity* and *degree of intersection* for PL-chains in general position. They correspond, under the above correspondence to those already defined. He proved also their invariance under deformation, the correspondence just defined allows us to translate his results to the subanalytic case.

Theorem. *Let M be a $2n$-dimensional, compact, connected and oriented real analytic manifold with boundary, and let (A, c_A) and (B, c_B) be two n subanalytic chains such that the supports of their boundaries are disjoint and contained in bM.*

1. *There exists subanalytic n-chains $(A', c_{A'})$ and $(B', c_{B'})$ in general position and such that*

    a. *The classes defined by $\partial(A, c_A)$ and $\partial(A', c_{A'})$ in $H_n^{\mathrm{sing}}(bM \setminus \operatorname{supp} \partial(B, c_B), Z)$ are equal where H^{sing} denotes singular homology.*

    b. *$\partial(B, c_B) = \partial(B', c_{B'})$.*

    c. *The homology classes in $H_n(M - bM, Z)$ defined by (A, c_A) and $(A', c_{A'})$ are equal.*

    d. *$\operatorname{supp}(A', c_{A'})$ and $\operatorname{supp}(B', c_{B'})$ are arbitrarily near A and B re-*

spectively. We can choose $(A', c_{A'})$ *such that* $\partial(A', c_{A'}) = \partial(A, c_A)$.

2. *Any pair of chains which verify* 1.a, 1.b *and* 1.c *have the same degree of intersection.*

Definition. The *degree of intersection* of two n-chains (A, c_B) and (B, c_B) with disjoint boundaries, supported in bM is equal to $\deg[(A', c_{A'}) \cdot (B', c_{B'})]$ where $(A', c_{A'})$ and $(B', c_{B'})$ verify 1.a, 1.b and 1.c in the theorem.

2.4. *Integration on Subanalytic Chains* [D_1, D_2]

Let M be a real analytic Riemannian manifold. For any subanalytic subset $A \subset M$ of dimension p, the p-dimensional Hausdorff measure of A is locally finite and the $(p+r)$-Hausdorff measure is zero $\forall r \in \mathbb{R}, r > 0$. In particular, for any compact K of M, the p-Hausdorff measure of $A \cap K$ is finite. If A is smooth, any orientation of A defines a unique class in $H_p(A, Z)$ which is a generator if A is also connected. It follows that if $c_A \in H_p(A, Z)$ is an orientation class, then $\int_{(A, c_A)} \omega < \infty$ if $\mathrm{supp}(\omega)$ is compact, ω a smooth p-form on M.

Let (A, c_A) be a p-subanalytic chain in M. Let $\{A_\alpha^r\}$, $\dim A_\alpha^r = r$ be a subanalytic stratification of A, such that all A_α^r are connected and orientable (it exists, for example a triangulation). We have then $H_r(A_\alpha^r, Z) \simeq Z$. For each α, we choose a generator $c_\alpha \in H_p(A_\alpha^p, Z)$. Then $c_A \mid_{A_\alpha^p} = k_\alpha \cdot c_\alpha$ and the chain (A, c_A) is equivalent to the chain $\sum_\alpha k_\alpha \cdot (A_\alpha^p, c_\alpha)$. For any p-smooth form ω of M, we define $\int_{(A, c_A)} \omega = \sum_\alpha k_\alpha \cdot \int_{(A_\alpha^p, c_\alpha)} \omega$.

This definition does not depend on the choice of the stratification. If (A, c_A) and (B, c_B) are equivalent, then $\int_{(A, c_A)} \omega = \int_{(B, c_B)} \omega$.

Stokes Theorem. *For any subanalytic p-chain and any smooth $(p-1)$-form ω*

$$\int_{(A, c_A)} d\omega = \int_{\partial(A, c_A)} \omega.$$

Let M and N be two real analytic manifolds, c_M an orientation class of M, $I \subset \mathbb{R}$ a segment, oriented as usual, (A, c_A) a p-dimensional subanalytic, relatively compact chain in $I \times M$ and $f : I \times M \to N$ a real analytic map. For each $t \in I$, c_M determines an orientation class c_t of $M_t = \{t\} \times M$. Assume that $\forall t \in I$ the chains (A, c_A) and (M_t, c_t) are in general position.

The orientations of I and M determine an orientation of $I \times M$, so the intersection chain $(A, c_A) \cdot (M_t, c_t)$ is well defined. Denote it by $(A, c_A)_t$. Let ω be a smooth form in N and $a : I \to \mathbb{R}$ a smooth function.

Theorem.

(i) *The function* $h(t) = \int_{(A, c_A)_t} f^*(\omega)$ *is continuous,*

(ii) $\int_{(A,c_A)} a(t)dt \wedge f^*(\omega) = \int_I a(t) \cdot h(t)dt.$

3. Multiplicity of Intersection of Lagrangian Cycles

3.1. *Lagrangian Subsets*

Let M be a real analytic manifold, of even dimension $2n$ and w be a symplectic analytic form on M, that is a closed analytic 2-form such that $\forall m \in M$ and $\forall v \in T_m M$ if $w(v, \ ) : T_m M \to \mathbb{R}$, $u \mapsto w(v,u)$ is the zero function, then $v = 0$.

Definition. A subanalytic subset $A \subset M$ is *Lagrangian* if it contains an open, smooth, dense subset A^* such that $w \mid_{A^*} \equiv 0$ and if $\dim A = n = \frac{1}{2} \dim M$.

The following proposition is proved using Whitney stratifications.

Proposition. *Let $A \subset M$ be a Lagrangian subset and let $S_{\bar{A}}$ be an arbitrary stratification of $\bar{A}$. Then $\forall S \in S_{\bar{A}}$, $w \mid_S \equiv 0$.*

As usual, we associate to the symplectic structure w the isomorphism $I : T^* M \to TM$ defined by the condition $w(I(v), u) = v(u)$. To any smooth function $H : U \to \mathbb{R}$, $U \subset M$ open, we associate the Hamiltonian vector field $IdH : U \to TM \mid_U$, it verifies $\forall v \in TM \mid_U$, $w(IdH, v) = dH(v)$.

Let $\mathcal{V} : U \to TM \mid_U$ be an analytic vector field. An *integral* (or integral flow) to $\mathcal{V}$ is a real analytic map $f : (-a, a) \times V \to U$, $a > 0$, $V \subset U$ open, such that

i. $f_* \left[\left(\frac{d}{dt}, 0 \right) \big|_{(t,v)} \right] = \mathcal{V}(f(t,v)).$

ii. $f(t_1, f(t_2, v)) = f(t_1 + t_2, v)$ whenever t_1, t_2, $t_1 + t_2$ are in $(-a, a)$ and $f(t_2, v) \in V$.

iii. For each $t \in (-a, a)$, the map $f_t : V \to U$, $f_t(x) = f(t, x)$ is injective and of maximal rank and $f(0, x) = x$.

If $K \subset U$ is compact then on every sufficiently small open neighborhood $V_K \subset U$ of K integrals to $\mathcal{V}$ exist. They are essentially unique, that is, if

$$f : (-a, a) \times V_1 \to U \quad \text{and}$$

$$g : (-b, b) \times V_2 \to U \quad \text{are integrals to } \mathcal{V} \text{ then}$$

$f = g$ on the intersection of their domains. If the vector field $\mathcal{V}$ has compact support then it admits a global integral $f : \mathbb{R} \times U \to U$. We will also need the case with parameters: T a *compact* analytic manifold (eventually with boundary), $\mathcal{V} : \Gamma \times U \to TM|_U$ an analytic family of vector fields over U. As before, for any compact $K \subset U$ there exist open neighborhoods V_K of

K in U and real analytic maps $F : \Gamma \times (-a, a) \times V_K \to U$ such that $\forall s \in \Gamma$, $f^s : (-a, a) \times V_K \to U$, $f^s(t, x) = F(s, t, x)$ is an integral to the vector field $\mathcal{V}_s : U \to TM|_U$, $\mathcal{V}_s(x) = \mathcal{V}(s, x)$.

All these results are immediate consequences of standard theorems in ordinary differential equations.

Parametrized families of vector fields will appear either as Hamiltonian vector fields IdH_s associated to an analytic family of functions or as Hamiltonian vector fields $I_s dH$ associated to a unique function H via an analytic family of isomorphisms $I_s : T^*M \to TM$ deduced from an analytic family of symplectic forms w_s.

The following two propositions are standard [A].

Proposition 1. *A Hamiltonian flow preserves the simplectic form,* i.e.

$$w\big(f_{t,*}(u), f_{t,*}(v)\big) = w(u, v) \quad \forall (u, v) \in TM \times TM.$$

Proposition 2. *The maps $f_t : V \to U$ preserve the "level surfaces",* that is

$$f_t(H^{-1}(c) \cap V) \subset H^{-1}(c) \cap U.$$

The next two are proved with the help of suitable Whitney stratifications. The last one is important in the proof of the main deformation theorem: it assures that (the proposed construction of) a 3-cell does not degenerate into a 2-cell.

Proposition 3. *Let A be a smooth subanalytic subset of M. If $w|_A \equiv 0$, then A contains no non-degenerated trajectory integral to an Hamiltonian vector field; that is, if $f : \overline{I} \times V \to M$ is an integral to IdH then $f(\overline{I} \times \{x\}) \cap A$ is finite $\forall x \in V$. In particular if IdH never vanishes, then $f(J \times \{x\}) \subset A$, for any interval $J \subset I$, implies that J is reduced to a point.*

Proposition 4. *Let $A \subset M$ be a Lagrangian subanalytic subset, $S_{\overline{A}}$ a stratification of $\overline{A}$ and $K \subset \overline{A}$ a compact subset over which IdH never vanishes and V_K a sufficiently small neighborhood of K. Let $f : I \times V_K \to U$ be an integral to IdH. For ε sufficiently small the restriction of f to $(-\varepsilon, \varepsilon) \times (S \cap V_K)$, $\forall S \in S_{\overline{A}}$, has everywhere maximal rank $1 + \dim S$.*

3.2. *Deformation Theorems*

Let U be an open subset of M, $H : U \to [0, T)$, $T \in \mathbb{R}_+ \cup \{\infty\}$, be a proper analytic function and A and B be two subanalytic subsets of U.

Notation. For any compact K, $W(K)$ denotes an open neighborhood of K; $I_a = (-a, a)$.

Definition. A point $c \in [0, T)$ is an *H-bad point* for (A, B) if, at least one of the following three conditions is verified:

(i) Given an integral $f : I_\varepsilon \times W(H^{-1}(c)) \to U$ to the Hamiltonian vector field IdH, there exists a sequence $(t_i)_{i \in \mathbb{N}}, t_i \neq 0, t_i \to 0$ such that

$$f(\{t_i\} \times A) \cap B \cap H^{-1}(c) \neq \phi \quad \forall i \in \mathbb{N}.$$

Recall that, by Proposition 2 in 3.1, $f(\{t_i\} \times A) \cap B \cap H^{-1}(c) = f(\{t_i\} \times [A \cap H^{-1}(c)]) \cap B$.

(ii) c is a critical value of H.

(iii) $\dim A \cap H^{-1}(c) > \dim A - 1$ or $\dim B \cap H^{-1}(c) > \dim B - 1$.

If we replace condition (i) by

(i′) There exists two sequences $(t_i)_{i \in \mathbb{N}}, (c_i)_{i \in \mathbb{N}}, t_i \neq 0, t_i \to 0, c_i \to c$ such that

$$f(\{t_i\} \times A) \cap B \cap H^{-1}(c_i) \neq \phi$$

we obtain the definition of *H-wild point* for (A, B).

Theorem 1. *If A and B are Lagrangian then the set of H-bad points for (A, B) is discrete in $[0, T)$.*

Corollary 1. *The set of H-wild points for (A, B) is discrete in $[0, T)$.*

Remark. Let $S_{\bar{A}}$ and $S_{\bar{B}}$ be Whitney stratifications of $\overline{A}, \overline{B}$. The critical values taken by $H|_S$ over $S \cap H^{-1}([0, \alpha]), \forall S \in S_{\bar{A}} \cup S_{\bar{B}}, \alpha \in \mathbb{R}_+$, is finite (see §1). So it is enough to consider bad points (wild points) which are not critical values. This is useful in some technical steps in the proofs.

This is the main deformation theorem. The following variant, whose proof is analogous, is necessary in order to define intrinsically local multiplicities of intersection.

Theorem 2. *Let h_t be an analytic family of analytic Riemannian metrics on TM, parameterized by a compact analytic manifold T (eventually with boundary). Let x_0 be a point of M and $H_t(x) = \text{dist}_t(x, x_0)$, where dist_t is the distance function in M deduced from the Riemannian metric h_t. There exists a neighborhood V of x_0 such that $d(t, x) = \text{dist}_t(x, x_0)$ is an analytic function on $T \times V$ and*

i. *$\exists \varepsilon_0 > 0$ such that no point in $(0, \varepsilon_0)$ is a H_t-bad point for (A, B), for any $t \in T$.*

ii. *Let $W(x_0) \subset V$ be an open neighborhood of x_0 and $F : T \times I_a \times W(x_0) \to V$ be an analytic family of integrals to IdH_t. For every $\varepsilon \in (0, \varepsilon_0)$ there exists $\delta_\varepsilon > 0$ such that*

$$F\left[T \times I_{\delta_\varepsilon}^* \times (A \cap W(x_0))\right] \cap B \cap H^{-1}(\varepsilon) = \phi$$

where

$$I_{\delta_\varepsilon}^* = (-\delta_\varepsilon, \delta_\varepsilon) \setminus \{0\}.$$

3.3. *Degree and Multiplicity of Intersection*

Let (A, c_A) and (B, c_B) be two subanalytic Lagrangian cycles of middle dimension n, that is, A and B are Lagrangian. Let $H : U \to [0, T)$ be a proper real analytic map. Then for $c \in [o, T)$, $H_c = H^{-1}([0, c))$ is a relatively compact subanalytic open subset of U. Whenever c is not a critical value of H, $\overline{H}_c$ is a compact manifold with boundary bH_c. Let $(A, c_A)_c$ and $(B, c_B)_c$ be respectively the chains $(A \cap H_c, c_{A|H_c})$ and $(B \cap H_c, c_B \mid H_c)$. Their boundaries $\partial(A, c_A)_c = (b(A \cap H_c), \partial(c_{A|H_c}))$, $\partial(B, c_B)_c = (b(B \cap H_c), \partial(c_{B|H_c}))$ are supported on subsets of ∂H_c.

Whenever they are disjoint, their degree of intersection is well defined (2.2). But this is not always the case. The deformation Theorem 1 allows us to define $\deg(A, c_A)_c \cdot (B, c_B)_c$ when c is *not* a bad point, that is, for all points in $[0, T)$ except a discrete subset.

Let $f : I \times W(\overline{H}_c) \to U$ be an integral to IdH. If c is not a bad point, by Theorem 1, there exists $\varepsilon_c > 0$ such that $\forall \varepsilon \in (0, \varepsilon_c)$, $f(\{\varepsilon\} \times (A \cap W(\overline{H}_c)) \cap B \cap bH_c = \phi$. It follows, from the Theorem in 2.3, that the chains $f_{\varepsilon,*}((A, c_A)_c) = (f_\varepsilon(A \cap H_c), f_{\varepsilon,*}(c_{A|H_c}))$ have, $\forall \varepsilon \in (0, \varepsilon_c)$, the same degree of intersection with $(B, c_B)_c$. We conclude that the following is a good definition of $\deg(A, c_A)_c \cdot (B, c_B)_c$.

Definition 1. Let $c \in [0, T)$ be not a bad point for (A, B), $\deg_\pm(A, c_A)_c \cdot (B, c_B)_c = \deg[[f_{\pm\varepsilon,*}(A, c_A)_c] \cdot (B, c_B)_c]$ for $\varepsilon > 0$ small enough.

Remark 1. In general there is no reason for $\deg_+$ and $\deg_-$ to be equal.

Remark 2. $\deg_\pm(B, c_B) \cdot (A, c_A)$ is calculated by deforming (B, c_B). As $f(\{\varepsilon\} \times B) \cap A = f(\{0\} \times B) \cap f(\{-\varepsilon\} \times A) = B \cap f(\{-\varepsilon\} \times A)$ and intersection of cycles is anticommutative, it follows that

$$\deg_+(B, c_B)_c \cdot (A, c_A)_c = (-1)^{n \cdot n} \deg_-(A, c_A) \cdot (B, c_B)_c.$$

Let $x_0 \in A \cap B$, h be an arbitrary Riemannian metric and $H : V \to \mathbb{R}$, $x \mapsto \text{dist}(x, x_0)$ the associated distance function, where V is an open neighborhood of x_0 in M. Using Corollary 1 and Theorem 2 instead of Theorem 1, analogous considerations show that $\deg_\pm(A, c_A)_c \cdot (B, c_B)_c$ is constant on c, for c small enough and independent of the analytic Riemannian metric h. So the following definition of *multiplicity of intersection* of (A, c_A) and (B, c_B) at x_0 makes sense.

Definition 2.

$$\text{mult}_{x_0}^+[(A, c_A), (B, c_B)] = \lim_{c \to 0} \deg_+(A, c_A)_c \cdot (B, c_B)_c$$

$$\text{mult}_{x_0}^-[(A, c_A), (B, c_B)] = \lim_{c \to 0} \deg_-(A, c_A)_c \cdot (B, c_B)_c.$$

Again $\text{mult}_{x_0}^+[(A, c_A), (B, c_B)] = (-1)^{n \cdot n} \text{mult}_{x_0}^-[(B, c_B)_c \cdot (A, c_A)_c]$.

There is no reason for $\text{mult}_{x_0}^+$ to be equal to $\text{mult}_{x_0}^-$. But both coincide in the complex case, as is explained below.

3.4. *Complex Case*

Let M be a complex manifold, $M_{\mathbb{R}}$ the underlying real analytic manifold and w an holomorphic symplectic form on M, then $\dim_{\mathbb{C}} M$ is necessarily even. The main example for us here is the holomorphic cotangent bundle T^*M, equipped with its canonical symplectic form [B]. Complex analytic subsets are always oriented by the fundamental class associated to their complex structure.

Notation. TM: the holomorphic tangent bundle to M, $(TM)_{\mathbb{R}}$: the real underlying vector bundle to TM, $TM_{\mathbb{R}}$: the real tangent to the real manifold $M_{\mathbb{R}}$, $J : TM_{\mathbb{R}} \to TM_{\mathbb{R}}$: the automorphism associated to the complex structure of M, $J^2 = -I_{TM_{\mathbb{R}}}$. There is a canonical isomorphism $\eta : (TM)_{\mathbb{R}} \to TM_{\mathbb{R}}$. As usual we define an action of $\mathbb{C}$ on $TM_{\mathbb{R}}$: if $v \in T_m M_{\mathbb{R}}$, $(a + ib) \cdot v = a \cdot v + bJ(v)$. It defines on $TM_{\mathbb{R}}$ a structure of complex vector bundle, denoted by $(TM_{\mathbb{R}})_{\mathbb{C}}$. The isomorphism η induces an isomorphicm $\eta_{\mathbb{C}} : TM \to (TM_{\mathbb{R}})_{\mathbb{C}}$. From now on the pairs vector bundles $(TM)_{\mathbb{R}}$ and $TM_{\mathbb{R}}$ and TM and $(TM_{\mathbb{R}})_{\mathbb{C}}$ will be identified. It follows that, given any holomorphic p-form u on M, $Re(u)$ is a p-form on $M_{\mathbb{R}}$. In particular, $\forall \theta \in [0, 2\pi]$, $w_\theta = Re(e^{i\theta} \cdot w)$ is a real analytic symplectic form on $M_{\mathbb{R}}$ and, obviously, $\{w_\theta\}$ is an analytic family.

Let X be a complex analytic subset of M and X^* the complex submanifold of its smooth points, X is *Lagrangian* if $w|_{X^*} \equiv 0$ and X is of pure dimension $n = \frac{1}{2} \dim_{\mathbb{C}} M$. Lagrangian subsets for w are also, as real analytic subsets of $M_{\mathbb{R}}$, Lagrangian for w_θ, $\forall \theta \in [0, 2\pi]$.

Let (A, c_A) and (B, c_B) be two complex analytic Lagrangian cycles. We now know that $\forall x \in A \cap B$ and $\forall \theta \in [0, 2\pi]$, the multiplicities of intersection $\text{mult}_x^{\pm}[(A, c_A), (B, c_B), w_\theta] = \text{mult}_{x, \theta}^{\pm}[(A, c_A), (B, c_B)]$ are well defined. Theorem 3 below implies that $\forall \theta \in [0, 2\pi]$ they are equal. We will write simply $\text{mult}_x^{\pm}[(A, c_A), (B, c_B)]$.

Corollary. $\text{mult}_x^+[(A, c_A), (B, c_B)] = \text{mult}_x^-[(A, c_A), (B, c_B)]$.

Proof. Let $F : [0, 2\pi] \times I_\varepsilon \times U \to M$ be an integral to the analytic family of vector fields $I_\theta dH$, where H is the distance function to x, deduced from some analytic Riemannian metric on $M_\mathbb{R}$ and $I_\theta : T^*M_\mathbb{R} \to TM_\mathbb{R}$ is the symplectic isomorphism associated to the symplectic form w_θ. As $w_0 = -w_\pi$, $I_0 dH = -I_\pi dH$, so $F(0, t, x) = F(\pi, -t, x)$. It follows that $\text{mult}_{x,0}^+ = \text{mult}_{x,\pi}^- = \text{mult}_{x,0}^-$.   □

Theorem 3. (1) *There exists $c_0 \in \mathbb{R}$, $c_0 > 0$ such that no point in $(0, c_0)$ is a bad point for A, B, for the symplectic structure w_θ, $\forall \theta \in [0, 2\pi]$.*
(2) *$\forall c \in (0, c_0)$, there exists $\varepsilon(c) > 0$ such that*

$$F([0, 2\pi] \times (-\varepsilon(c), \varepsilon(c))^* \times A) \cap B \cap H_c^{-1} = \phi$$

where $(-\varepsilon(c), \varepsilon(c))^* = (-\varepsilon(c), \varepsilon(c)) \setminus \{0\}$.

Remark. (2) does not follow from (1): that c is not a bad point for any $\theta \in [0, 2\pi]$ only says that for each θ, $\exists t_\theta$ such that

$$f(\{\theta\} \times (0, t_\theta) \times A) \cap B \cap H^{-1}(c) = \phi$$

but it could happen that $t_{\theta_j} \to 0$ when $\theta_j \to \theta_0$ for some sequence $\{\theta_j\}$ and some θ_0. To show that this is not the case requires an analogous but distinct proof.

4. Examples and Applications

4.1. Let w be a standard symplectic form on $\mathbb{C}^{2N}$. Any two Lagrangian subspaces have multiplicity of intersection at $0 \in \mathbb{C}^{2N}$ equal to 1. In $\mathbb{R}^{2N}$, the multiplicity of intersection at $0 \in \mathbb{R}^{2N}$ of two oriented Lagrangian subspaces is ± 1, it depends on their orientations. This is easily proved using $H = \sum x_i^2 + y_i^2$, $w = \sum dx_i \wedge dy_i$.

4.2. Let M be a complex manifold, $X \subset M$ a complex analytic subset, X^* the complex submanifold of its smooth points and $T_X^* M$ the conormal to X defined as $\overline{T_{X^*}^* M}$ where

$$T_{X^*}^* M = \{(x, v), \ x \in X^*, \ v \in T_x^* M \text{ s.t. } v(T_x X^*) = 0\}.$$

The conormal to X is a conic complex analytic subvariety of T^*M of middle dimension. It is Lagrangian for the standard symplectic structure of T^*M. We write also $T_X^* M$ for the cycle $(T_x^* M, c_X)$, c_X the canonical fundamental class of $T_X^* M$ [B-H].

Proposition. $(-1)^{\dim X} \cdot \text{mult}(T_X^* M, T_M^* M, (x, 0)) = Eu_X(x)$ *the Euler obstruction at x.*

See [M, D_1, D_4] for the definition of Eu_X and different evaluations in terms of multiplicities of intersection of suitable algebraic classes in $\hat{X}_x$

(the fiber over x of the Nash blow-up $\pi : \hat{X} \to X$) and also in terms of Euler characteristics of local plane sections of X near x.

4.3. *Constructible Complexes and Holonomic Modules* [D_4]

For any complex $\mathbb{L}^\bullet$, we denote by $\mathbb{L}^\bullet[-1]$ the complex $\mathbb{L}^\bullet$ shifted by -1. Let $\mathbb{K}^\bullet$ be a constructible complex, that is, a bounded complex of sheaves over a complex manifold M, whose local cohomology sheaves have finite dimensional fibers and are locally constant over the strata of a complex analytic stratification $S_{\mathbb{K}^\bullet}$. Let $\mathbb{H}^*(V, \mathbb{K}^\bullet|_V)$ be the hypercohomology groups of the complex $\mathbb{K}^\bullet|_V$, $V \subset M$ open. Denote by $\chi(\mathbb{K}^\bullet|_V)$ the Euler characteristic $\sum_i (-1)^i \dim \mathbb{H}^i(V, \mathbb{K}^\bullet|_V)$. Whenever V is relatively compact and has smooth boundary bV, transverse to the strata of $S_{\mathbb{K}^\bullet}$, the hypercohomology groups are finite and then so is $\chi(\mathbb{K}^\bullet|_V)$. Let $\chi(\mathbb{K}_x^\bullet) = \sum_i (-1)^i \dim \mathcal{H}^i(\mathbb{K}_x^\bullet)$. It is not difficult to verify, using stratifications, that $\chi(\mathbb{K}_x^\bullet) = \chi(\mathbb{K}^\bullet|_{B_\varepsilon}(x))$ for $\varepsilon << 1 \varepsilon > 0$.

Let $x \in M$, U_x be a neighborhood of x and $f : U_x \to \mathbb{C}$ an holomorphic function. The complex $\Phi(\mathbb{K}^\bullet, f)_x$ of *vanishing cycles at x in the direction f* is defined in [De]. Its Euler characteristic $\chi\Phi(\mathbb{K}^\bullet, f)_x$ is finite.

There exist complex analytic stratifications S_M of M such that

i. The local cohomology sheaves $\mathcal{H}^i(\mathbb{K}^\bullet)$ are locally constant on all strata.

ii. For any $S \in S_M$, let $x \in S$, $v \in [T_S^* M \setminus \bigcup_{S' \neq S} \overline{T_{S'}^* M}]$, and f be an holomorphic germ at x, sufficiently general among those which verify $f(x) = 0$, $\partial f(0) = v$. The number $\chi\Phi(\mathbb{K}^\bullet, f)[-1]$ is independent of f, x and v. We will denote it by $\chi\Phi(\mathbb{K}^\bullet, S)[-1]$.

Definition [D_4]. The *characteristic cycle* $Ch(\mathbb{K}^\bullet)$ of $\mathbb{K}^\bullet$ is the cycle

$$\sum_{S \in S_M} (-1)^{\dim M} \chi\Phi(\mathbb{K}^\bullet, S)[-1] \cdot [\overline{T_S^* M}],$$

where S_M is a stratification of M, which verifies i. and ii. above. The characteristic cycle is independent of the stratification S_M.

Let $U \subset M$ be an open subset and $H : U \to [0, T)$ be a proper analytic function and $\overline{H} : TM \to [0, T)$ be the composition $H \circ \pi$, $\pi : TM \to M$ the projection. The set $\mathcal{C}$ of points of $[0, T)$ which are bad points for any of the pairs $(\overline{T_S^* M}, T_M M)_{S \in S_M}$ is discrete (even if $\overline{H}$ is not proper). Let $\mathcal{C}'$ be the union of $\mathcal{C}$ and the critical values of H. It is also discrete in $[0, T)$.

Theorem 1. *For all $t \in [o, T) \setminus \mathcal{C}'$, $V_t = H^{-1}([0, t))$ is a relatively open subset of U, with smooth boundary bV_t, transverse to all strata $S \in S_M$ and*

$$\chi\mathbb{H}(V_t, \mathbb{K}^\bullet|_{V_t}) = (-1)^{\dim M} \deg([T_M^* M]|_{V_t}, Ch(\mathbb{K}^\bullet)|_{V_t}).$$

Theorem 2. $\chi(\mathbb{K}_x^\bullet) = (-1)^{\dim M} \mathrm{mult}_{(x,0)}([T_M^* M], Ch(\mathbb{K}^\bullet))$.

Theorem 3. $\chi\Phi(\mathbb{K}^\bullet, f)_x[-1] = (-1)^{\dim M}\mathrm{mult}_{(x,\partial f(x))}([\partial f(V_x)],$
$Ch(\mathbb{K}^\bullet))$ *where* U_x *is a neighborhood of* $x \in M$, $f : U_x \to \mathbb{C}$ *is holomorphic,* $V_x \subset U_x$ *is a sufficiently small subanalytic open neighborhood of* x, $\partial f : U_x \to T^*M \mid_{U_x}$ *is the holomorphic differential of* f *and* $[\partial f(V_x)]$ *is the chain defined by* $\partial f(V_x)$ *and its fundamental class.*

Let $\mathcal{M}$ be a Holonomic Module over M. Its de Rham resolution $DR^*(\mathcal{M}) = \Omega^*_M \otimes_{\mathcal{D}_M} \mathcal{M}$ is a constructible complex of sheaves [B].

Let $Ch(\mathcal{M})$ be the characteristic variety of $\mathcal{M}$ (see [B]). For each irreducible component Z of $Ch(\mathcal{M})$, let $b_Z(\mathcal{M})$ be the "multiplicity of $\mathcal{M}$ at Z" (see [K]). The cycle $Ch(\mathcal{M}) = \sum_Z b_Z(\mathcal{M}) \cdot [Z]$ (the sum is over the family of irreducible components of $Ch(\mathcal{M})$) is called the characteristic cycle of $\mathcal{M}$.

Theorem 4. [D₄] $Ch(DR^*\mathcal{M}) = Ch(\mathcal{M})$.

Define the local index $\chi_x(\mathcal{M})$ at x by $\chi_\mathcal{M}(x) = \sum(-1)^i \mathcal{E}xt_{\mathcal{D}_M}(\mathcal{M}, \mathcal{O}_M)$ where $\mathcal{O}_M$ and $\mathcal{D}_M$ are respectively the sheaf of holomorphic functions and the sheaf of holomorphic differential operators of finite order on M. Then $\chi_\mathcal{M}(x) = \chi[(DR^*\mathcal{M})_x]$.

Corollary. $\chi_\mathcal{M}(x) = (-1)^{\dim M}\mathrm{mult}_{(x,0)}([T_M M], Ch(\mathcal{M}))$.

These results are reformulations in terms of multiplicities of intersection of the formulas for the index in [D₄].

4.4. Let $X \subset M$ be a complex analytic lagrangian subvariety of the complex symplectic manifold M. The notion of local multiplicity of intersection provides us with a function $\mathrm{mult}_X : X \to Z$, $\mathrm{mult}_X(x) = \mathrm{mult}_x([X], [X])$. It is a constructible function. Obviously, $\mathrm{mult}_X(x) = 1$ if x is a smooth point.

5. Sketch of the Proof of the Main Deformation Theorem

The notations are those used in the statement of the corresponding theorem.

The proof of Theorem 1 proceeds by the absurd: if the set of bad points has an accumulation point, then we construct a subanalytic 3-cell Θ in M such that $\int_{\partial\Theta} w \neq 0$. Stokes theorem says $\int_{\partial\Theta} w = \int_\Theta dw = \int_\Theta 0 = 0$. Absurd.

The proof of Theorem 2 is analogous, of course the 3-cell is different and lives in a different space. We integrate a suitable pull-back of w. The same for Theorem 3, (the 3-cell is in $S^1 \times \mathbb{R} \times M$) except that we now integrate a pull-back $p^*(Re(e^{i\theta} \cdot w))$ of the form $Re(e^{i\theta} \cdot w)$ which is not closed. So we have to prove directly that $\int_\Theta dp^*(Re(e^{i\theta} \cdot w)) = 0$. The problem is reduced, via the last theorem in 2.4, to an elementary differential equation.

Sketch of the Proof of Theorem 1.

A. Let $\{c_i\}_{c \in \mathbb{N}}$ be a sequence of H-bad points for A, B, which converges to c_0 in $[0, T)$. We can assume $c_i > c_{i+1} > \cdots > c_0$. Then, for each c_i, there exists a sequence $\{t_j\}_{j \in \mathbb{N}} \subset \mathbb{R}^*$, which converges to 0 and such that

$$f(\{t_j\} \times [A \cap H^{-1}(c_i)]) \cap B \neq \phi.$$

From the subanalytic theory it follows

(i) $\dim[f(I_\varepsilon \times A) \cap B \cap H^{-1}(c_i)] \geq 1$ $I_\varepsilon = (-\varepsilon, \varepsilon)$

(ii) $\dim[f(I_\varepsilon \times A) \cap B \cap H^{-1}[(c_0 - \delta, c_0 + \delta)^*]] \geq 2$ $(r, s)^* = (r, s) \setminus \{0\}$, $\delta > 0$.

Let $\overline{f} : I \times V \to I \times U$ be defined by $\overline{f}(t, x) = (t, f(t, x))$. From (ii) it is not difficult to deduce the following lemma.

Lemma. Let $Y = \overline{f}(I_\varepsilon \times A) \cap (I_\varepsilon \times B) \cap (I_\varepsilon \times H^{-1}(c_0 - \delta, c_0 + \delta))$. Then

a. $\dim[\overline{Y} \cap (\{0\} \times A)] \geq 1$.

b. $H(\overline{Y} \cap (\{0\} \times A))$ contains an interval $(c_0, c_0 + \mu)$ or $(c_0 - \mu, c_0)$, $\mu > 0$.

B. The following proposition is proved with the help of the results in §1 (in particular Theorem 2), and in §3.1:

Proposition. There exists a subanalytic map $\eta : [0, 1] \times [0, 1] \to I_\varepsilon \times A$, $\eta = (\eta_1, \eta_2)$, such that

i. $\varphi([0, 1] \times [0, 1]) \subset f(I_\varepsilon \times A) \cap B \cap H^{-1}([a, b])$, where $\varphi = f \circ \eta$, $[a, b] \subset (c_0 - \delta, c_0 + \delta)$.

ii. $\varphi(\{0\} \times [0, 1]) \subset H^{-1}(a)$
$\varphi(\{1\} \times [0, 1]) \subset H^{-1}(b)$
$\eta([0, 1] \times \{0\}) \subset \{0\} \times A$
$\eta_1(s, t) > 0$ if $t > 0$.

iii. φ is a subanalytic homeomorphism onto its image, and restricted to each open face of $[0, 1] \times [0, 1]$ is smooth of maximal rank everywhere.

iv. The restriction of H to each open face of $\varphi([0, 1] \times [0, 1])$ except $\varphi(\{z\} \times (0, 1))$, $z = 0, 1$, is of maximal rank.

Let μ be the 2-cell $\eta([0, 1] \times [0, 1]) \subset I \times A$.

Let $h = \{(t, x) : \exists (t', x) \in \mu, t' \geq t, 0 \leq t\}$. It is a subanalytic 3-cell in $I \times A$ (the proof is easy with the help of Proposition 3 in 3.1). The faces of h are

$$e_a = \{(t,x) \in h : f(t,x) \subset H^{-1}(a)\}$$
$$e_b = \{(t,x) \in h : f(t,x) \subset H^{-1}(b)\}$$
$$\mu_0 = \{(t,x) \in h : t = 0\}$$
$$\ell = \{(t,x) : \exists t' \geq t, (t',x) \in \eta([0,1] \times 1)\},$$

then $bh = \mu \cup e_a \cup e_b \cup \mu_0 \cup \ell$.

As the orientation problems here are trivial we denote also by the same letters the subanalytic chains such that

$$\partial h = \mu + e_a + e_b + \mu_0 + \ell.$$

Let $\Theta = f_*(h)$, then $\partial\Theta = f_*(\partial h) = f_*(\mu) + f_*(\mu_0) + f_*(e_a) + f_*(e_b) + f_*(\ell)$.

$$\int_\mu w = 0 \quad \text{because } \mu \subset B$$

$$\int_{\mu_0} w = 0 \quad \text{because } \mu_0 \subset A$$

$$\int_{e_a} w = 0 \quad \text{because } IdH \text{ is a nonvanishing vector field on } e_a \text{ and for any}$$

other field of vectors v independent of IdH, as $e_a \subset H^{-1}(a)$, we have $v \in TH^{-1}(a)$, so $0 = dH(v) = w(IdH, v)$

$$\int_{e_b} w = 0 \quad \text{same proof.}$$

Claim. $\int_\ell w \neq 0$.

Condition iv. in the Proposition above implies the existence of a vector field $\tau : \ell \to T\ell$ such that $dH(\tau(x)) > 0 \; \forall x \in \ell$. Then $(IdH(x), \tau(x))$ is a basis of $T_x\ell$ and $w(IdH, \tau) = dH(\tau) > 0$ everywhere. $\qquad\square$

REFERENCES

[A] V. Arnold, *Méthodes mathématiques de la mecanique classique*, Editions MIR, Moscow, 1976.

[B] J.F. Bjork, *Rings of differential operators*, North-Holland, Amsterdam, 1979.

[B-H] A. Borel and A. Haefliger, *La classe d'homologie* ..., Bull. Soc. Math. Fr. **89** (1961), 461–513.

[De] P. Deligne, Exposé XIV in *Groupes de Monodromie en géométrie algebrique*, Lecture Notes in Math. No. 340, Springer-Verlag, 116–164.

[D$_1$] A.S. Dubson, These de Doctorat d'Etat, Université de Paris VII, Paris, 1982.

[D$_2$] A.S. Dubson, *Théorie d'intersection* ..., CRAS **286** (2 may 1978), 755–758.

[D$_3$] A.S. Dubson, *Majorations Holder* ..., CRAS **286** (17 April 1978), 667–670.

[D$_4$] A.S. Dubson, *Formule pour l'indice* ..., CRAS **298**, Serie I, No. 6, 1984, 113–116; *Formule pour les cycles évanescents*, CRAS **299**, Serie I, 1984, 181–184.

[H] H. Hironaka, *Triangulations of algebraic sets*, Proc. Symp. Pure Math., Amer. Math. Soc. **29** (1975), 165–185; *Subanalytic sets*, in *Number Theory, Algebraic Geometry and Commutative Algebra*, Kinokuniya Publications, 1973, 453–493.

[K] M. Kashiwara, *Systèmes d'équations microdifférentielles*, Birkhauser, Boston, 1983.

[L] S. Lefschetz, *Topology*, Chelsea Publishing Company, New York, 1965.

[M] R.D. MacPherson, *Chern classes for singular varieties*, Annals of Math., Vol. 100, No. 2, Sept. 1974, 423–432.

Alberto S. Dubson
CONICET
Instituto Argentino de Matemática
Viamonte 1636 – 1er.crp. 1er. piso
1055 Buenos Aires, ARGENTINA

Progress in Mathematics

Edited by:

J. Oesterlé
Département de Mathématiques
Université de Paris VI
4, Place Jussieu
75230 Paris Cedex 05, France

A. Weinstein
Department of Mathematics
University of California
Berkeley, CA 94720
U.S.A.

Progress in Mathematics is a series of books intended for professional mathematicians and scientists, encompassing all areas of pure mathematics. This distinguished series, which began in 1979, includes authored monographs and edited collections of papers on important research developments as well as expositions of particular subject areas.

We encourage preparation of manuscripts in some form of TeX for delivery in camera-ready copy which leads to rapid publication, or in electronic form for interfacing with laser printers or typesetters.

Proposals should be sent directly to the editors or to: Birkhäuser Boston, 675 Massachusetts Avenue, Cambridge, MA 02139, U. S. A.

A complete list of titles in this series is available from the publisher.